Fuzzy Stochastic Optimization

Shuming Wang • Junzo Watada

Fuzzy Stochastic Optimization

Theory, Models and Applications

Shuming Wang
Waseda University
Hibikino, Wakamatsu-ku
Kitakyushu-City 2-7
Fukuoka, Japan

Junzo Watada
Waseda University
Hibikino, Wakamatsu-ku
Kitakyushu-City 2-7
Fukuoka, Japan

ISBN 978-1-4899-9273-4 ISBN 978-1-4419-9560-5 (eBook)
DOI 10.1007/978-1-4419-9560-5
Springer New York Heidelberg Dordrecht London

Softcover reprint of the hardcover 1st edition 2012

Printed on acid-free paper

Springer is part of Springer Science+Business Media (www.springer.com)

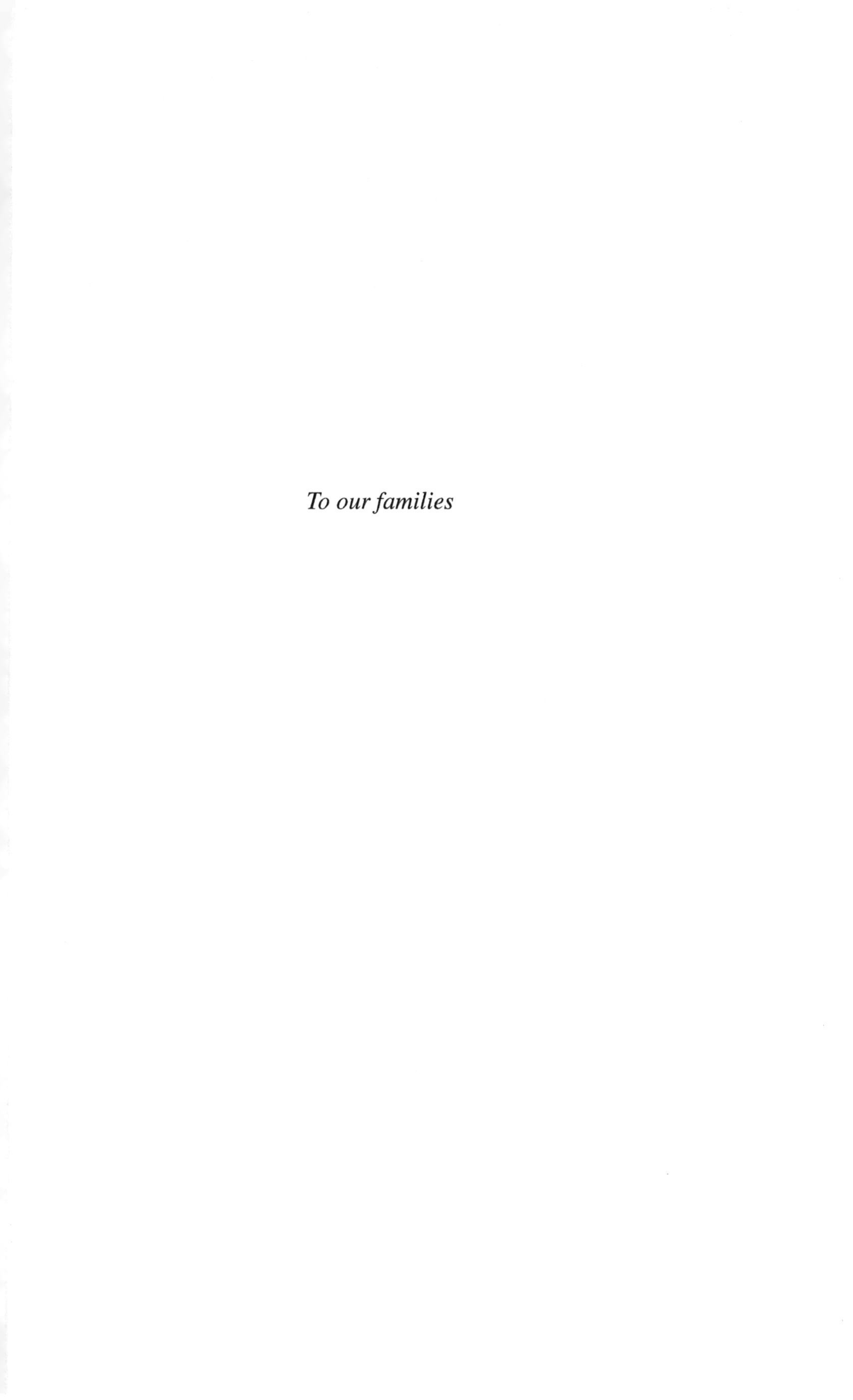

To our families

Preface

Randomness and fuzziness (or vagueness) are two major sources of uncertainty in the real world. In practical applications in areas of industrial engineering, management, and economics, chances are pretty good that decision makers are being confronted with information that are simultaneously probabilistically uncertain and fuzzily imprecise, and an optimization (decision making) has to be performed under such a twofold uncertain environment of a co-occurrence of randomness and fuzziness.

Fuzzy random variable originally presented by H. Kwakernaak is a tailor-made mathematical tool to describe the twofold or hybrid uncertainty. It owns a twofold distribution structure being able to carry a joint integrality of the simultaneous probabilistic and fuzzy information which goes beyond the juxtaposition of information contained in random variable in probability theory and fuzzy variable in theory of fuzzy set and possibility. Naturally, it is regarded as a generalization for both random variable and fuzzy variable. Being capable of modeling the randomness and fuzziness as an ensemble, the fuzzy random variable has become the part and parcel in the optimization under integrated uncertainties of randomness and fuzziness (fuzzy stochastic optimization).

From both theoretical and practical perspectives, this book aims to present a self-contained, systematic, and up-to-date description of fuzzy stochastic optimization on the basis of the fuzzy random variable being a core mathematical vehicle to model the integrated fuzzy random uncertainty. It goes along a direction from theoretical aspects of the fuzzy random variable to fuzzy stochastic optimization models and their real-life case studies.

We now describe in detail the structure of the book. After an introduction in Chap. 1 which outlines the development of theory of fuzzy random variables, and of fuzzy stochastic optimization models, the core work consists of the next eight chapters that form relatively independent three parts: theory, models, and real-life applications.

Part I: Theory

In this first part, we present mathematical foundations of fuzzy random variable. Chapter 2 introduces necessary preliminaries on fuzzy random variable and then presents analytical properties of fuzzy random variable in different aspects. In Chap. 3, we discuss two renewal processes, i.e., fuzzy stochastic renewal process and fuzzy stochastic renewal reward process, and a fuzzy random elementary renewal theorem and a fuzzy random renewal reward theorem are described in depth.

Part II: Models

In the second part we present a series of fuzzy stochastic optimization models using fuzzy random variables to model uncertain parameters. Chapter 4 discusses two redundancy allocation models with fuzzy random lifetimes for system reliability optimization in different objectives of reliability maximization and cost minimization. In Chap. 5, we discuss a two-stage facility location selection model with fuzzy random variable cost and client demand in which the capacity is fixed and the objective is to maximize the expected profit. In this two-stage problems, given each first-stage decision (location decision) and each realization of fuzzy random parameter, we have a number of second-stage optimization subproblems to solve to determine the objective value at the first-stage decision. In Chap. 6 we establish a genetic risk optimization model in the fuzzy random environment, namely, two-stage fuzzy stochastic programming with Value-at-Risk (VaR). In this chapter, the notation of VaR metric is introduced into the fuzzy random case, and a fuzzy random VaR criterion is defined to form an objective of VaR minimization in the context of a two-stage model. Applying the generic fuzzy stochastic VaR model to the facility location selection with variable capacity, in Chap. 7 we build a two-stage fuzzy stochastic facility location model with VaR objective and variable capacity, in which both the location and capacity are decision variables. Owing to the solution difficulties that above fuzzy stochastic models bear, in general, they cannot be surmounted by classic mathematical programming approaches, the models are therefore solved with the aid of hybrid approaches that fuse some improved metaheuristic algorithms and approximate vehicles.

Part III: Real-Life Applications

The third part consists of two real-life applications of the optimization models discussed in Part II. Chapter 8 presents a case study on a dam control system design problem which is an application of the system reliability optimization models built in Chap. 4. In Chap. 9 we present a case study on location selection problem for frozen food plants to which the models in Chaps. 5 and 6 are applied.

We are grateful to the following scholars and friends for the insightful discussions and suggestions they provided during the creation of the book. They include Professor Witold Pedrycz of the University of Alberta, Canada; Professor Yan-Kui Liu of Hebei University, China; Professor Lakhmi C. Jain of the University of South Australia, Australia; Professor Hiroaki Ishii of Osaka University, Japan; Professor Baoding Liu of Tsinghua University, China; Professor Vyacheslav Kalashnikov of the Monterrey Institute of Technology, Mexico; Professor Jaeseok Choi of Gyeongsang National University, Korea; Professor Jeng-Shyang Pan of the National Kaohsiung University of Applied Sciences; Professor Huey-Ming Lee of the Chinese Culture University; and Professor Berlin Wu of the National Chengchi University, Taiwan.

Also, the first author would like to thank the financial support from the Research Fellowship of the Japan Society for the Promotion of Science (JSPS) for Young Scientists, and the Research Fellowship of the "Ambient SoC Global COE Program of Waseda University" of the Ministry of Education, Culture, Sports, Science and Technology (MEXT), Japan.

Finally, we wish to thank Dr. Brett Kurzman and Ms. Elizabeth Dougherty at Springer for their much appreciated editorial work throughout the process of publishing this book.

Beijing, China
Kitakyushu, Japan

Shuming Wang
Junzo Watada

Contents

Frequently Used Acronyms

$\mathscr{A}$	Ample field
ξ	Fuzzy random vector or fuzzy random variable
Σ	σ-algebra
$(\Omega, \Sigma, \mathrm{Pr})$	Probability space
$(\Gamma, \mathscr{A}, \mathrm{Pos})$	Possibility space
Cr	Credibility measure
Ch	(mean) Chance measure
Pr	Probability measure
$E[\cdot]$	Expected value operator
Pos	Possibility measure
Nec	Necessity measure
$\Re$	Set of real numbers
$\Re^n$	n-dimensional Euclidean space
$I_{\{\cdot\}}$	Indicator function of set $\{\cdot\}$
$\vee$	Maximum operator
$\wedge$	Minimum operator
$\mu_X(\cdot)$	Membership function of fuzzy variable X
$\emptyset$	Empty set
∞	Infinity
i.i.d.	Independent and identically distributed
lim sup	Limit superior
$\mathscr{F}_v^n$	A collection of n-dimensional fuzzy vectors
$G_\xi(\cdot)$	Distribution function of fuzzy random variable ξ
$\top$	Triangular norm
$\perp$	Triangular conorm
$f^{[-1]}$	Pseudo-inverse of f
Π	Possibility function
$\mathrm{co}(g)_\Xi$	Convex hull of function g on its support Ξ

$N(t)$	Fuzzy random renewal variable
$C(t)$	Fuzzy random total reward
x	Decision vector
$\mathscr{L}$	Loss variable
$\mathbb{N}$	Set of natural numbers
$\mathbb{Z}$	Set of integers
ACO	Ant Colony Optimization
GA	Genetic Algorithm
PSO	Particle Swarm Optimization
TS	Tabu Search
FRS	Fuzzy random solution
RS	Random solution
VFI	Value of fuzzy information
VaR	Value-at-risk

Chapter 1
Introduction

Randomness and fuzziness (or vagueness) are two major sources of uncertainty in the real world. Randomness relates to the stochastic variability of all possible outcomes of a situation and can be perfectly and mathematically described by probability theory with random variable. Fuzziness, on the other hand, stems from the imprecision of subjective human knowledge and exists objectively with a variety of manifestations in numbers of situations such as data capture and process, blurred boundaries of the parameters, expertise applications, and lack of precise knowledge. Fuzzy variable in the context of theory of fuzzy set and possibility (see [114, 117, 118, 171, 172]) is widely accepted as an effective mathematical approach to model the fuzzy uncertainty.

On the basis of probability theory and statistics, stochastic programming (see [13, 60]) is deputed a conventional approach to the optimization problems under uncertainty. In line with the probability theory, stochastic programming assumes all the uncertain parameters to be random variables, and the computation and the solution are overly dependent on the probability distributions identified for the parameters.

As we all know that any (reliable) probability distribution of a random variable should be determined via the statistic analysis and inference based on adequate crisp data. However, in many practical situations, such an essential ingredient (adequate crisp data) is not available. For instance, it is sometimes not easy to acquire a sufficient number of crisp statistic data, and the imprecise expert estimation is utilized as a compensation to the scarce precise information, or the captured data themselves are not precise but more or less imprecise or fuzzy (see [10, 17, 37, 69, 81, 82, 113, 115, 116, 120, 123, 181]).

In such circumstances, since the parameter distribution intrinsically embraces both randomness and fuzziness, random variable apparently is not the best way to handle this sort of hybrid or twofold uncertainty. Therefore the stochastic programming techniques are not as suitable to cope with the optimization problems in such a twofold uncertain environment as they are in the single-fold stochastic case. In fact, if we impose the farfetched stochastic programming as the approach to the problem with randomness and fuzziness simultaneously, then we have to ignore the fuzziness to make an uncertainty reduction (simplification). Or more

S. Wang and J. Watada, *Fuzzy Stochastic Optimization: Theory, Models and Applications*,
DOI 10.1007/978-1-4419-9560-5_1,

concretely, we have to assume all stubbornly imprecise elements are known with precision by assigning some fixed values to create the required "perfect" probability distribution so as to satisfy the working condition of stochastic approach. Oblivious of the fuzziness, in this way we will be inevitably running a risk of "Garbage In, Garbage Out (GIGO)[1]" which may spoil the whole decision making.

Likewise, in light of that the fuzzy variable fails to describe the stochastic variability which is essentially distinct from the vagueness, fuzzy programming [88, 94, 103] with the fuzzy variable as its part and parcel also falls much short of reality when it is employed to deal with the decision making under the hybrid uncertainty of simultaneous randomness and fuzziness.

The concept of a fuzzy random variable was originally discussed by Kwakernaak [72], it owns a powerful distribution structure being able to carry a joint integrality of the simultaneous probabilistic and fuzzy information which goes beyond the juxtaposition of information contained in random variable and fuzzy variable. Mathematically, a fuzzy random variable is defined as a measurable mapping from a probability space to a collection of fuzzy variables; or roughly speaking, it can be regarded as a "random variable" that takes on fuzzy or imprecise values (see Fig. 1.1). Naturally, random variable and fuzzy variable are just two special cases in the sense of fuzzy random variable.

Being capable of characterizing mathematically the randomness and fuzziness as an ensemble, the fuzzy random variable facilitates modeling the optimization problems that must base decisions and analysis on information that is simultaneously probabilistically uncertain and fuzzily imprecise. It is deemed as the core mathematical vehicle in fuzzy stochastic optimization just like the role the random variable plays in the stochastic optimization.

Over the past three decades, the theory of fuzzy random variables which to a large extent serves as the mathematical foundation for fuzzy stochastic optimization has been developed in breadth and depth via a large number of studies:

(a) Following the original ideas of Kwakernaak [72, 73], several variants as well as extensions of fuzzy random variable have been presented by other researchers. For instance, the mathematical concept for fuzzy random variable was further formalized in [69] and was defined under some different measurability conditions from those in [72, 73]. Some constructive definitions of fuzzy random variables were discussed in [101]. Recently, a modified fuzzy random variable with an optimization-oriented measurability condition was introduced by [89] for the purpose of modeling fuzzy random decision-making problems. More extensive discussions can be found in [38, 131, 143, 173–177].

[1] "Garbage In, Garbage Out" (abbreviated to GIGO) is a phrase in the fields of computer science and information communication technology. It is used primarily to call attention to the fact that computers will unquestioningly process the most nonsensical of input data and produce nonsensical output. It was popular in the early days of computing, but it applies even more today to describe failures in human decision making due to imprecise, incomplete, or faulty data (see [12]).

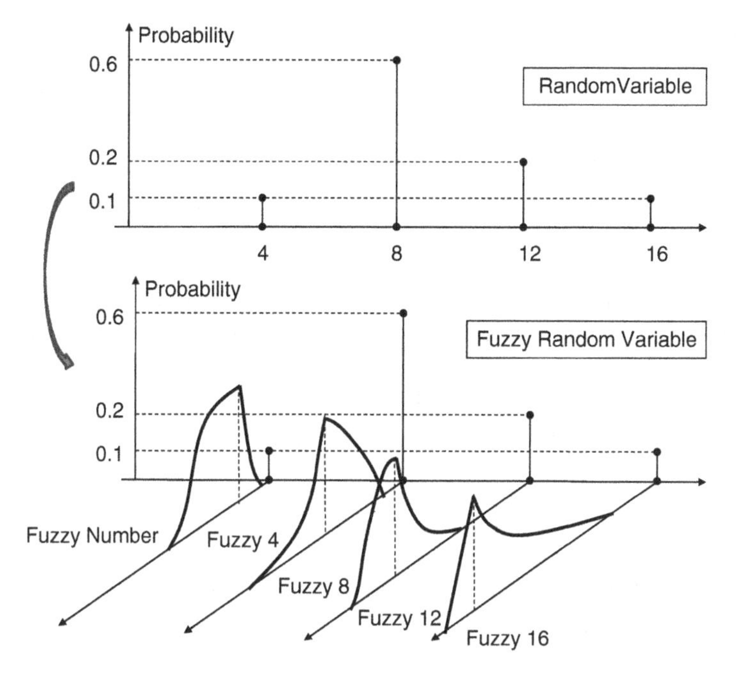

Fig. 1.1 Fuzzy random variable: a "random" variable assuming fuzzy realizations

(b) A variety of theoretical aspects of the fuzzy random variable have been examined. For example, measurability criteria (see [1, 35, 93, 146]), independence (see [52, 89, 146]), variance formulas with particular distribution types (see [45, 67]), analytical properties of distribution functions and critical value functions (see [147, 164]), convergence criteria and relations (see [58, 87, 98]), limit theorems and laws of large numbers (see [68, 79, 153, 165]), convexity analysis (see [95, 134]), and so on. Some important theoretical results of fuzzy random variable are discussed in Chap. 2.

(c) As the extensions of stochastic processes in fuzzy random environments, some types of fuzzy stochastic processes have been established by investigating the properties of a collection of fuzzy random variables. Along this direction, a fuzzy random renewal reward process was initially studied by [121], in which the rewards are assumed as independent and identically distributed (i.i.d.) fuzzy random variables, and the limit value for long-run reward rate was derived in probability. In [53], a fuzzy random renewal process was discussed, and the limit of long-run renewal rate in probability was obtained. A Blackwell's renewal theorem and a Smith's key renewal theorem for fuzzy random interarrival times were proved in [180], and a fuzzy random delayed renewal process was discussed in [80]. Recently, a fuzzy random renewal process [148] and a fuzzy random renewal reward process [150] were intensively studied on the basis of mean chance measure and expected value of fuzzy random variable, and a fuzzy random elementary renewal theorem and a fuzzy random renewal reward theorem were established which prove parallel to the stochastic elementary renewal and renewal reward theorems in the stochastic case, respectively. Some other fuzzy stochastic processes can be found in [4, 78]. The fuzzy stochastic renewal processes are discussed in depth in Chap. 3.

The incessant development of the theory of fuzzy random variable gives a constant boost to the fuzzy stochastic optimization. Ever since the initial work [142] which is a fuzzy stochastic linear programming with fuzzy random variable coefficients for an economic problem of polluted water treatment, an increasing number of fuzzy stochastic optimization models with fuzzy random variable parameters have been discussed.

Most fuzzy stochastic optimization models can be roughly sorted into two categories, i.e., single-stage problems and multistage problems. The former covers most fuzzy stochastic models such as fuzzy stochastic inventory models [6, 24, 32, 166], fuzzy stochastic regression analysis [16, 115, 159–161], portfolio selection models in fuzzy random environments [50, 133], fuzzy stochastic data envelopment analysis (DEA) model [124], system reliability optimization models with fuzzy random lifetimes [149, 151], and many abstract programming models with fuzzy random variable parameters [2, 47, 84, 85, 90, 91, 105, 106, 169]. As typical models of this type, two fuzzy stochastic system reliability optimization models are described in detail in Chap. 4.

The multistage problems, generally speaking, are much more difficult in solution than the single-stage problems, since given each solution, each realization of fuzzy random parameters corresponds each optimization subproblem to solve. That is partly why the existing fuzzy stochastic multistage models are found much less than the single-stage models. Among the limited number of fuzzy stochastic multistage models, a two-stage fuzzy stochastic programming was first formulated with an expected value objective and its approximation properties were discussed [96]. A two-stage fuzzy random facility location model with fixed capacity was built in [154] and its properties as well as hybrid metaheuristic algorithm were discussed. Introducing the Value-at-Risk (VaR) metric into fuzzy stochastic optimization, a two-stage fuzzy stochastic programming model with VaR was proposed and thoroughly discussed by [155, 156]. Furthermore, applying the VaR model [156] to the facility location selection with variable capacity in a fuzzy random environment, a VaR-based fuzzy stochastic facility location model was established by [158]. Three out of four chapters in Part II are devoted to the two-stage fuzzy stochastic optimization models, where the course-based two-stage fuzzy random location model is discussed in Chap. 5, the two-stage fuzzy stochastic programming with VaR is discussed in Chap. 6, and the VaR-based two-stage fuzzy stochastic location model is described in Chap. 7.

In addition, since the uncertain parameters are fuzzy random variables which owe the complex twofold distribution structures, fuzzy stochastic optimization models are a family of complex optimization problems which in general are even more difficult than the stochastic optimization models. In many cases (e.g., continuous fuzzy random parameter case), the objective functions or constraints consisting of fuzzy random parameters cannot be expressed analytically, which makes the general nonlinear programming approaches not applicable to the fuzzy stochastic problems (detailed descriptions for model difficulties can be found in Chap. 4 for a single-stage problem case and Chaps. 5–7 for two-stage problem cases). As a consequence, a variety of metaheuristic algorithms such as genetic algorithms (GA) [39, 109], particle swarm optimization (PSO) [54, 63, 64, 132], and tabu search (TS) [40–42], as an important and effective group of optimizers, have been applied to tackle some fuzzy stochastic optimization problems. Different types of metaheuristic-based solution approaches are designed to handle the different problems in Chaps. 4–7 in accordance with the problem structure.

As for the real-life applications of fuzzy stochastic optimization models, few can be found in the literature; it is only because of the solution difficulties as mentioned above, but it is also due to the complicated process of determining the distributions of fuzzy random parameters. See Chaps. 8 and 9 for a full description of two real-life case studies.

Part I
Theory

Chapter 2
Fuzzy Random Variable

In this chapter, we review some necessary basics of fuzzy random variable, and then lay out the description of the analytical aspects of the fuzzy random variable.

In Sect. 2.1, we first give the definitions of probability measure, random variable as well as its mathematical expectation in probability theory, and parallel, the definitions of possibility and credibility measures, fuzzy variable as well as its mathematical expectation in possibility theory. Then, we introduce the formal definition of fuzzy random variable and some relating concepts including mean chance measure, distribution functions, expectations, and convergence mode.

In Sect. 2.2, we discuss some different semicontinuity and continuity conditions for distribution functions of functions of fuzzy random variable. Some of the results are shown useful to the analysis of optimization models in Part II.

In Sect. 2.3, we introduce the concept of triangular norm (t-norm) and give some theoretical results on $\top$-independent fuzzy variables. Furthermore, we discuss the independence condition for a fuzzy random vector based on $\top$-independence.

In Sect. 2.4, we discuss some limit theorems for the sum of fuzzy random variables by considering a sort of conditions based on continuous Archimedean t-norms. Then, on the basis of the limit theorems, we give two laws of large numbers for fuzzy random variable sequences in different convergence modes. Many of the results prove useful and pivotal to the fuzzy random renewal processes discussed in Chap. 3.

2.1 Basic Concepts

For readers' convenience, this subsection is intended to dabble on some basic and necessary concepts of random variable, fuzzy variable, and fuzzy random variable that will be frequently used in later parts of the book. However, these preliminaries are presented as concisely as possible for the sake of easy reading.

S. Wang and J. Watada, *Fuzzy Stochastic Optimization: Theory, Models and Applications*,
DOI 10.1007/978-1-4419-9560-5_2,

2.1.1 Random Variable

Let us start from probability space. A probability space is defined as a triplet $(\Omega, \Sigma, \Pr)$, where $\Omega = \{\omega_\tau, \tau \in \mathscr{T}\}$ is a sample space; $\mathscr{T}$ is an index set, Σ is a σ-algebra of subsets of Ω, that is, the set of all possible "interesting" events; and a probability measure on Ω, denoted by Pr, satisfies:

(1) $\Pr(\emptyset) = 0$
(2) $\Pr(\Omega) = 1$
(3) $\Pr\left(\bigcup_{i=1}^{\infty} A_i\right) = \sum_{i=1}^{\infty} \Pr(A_i)$, for any countable and mutually disjoint events $A_i \in \Sigma, i = 1, 2, \cdots$

In probability theory, a random variable Y is a measurable function from a probability space $(\Omega, \Sigma, \Pr)$ to the set of real numbers $\Re$, that is, for any $t \in \Re$, we have $\{\omega \in \Omega \mid Y(\omega) \leq t\} \in \Sigma$. In this case, as a function of $\omega \in \Sigma$, Y is also called Σ-measurable. In addition, any event to happen with probability 1 is said to happen *almost surely* (a.s.)., or, the event can be said to happen for almost every $\omega \in \Omega$.

Furthermore, the expected value of a random variable Y can be defined generally by the *integration of a measurable function* (see [11,43]) as

$$E[Y] = \int_{\Omega} Y(\omega) \Pr(\mathrm{d}\omega),$$

which is equivalent to

$$E[Y] = \int_{-\infty}^{\infty} x p(x) \mathrm{d}x,$$

in the case that Y is a continuous random variable with density function $p(x)$, or equivalent to

$$E[Y] = \sum_{i=1}^{\infty} y_i \Pr\{Y = y_i\},$$

if Y is a discrete random variable assuming values $y_i, i = 1, 2, \cdots$.

A random variable Y is said to be integrable if $E[|Y|] < \infty$. As a useful result on random variables, the following famous *dominated convergence theorem* of a sequence of integrations (expectations) will be utilized frequently in later contents.

Theorem 2.1. *Let Y_n and Z be integrable random variables on $(\Omega, \Sigma, \Pr)$. If $|Y_n| \leq Z$ and $Y_n \to Y$, almost surely, then*

$$\lim_{n \to \infty} \int_{\Omega} Y_n(\omega) \Pr(\mathrm{d}\omega) = \int_{\Omega} \lim_{n \to \infty} Y_n(\omega) \Pr(\mathrm{d}\omega) = \int_{\Omega} Y(\omega) \Pr(\mathrm{d}\omega).$$

2.1.2 Fuzzy Variable

In contrast with probability measure and random variables in probability theory which are the mathematical tools to deal with the stochastic variability, credibility measure and fuzzy variable in possibility theory (see [29, 172]) are utilized to depict the imprecise or fuzzy uncertainty.

Given a universe Γ, an ample field [162] $\mathscr{A}$ on Γ is a class of subsets of Γ that is closed under arbitrary unions, intersections, and complementation in Γ. Let Pos be a set function defined on the ample field $\mathscr{A}$. Pos is said to be a possibility measure if it satisfies the following conditions:

(1) $\mathrm{Pos}(\emptyset)=0$.
(2) $\mathrm{Pos}(\Gamma)=1$.
(3) $\mathrm{Pos}\left(\bigcup_{i\in\mathscr{I}} A_i\right)=\sup_{i\in\mathscr{I}}\mathrm{Pos}(A_i)$ for any subclass $\{A_i \mid i\in\mathscr{I}\}$ of $\mathscr{A}$, where $\mathscr{I}$ is an arbitrary index set.

The triplet $(\Gamma,\mathscr{A},\mathrm{Pos})$ is called a possibility space [29]. In addition, the dual measure of Pos, named necessity measure (denoted by Nec), is defined by

$$\mathrm{Nec}\{A\}=1-\mathrm{Pos}\{A^c\}, \quad A\in\mathscr{A},$$

where A^c is the complement of A.

Based on possibility measure, a self-dual set function Cr, called *credibility measure*, is formally defined as follows:

Definition 2.1 ([88]). Let Γ be a universe, $\mathscr{A}$ an ample field on Γ. The credibility measure, denoted Cr, is defined as

$$\mathrm{Cr}(A)=\frac{1}{2}\Big(1+\mathrm{Pos}(A)-\mathrm{Pos}(A^c)\Big), \quad A\in\mathscr{A}. \tag{2.1}$$

Apparently, the credibility measure can be written equivalently as

$$\mathrm{Cr}(A)=\frac{1}{2}\Big(\mathrm{Pos}(A)+\mathrm{Nec}(A)\Big), \quad A\in\mathscr{A}.$$

A credibility measure has the following properties (see [87]):

(1) $\mathrm{Cr}(\emptyset)$=0, and $\mathrm{Cr}(\Gamma)=1$.
(2) Monotonicity: $\mathrm{Cr}(A)\le\mathrm{Cr}(B)$ for all $A,B\subset\Gamma$ with $A\subset B$.
(3) Self-duality: $\mathrm{Cr}(A)+\mathrm{Cr}(A^c)=1$ for all $A\subset\Gamma$.
(4) Subadditivity: $\mathrm{Cr}(A\cup B)\le\mathrm{Cr}(A)+\mathrm{Cr}(B)$ for all $A,B\subset\Gamma$.

Let $(\Gamma,\mathscr{A},\mathrm{Pos})$ be a possibility space. A fuzzy vector is a map $X=(X_1,X_2,\cdots,X_n)$ from Γ to $\Re^n$ such that $\{\gamma\in\Gamma \mid X(\gamma)\le t\}\in\mathscr{A}$ for every $t\in\Re^n$. As $n=1$, it is called a fuzzy variable. In possibility theory, the membership function of a fuzzy vector X can be expressed through the possibility measure as

$$\mu_X(t)=\mathrm{Pos}\{\gamma \mid X(\gamma)=t\} \tag{2.2}$$

for every $t=(t_1,t_2,\cdots,t_n)\in\Re^n$.

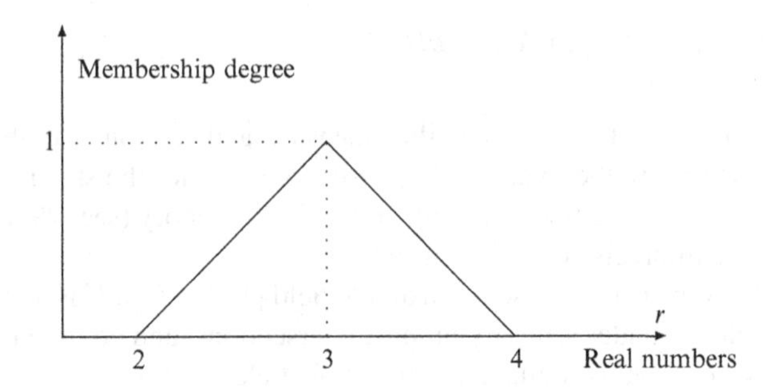

Fig. 2.1 The membership function of fuzzy variable $X = (2,3,4)$

Triangular fuzzy variable is one of most common sort of fuzzy variables, which is denoted as $X = (a,b,c)$ with membership degree

$$\mu_X(t) = \begin{cases} \dfrac{t-a}{b-a}, & \text{if } a \le t \le b \\ \dfrac{c-t}{c-b}, & \text{if } b \le t \le c \\ 0, & \text{otherwise.} \end{cases}$$

In some applications, a triangular fuzzy variable $X = (a,b,c)$ is also represented as by its mode $X^C = b$ which is the value assuming the membership degree 1, and both its left and right spreads, i.e., $X^L = b - a$ and $X^R = c - b$, respectively. That is,

$$X = \left(X^C, X^L, X^R\right) = (b, b-a, c-b).$$

Such representation is to prove useful when dealing with fuzzy statistics in real applications (Chap. 9).

Example 2.1. For a triangular fuzzy variable $X = (2,3,4)$ whose membership function (see Fig. 2.1) is given by

$$\mu_X(t) = \begin{cases} t-2, & 2 \le t < 3 \\ 4-t, & 3 \le t < 4 \\ 0, & \text{otherwise,} \end{cases} \tag{2.3}$$

the credibility of $X \ge r$ (see Fig. 2.2) is calculated as follows:

$$\mathrm{Cr}\{X \ge r\} = \begin{cases} 1, & r \le 2 \\ \dfrac{4-r}{2}, & 2 < r \le 4 \\ 0, & \text{otherwise.} \end{cases} \tag{2.4}$$

Similarly, we can also get the credibility of $\xi < r$ by

$$\mathrm{Cr}\{\xi < r\} = 1 - \mathrm{Cr}\{\xi \ge r\} \text{ (Fig. 2.2).}$$

Fig. 2.2 The credibility values of events $X \geq r$ and $X < r$

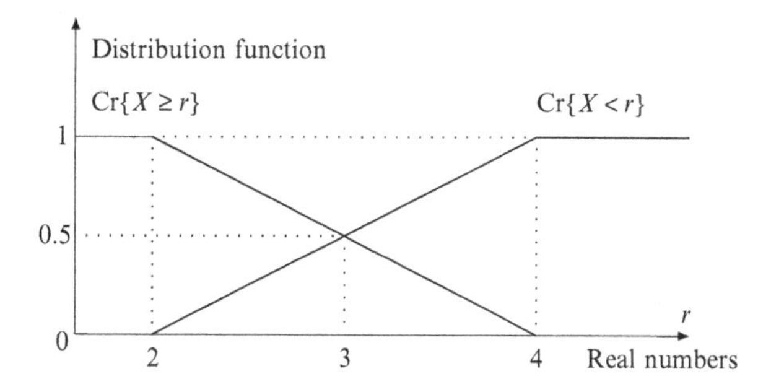

Definition 2.2 ([88]). Let X be a fuzzy variable. The expected value of X is defined as

$$E[X] = \int_0^{\infty} \mathrm{Cr}\{X \geq r\}\mathrm{d}r - \int_{-\infty}^{0} \mathrm{Cr}\{X \leq r\}\mathrm{d}r, \tag{2.5}$$

provided that one of the two integrals is finite.

Example 2.2. For any triangular fuzzy variable $X = (a,b,c)$, the expected value of X is

$$E[X] = \frac{a+2b+c}{4}.$$

2.1.3 Fuzzy Random Variable

The concept of fuzzy random variable was initially introduced by Kwakernaak [72] in 1978 to model the hybrid uncertainty which embraces randomness and fuzziness simultaneously. Since then, the original idea of fuzzy random variable has been enhanced, and its extensions have been presented in the literature (see [69, 89, 101, 104, 122]). Among them in [89], on the viewpoint of optimization, a new version was distilled from intensive comparison and analysis among the characterizations of different definitions on fuzzy random variable, and depending on the applications, it has proved a suitable basis of optimization techniques and numerical simulation in fuzzy random systems.

Let us assume that $(\Omega, \Sigma, \mathrm{Pr})$ is a probability space, and $\mathscr{F}_v^n$ is a collection of fuzzy vectors defined on a possibility space $(\Gamma, \mathscr{A}, \mathrm{Pos})$, then the concept of fuzzy random vector is defined as follows:

Definition 2.3 ([89]). A fuzzy random vector is a map $\xi = (\xi_1, \xi_2, \cdots, \xi_n) : \Omega \to \mathscr{F}_v^n$ such that for any closed subset $F \subset \Re^n$, $\mathrm{Pos}\{\gamma \mid \xi(\omega,\gamma) \in F\}$ is a Σ-measurable function of $\omega \in \Omega$, i.e., for any $t \in [0,1]$, we have

$$\{\omega \in \Omega \mid \mathrm{Pos}\{\gamma \mid \xi(\omega,\gamma) \in F\} \leq t\} \in \Sigma.$$

In the case of $n = 1$, ξ is called a fuzzy random variable.

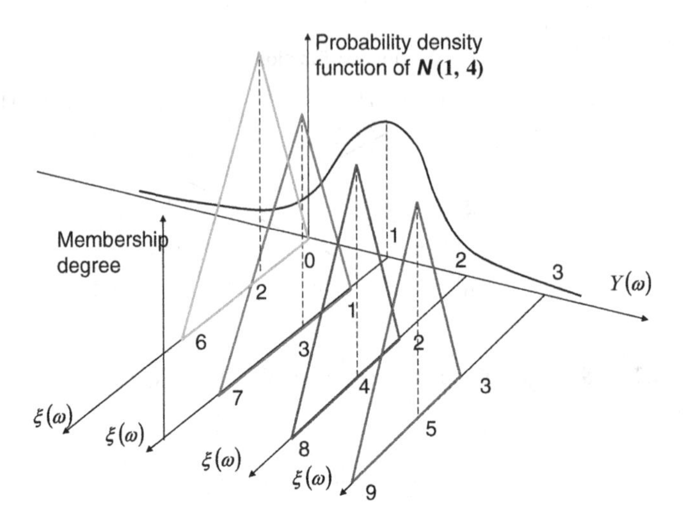

Fig. 2.3 Distribution of fuzzy random variable $\xi(\omega) = (Y(\omega), Y(\omega)+2, Y(\omega)+6)$

From the above definition, a fuzzy random variable can be roughly interpreted as a "generalized" random variable taking on values that could be fuzzy but not limited to crisp ones. That is, more precisely, given a fuzzy random variable ξ on $(\Omega, \Sigma, \mathrm{Pr})$, the realization $\xi(\omega)$ for each $\omega \in \Omega$ is a fuzzy variable which is more general than a crisp number (the latter is no more than a special case of the former).

Remark 2.1. It has been showed in the literature that both random variable and fuzzy variable are the special cases of fuzzy random variable in the sense of Definition 2.3 (refer to [89]).

Example 2.3. Let Y be a random variable defined on probability space $(\Omega, \Sigma, \mathrm{Pr})$. We call ξ a triangular fuzzy random variable (see [89]), if for every $\omega \in \Omega$, $\xi(\omega)$ is a triangular fuzzy variable defined on some possibility space $(\Gamma, \mathscr{A}, \mathrm{Pos})$. For example,

$$\xi(\omega) = (Y(\omega), Y(\omega)+2, Y(\omega)+6)$$

is a triangular fuzzy random variable. In Fig. 2.3, the distribution of the above triangular fuzzy random variable ξ is provided, where the random parameter Y is a normal distributed random variable, i.e., $Y \sim \mathscr{N}(1,4)$.

Example 2.4. We say ξ is a normal fuzzy random variable (see [89]), denoted by $\mathscr{N}_{FR}(Y,b)$, where $b > 0, Y$ is a random variable, if for every $\omega \in \Omega$, the membership function of $\xi(\omega)$ is

$$\mu_{\xi(\omega)}(r) = EXP\left(\frac{-(r - Y(\omega))^2}{b}\right),$$

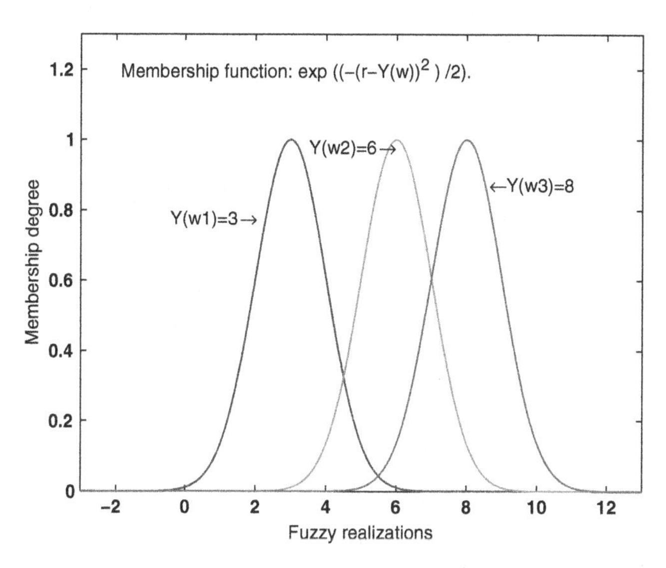

Fig. 2.4 Distribution of fuzzy random variable $\mathscr{N}_{FR}(Y,2)$

where $EXP(\cdot) = e^{(\cdot)}$ is the exponential function. The following Fig. 2.4 illustrates the distribution of $\mathscr{N}_{FR}(Y,b)$, where $b=2$, and Y is a discrete random variable assuming three values ω_1, ω_2, and ω_3.

From Definition 2.3, it is not difficult to prove that $C(\omega) \triangleq \mathrm{Cr}\{\xi(\omega) \in B\}$ as a function of $\omega \in \Sigma$ is Σ-measurable, or equivalently, it is a random variable. In order to measure an event induced by a fuzzy random variable ξ, the mean chance measure is given as follows:

Definition 2.4 ([91]). Let ξ be a fuzzy random variable, and B a Borel subset of $\Re$. The mean chance of an event $\xi \in B$ is defined as

$$\mathrm{Ch}\{\xi \in B\} = \int_{\Omega} \mathrm{Cr}\{\xi(\omega) \in B\}\,\mathrm{Pr}(\mathrm{d}\,\omega), \tag{2.6}$$

where $\mathrm{Cr}\{\cdot\}$ is the credibility measure in (2.1), and $\mathrm{Pr}\{\cdot\}$ is the probability measure.

The "mean chance" in this book is often simplified as "chance" without any confusion. From the above definition, the chance of an event $\xi \in B$ induced by a fuzzy random variable ξ is exactly the mean value (in the sense of probability) of all the credibility values of the corresponding fuzzy events $\xi(\omega) \in B$ relating to the fuzzy realizations $\xi(\omega)$ at different $\omega \in \Omega$. From the definition (2.6), the chance can also be equivalently written as

$$\mathrm{Ch}\{\xi \in B\} = \int_0^1 \mathrm{Pr}\{\omega \in \Omega \mid \mathrm{Cr}\{\xi(\omega) \in B\} \geq r\}\,\mathrm{d}r.$$

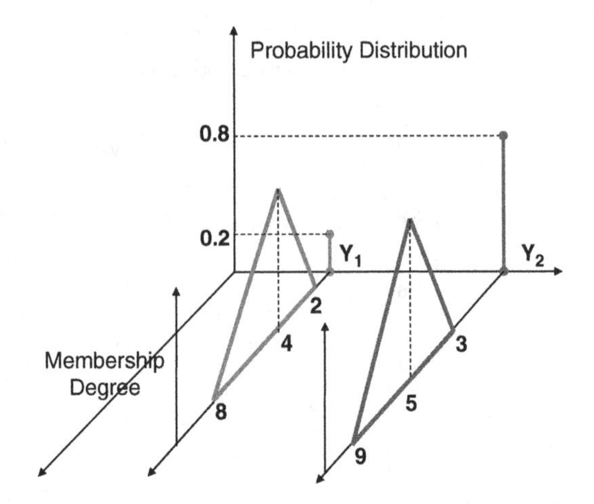

Fig. 2.5 Distribution of fuzzy random variable ξ in Example 2.5

Example 2.5. For the triangular fuzzy random variable ξ defined in Example 2.3, suppose that Y is a discrete random variable, which takes on values $Y_1 = 2$ with probability 0.2, and $Y_2 = 3$ with probability 0.8 (see Fig. 2.5). Try to determine the mean chance of fuzzy random event $\{\xi \leq 6\}$.

From the distribution of a random variable Y, we see that the fuzzy random variable ξ assumes values which are triangular fuzzy variables $\xi(Y_1) = (2,4,8)$ with probability 0.2, and $\xi(Y_2) = (3,5,9)$ with probability 0.8.

For the triangular fuzzy variable $\xi(Y_1) = (2,4,8)$, since its membership function is

$$\mu_{\xi(Y_1)}(t) = \begin{cases} \dfrac{t-2}{2}, & 2 < t \leq 4 \\ \dfrac{8-t}{4}, & 4 < t \leq 8 \\ 0, & \text{otherwise}, \end{cases} \tag{2.7}$$

from which we can compute that $\text{Cr}\{\xi(Y_1) \leq 6\} = 0.75$, with probability 0.2. Similarly, we get $\text{Cr}\{\xi(Y_2) \leq 6\} = 0.625$, with probability 0.8. From (2.6), we can calculate $\text{Ch}\{\xi \leq 6\} = 0.65$.

Based on mean chance Ch, the distribution functions of a fuzzy random variable are given as follows:

Definition 2.5 ([87]). Let ξ be a fuzzy random variable. The distribution functions of ξ are defined by

$$G_{\xi_L}(x) = \text{Ch}\{\xi \leq x\}, \quad x \in \Re \tag{2.8}$$

and

$$G_{\xi_U}(x) = \text{Ch}\{\xi \geq x\}, \quad x \in \Re. \tag{2.9}$$

Example 2.6. Consider the triangular fuzzy random variable ξ given in Example 2.5. Let us determine the distribution function $G_{\xi_L}(x)$ of ξ.

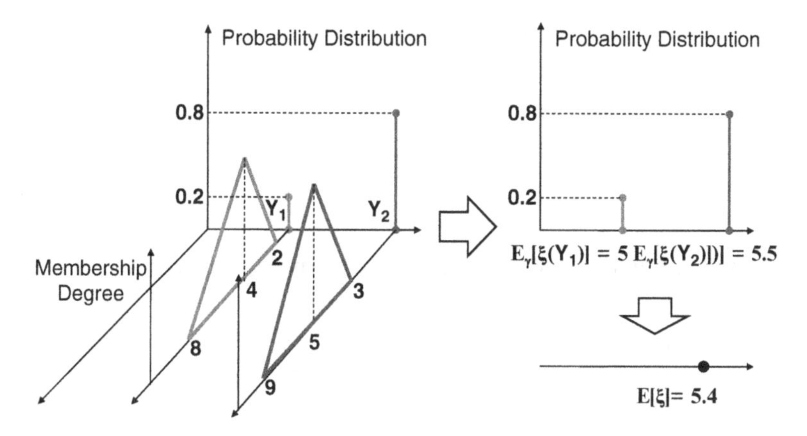

Fig. 2.6 Calculating the expected value of fuzzy random variable ξ in Example 2.5

Hence, we obtain

$$\int_0^{\infty} \mathrm{Cr}\{\xi(Y_1) \geq r\}\,\mathrm{d}r = 5.$$

Similarly, we compute

$$\int_0^{\infty} \mathrm{Cr}\{\xi(Y_2) \geq r\}\,\mathrm{d}r = 5.5.$$

Thus, by the definition in (2.11), the expected value of ξ is

$$E[\xi] = 0.2 \times 5 + 0.8 \times 5.5 = 5.4.$$

Figure 2.6 gives a graphic interpretation of the above calculation from which we can clearly see a three-to-one dimension-reduction process.

Since $\mathrm{Cr}\{\cdot\}$ owns the property of self-duality, it is easy to get that the mean chance measure $\mathrm{Ch}\{\cdot\}$ is also self-dual, $\mathrm{Ch}\{\xi \in B\} = 1 - \mathrm{Ch}\{\xi \in B^c\}$ for any Borel subset B of $\Re$. What's more, making use of the mean chance measure, the expected value (2.11) is equivalent to the following form (see [91]):

$$E[\xi] = \int_0^{\infty} \mathrm{Ch}\{\xi \geq r\}\,\mathrm{d}r - \int_{-\infty}^{0} \mathrm{Ch}\{\xi \leq r\}\,\mathrm{d}r. \tag{2.13}$$

As the last key concept in the preliminaries, we give the following convergence mode in chance for a sequence of fuzzy random variables which will be used frequently in the subsequent contents throughout the theoretical part of the book.

Definition 2.7 ([98]). A sequence $\{\xi_n\}$ of fuzzy random variables is said to converge in chance Ch to a fuzzy random variable ξ, denoted $\xi_n \xrightarrow{\text{Ch}} \xi$, if for every $\varepsilon > 0$,

$$\lim_{n\to\infty} \text{Ch}\{|\xi_n - \xi| \geq \varepsilon\} = 0.$$

2.2 Continuity Theorems of Distribution Functions

This section is intended to introduce some continuity theorems for distribution functions of functions of fuzzy random variables, i.e., $f(\xi)$, where ξ is a fuzzy random variable on probability space $(\Omega, \Sigma, \text{Pr})$ and f is a continuous and monotone real function. Several sufficient conditions for the right continuity, left continuity, and continuity of distribution functions $G_{f(\xi)_L}$ and $G_{f(\xi)_U}$ will be derived, respectively.

Given $\omega \in \Omega$, we know that realization $\xi(\omega)$ is a fuzzy variable, and $\mu_{\xi(\omega)}$ is its membership function. Before the introduction of the major results, we denote here two conditions for $\mu_{\xi(\omega)}, \omega \in \Omega$ as below for the sake of a more clear elaboration.

C1. $\limsup_{r\to x+} \mu_{\xi(\omega)}(r) \leq \sup_{r\leq x} \mu_{\xi(\omega)}(r)$ for all $x \in \Re$.

C2. $\limsup_{r\to x-} \mu_{\xi(\omega)}(r) \leq \sup_{r\geq x} \mu_{\xi(\omega)}(r)$ for all $x \in \Re$.

In the conditions C1 and C2 mentioned above,

$$\limsup_{r\to x+} f(r) \quad \text{and} \quad \limsup_{r\to x-} f(r)$$

represent the right and left limits superior of the real-valued function $f(r)$ at $x \in \Re$, respectively, and a brief explanation on those concepts in real analysis is provided in Appendix A.

2.2.1 *Semicontinuity Conditions*

To begin with, let us focus on the semicontinuity conditions for distribution function $G_{f(\xi)_L}(x)$.

Theorem 2.2 ([147]). *Let ξ be a fuzzy random variable, and f a nondecreasing and continuous function. If for almost every $\omega \in \Omega$, fuzzy variable $\xi(\omega)$ satisfies condition C1, then distribution function $G_{f(\xi)_L}(x)$ is a right continuous function.*

Proof. To prove the right continuity of $G_{f(\xi)_L}(x) = \text{Ch}\{f(\xi) \leq x\}$, it suffices to show

$$\lim_{n\to\infty} \text{Ch}\{f(\xi) \leq x + \varepsilon_n\} = \text{Ch}\{f(\xi) \leq x\}$$

for any sequence $\{\varepsilon_n\}$ with $\varepsilon_n \downarrow 0$.

We first prove that under the condition C1, credibility function $\mathrm{Cr}\{\xi(\omega)\leq x\}$ is right continuous almost surely. In fact, on one hand, since for almost every $\omega\in\Omega$,

$$\begin{aligned}\limsup_{r\to x+}\mu_{\xi(\omega)}(r) &= \inf_{\delta>0}\sup_{0<r-x<\delta}\mu_{\xi(\omega)}(r)\\ &\leq \sup_{r\leq x}\mu_{\xi(\omega)}(r)\\ &= \mathrm{Pos}\{\xi(\omega)\leq x\},\end{aligned}$$

one has for any $\varepsilon>0$, there is an $\eta>0$ such that

$$\begin{aligned}\sup_{0<r-x<\eta}\mu_{\xi(\omega)}(r)-\varepsilon &< \inf_{\delta>0}\sup_{0<r-x<\delta}\mu_{\xi(\omega)}(r)\\ &\leq \mathrm{Pos}\{\xi(\omega)\leq x\}.\end{aligned}$$

Thus, for all r with $0<r-x<\eta$, we have

$$\begin{aligned}\mathrm{Pos}\{x<\xi(\omega)\leq r\}-\varepsilon &\leq \sup_{0<r-x<\eta}\mu_{\xi(\omega)}(r)-\varepsilon\\ &< \mathrm{Pos}\{\xi(\omega)\leq x\},\end{aligned}$$

which implies

$$\begin{aligned}\mathrm{Pos}\{\xi(\omega)\leq x\} &\leq \mathrm{Pos}\{\xi(\omega)\leq r\}\\ &< \mathrm{Pos}\{\xi(\omega)\leq x\}+\varepsilon.\end{aligned}$$

It follows that $\mathrm{Pos}\{\xi(\omega)\leq x\}$ is a right continuous function.

On the other hand, noting that

$$\mathrm{Pos}\{\xi(\omega)>x\}=\sup_{t>x}\mu_{\xi(\omega)}(t)$$

is also a right continuous function with respect to (w.r.t.) x, as a consequence,

$$\mathrm{Cr}\{\xi(\omega)\leq x\}=\frac{1}{2}\Big(1+\mathrm{Pos}\{\xi(\omega)\leq x\}-\mathrm{Pos}\{\xi(\omega)>x\}\Big)$$

is a right continuous function.

Next, noting that f is a nondecreasing and continuous function, the inverse function f^{-1} is also nondecreasing and continuous. Since $\mathrm{Cr}\{\xi(\omega)\leq x\}$ is right continuous almost surely, we know the composite function $\mathrm{Cr}\{\xi(\omega)\leq f^{-1}(x)\}$ is also right continuous almost surely. Therefore, for almost every $\omega\in\Omega$,

$$\begin{aligned}&\lim_{n\to\infty}\mathrm{Cr}\{f(\xi(\omega))\leq x+\varepsilon_n\}\\ &=\lim_{n\to\infty}\mathrm{Cr}\{\xi(\omega)\leq f^{-1}(x+\varepsilon_n)\}\\ &=\mathrm{Cr}\{\xi(\omega)\leq f^{-1}(x)\}=\mathrm{Cr}\{f(\xi(\omega))\leq x\}.\end{aligned}$$

Furthermore, combining with $\mathrm{Cr}\{f(\xi(\omega)) \leq x+\varepsilon_n\} \leq 1$ for $n=1,2,\cdots$, it follows from the dominated convergence theorem (Theorem 2.1) that

$$\begin{aligned}
&\lim_{n\to\infty} \mathrm{Ch}\{f(\xi) \leq x+\varepsilon_n\} \\
&\quad = \lim_{n\to\infty} \int_{\Omega} \mathrm{Cr}\{f(\xi(\omega)) \leq x+\varepsilon_n\} \mathrm{Pr}(\mathrm{d}\,\omega) \\
&\quad = \int_{\Omega} \lim_{n\to\infty} \mathrm{Cr}\{f(\xi(\omega)) \leq x+\varepsilon_n\} \mathrm{Pr}(\mathrm{d}\,\omega) \\
&\quad = \int_{\Omega} \mathrm{Cr}\{f(\xi(\omega)) \leq x\} \mathrm{Pr}(\mathrm{d}\,\omega) \\
&\quad = \mathrm{Ch}\{f(\xi) \leq x\}.
\end{aligned}$$

This proves the right continuity of distribution function $G_{f(\xi)_L}(x)$. □

Corollary 2.1 ([147]). *If f is a nondecreasing and continuous function, and ξ is a fuzzy random variable such that for almost every $\omega \in \Omega$, fuzzy variable $\xi(\omega)$ is right continuous or upper semicontinuous, then distribution function $G_{f(\xi)_L}(x)$ is right continuous.*

Proof. For almost every $\omega \in \Omega$, if fuzzy variable $\xi(\omega)$ is right continuous or upper semicontinuous, i.e., membership function $\mu_{\xi(\omega)}$ is right continuous or upper semicontinuous, we can deduce the condition C1 is valid. Therefore, by Theorem 2.2, the corollary is valid. □

Theorem 2.3 ([147]). *Let ξ be a fuzzy random variable, and f a nonincreasing and continuous function. If for almost every $\omega \in \Omega$, fuzzy variable $\xi(\omega)$ satisfies condition C2, then distribution function $G_{f(\xi)_L}(x)$ is a right continuous function.*

Proof. Given condition C2, we first prove credibility function $\mathrm{Cr}\{\xi(\omega) \geq x\}$ is a left continuous almost surely. In fact, since

$$\begin{aligned}
\limsup_{r\to x-} \mu_{\xi(\omega)}(r) &= \inf_{\delta>0} \sup_{0<x-r<\delta} \mu_{\xi(\omega)}(r) \\
&\leq \sup_{r\geq x} \mu_{\xi(\omega)}(r) \\
&= \mathrm{Pos}\{\xi(\omega) \geq x\},
\end{aligned}$$

for any $\varepsilon > 0$, there exists $\eta > 0$ such that

$$\begin{aligned}
\sup_{0<x-r<\eta} \mu_{\xi(\omega)}(r) - \varepsilon &< \inf_{\delta>0} \sup_{0<x-r<\delta} \mu_{\xi(\omega)}(r) \\
&< \mathrm{Pos}\{\xi(\omega) \geq x\}.
\end{aligned}$$

Therefore, for all r with $0 < x - r < \eta$, we have

$$\begin{aligned}\text{Pos}\{r \le \xi(\omega) < x\} - \varepsilon &\le \sup_{0<x-r<\eta} \mu_{\xi(\omega)}(r) - \varepsilon \\ &< \text{Pos}\{\xi(\omega) \ge x\},\end{aligned}$$

which implies

$$\begin{aligned}\text{Pos}\{\xi(\omega) \ge x\} &\le \text{Pos}\{\xi(\omega) \ge r\} \\ &< \text{Pos}\{\xi(\omega) \ge x\} + \varepsilon.\end{aligned}$$

It follows that $\text{Pos}\{\xi(\omega) \ge x\}$ is left continuous w.r.t. x.

Combining that $\text{Pos}\{\xi(\omega) < x\} = \sup_{r<x} \mu_{\xi(\omega)}(r)$ is left continuous with respect to x, we have

$$\text{Cr}\{\xi(\omega) \ge x\} = \frac{1}{2}\Big(1 + \text{Pos}\{\xi(\omega) \ge x\} - \text{Pos}\{\xi(\omega) < x\}\Big)$$

is a left continuous function for almost every $\omega \in \Omega$.

Next, since f is continuous and nonincreasing, we know f^{-1} is also continuous and nonincreasing. By the left continuity of $\text{Cr}\{\xi(\omega) \ge x\}$, we get the composite function $\text{Cr}\{\xi(\omega) \ge f^{-1}(x)\}$ is a right continuous function w.r.t. x. Thus, for any sequence $\{\varepsilon_n\}$ with $\varepsilon_n \downarrow 0$ and almost every $\omega \in \Omega$,

$$\lim_{n\to\infty} \text{Cr}\{f(\xi(\omega)) \le x + \varepsilon_n\} = \text{Cr}\{f(\xi(\omega)) \le x\}.$$

Furthermore, since $\text{Cr}\{f(\xi(\omega)) \le x + \varepsilon_n\} \le 1$ for $n = 1, 2, \cdots$, similar to the proof in Theorem 2.2, by the dominated convergence theorem, we obtain

$$\lim_{n\to\infty} \text{Ch}\{f(\xi) \le x + \varepsilon_n\} = \text{Ch}\{f(\xi) \le x\}.$$

The right continuity of distribution function $G_{f(\xi)_L}(x) = \text{Ch}\{f(\xi) \le x\}$ is proved. □

Example 2.8. Suppose that ω is a discrete random variable taking two values ω_1 and ω_2, ξ is a fuzzy random variable with fuzzy realizations as follows (Fig. 2.7):

$$\mu_{\xi(\omega_1)}(x) = \begin{cases} x, & 0 \le x \le 1 \\ 1 - \dfrac{x}{2}, & 1 < x \le 2 \\ 0, & \text{otherwise,} \end{cases}$$

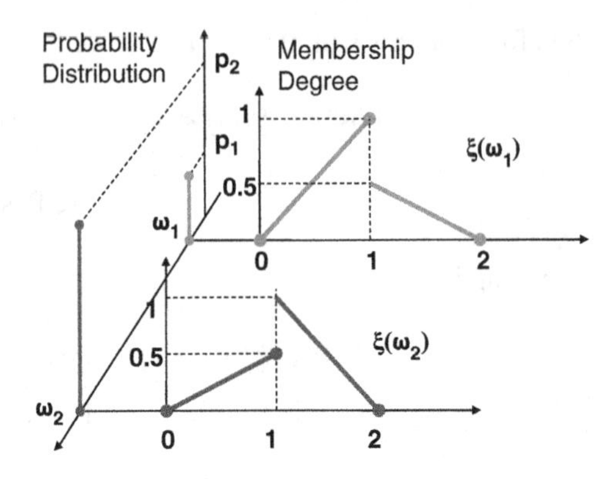

Fig. 2.7 The distribution of fuzzy random variable ξ in Example 2.8

and

$$\mu_{\xi(\omega_2)}(x) = \begin{cases} \frac{x}{2}, & 0 \leq x \leq 1 \\ 2-x, & 1 < x \leq 2 \\ 0, & \text{otherwise,} \end{cases}$$

and $f(x) = (1/2)^x, x \in \Re$. Try to verify the continuity of $G_{f(\xi)_L}(x)$.

First, we know $f(x) = (1/2)^x$ is a decreasing and continuous real function. Furthermore, we note that for $\omega = \omega_1$ and ω_2, membership functions $\mu_{\xi(\omega_1)}$ and $\mu_{\xi(\omega_2)}$ both are continuous except for $x = 1$. Since

$$\limsup_{r \to 1-} \mu_{\xi(\omega_1)}(r) = 1 = \sup_{r \geq 1} \mu_{\xi(\omega_1)}(r),$$

and

$$\limsup_{r \to 1-} \mu_{\xi(\omega_2)}(r) = \frac{1}{2} < 1 = \sup_{r \geq 1} \mu_{\xi(\omega_2)}(r),$$

condition C2 is satisfied. Therefore, by Theorem 2.3, $G_{f(\xi)_L}(x)$ is right continuous.

Corollary 2.2 ([147]). *If f is a nonincreasing and continuous function, and ξ is a fuzzy random variable such that for almost every $\omega \in \Omega$, fuzzy variable $\xi(\omega)$ is left continuous or upper semicontinuous, then distribution function $G_{f(\xi)_L}$ is right continuous.*

Proof. For almost every $\omega \in \Omega$, since membership function $\mu_{\xi(\omega)}$ is right continuous or upper semicontinuous, condition C2 is satisfied. By Theorem 2.3, the corollary is valid. □

Theorem 2.4 ([147]). *Assume that f is a continuous and nondecreasing function, and ξ is a fuzzy random variable such that for almost every $\omega \in \Omega$, fuzzy variable $\xi(\omega)$ is left continuous and lower semicontinuous. Then distribution function $G_{f(\xi)_L}(x)$ is left continuous.*

Proof. For almost every $\omega \in \Omega$, since fuzzy variable $\xi(\omega)$ is left continuous and lower semicontinuous, by Theorem B.1 in Appendix B, we obtain that credibility function $\mathrm{Cr}\{\xi(\omega) \leq x\}$ is left continuous. Noting that f is a continuous and nondecreasing function, we have composite function $\mathrm{Cr}\{\xi(\omega) \leq f^{-1}(x)\}$ is also left continuous. Therefore, for any sequence $\{\varepsilon_n\}$ with $\varepsilon_n \downarrow 0$

$$\begin{aligned}
&\lim_{n\to\infty} \mathrm{Cr}\{f(\xi(\omega)) \leq x-\varepsilon_n\} \\
&\quad = \lim_{n\to\infty} \mathrm{Cr}\{\xi(\omega) \leq f^{-1}(x-\varepsilon_n)\} \\
&\quad = \mathrm{Cr}\{\xi(\omega) \leq f^{-1}(x)\} = \mathrm{Cr}\{f(\xi(\omega)) \leq x\}.
\end{aligned}$$

Noting that

$$\mathrm{Cr}\{f(\xi(\omega)) \leq x-\varepsilon_n\} \leq 1,$$

for $n = 1, 2, \cdots$, by the dominated convergence theorem (see [43]), we have

$$\begin{aligned}
&\lim_{n\to\infty} \mathrm{Ch}\{f(\xi) \leq x-\varepsilon_n\} \\
&\quad = \lim_{n\to\infty} \int_\Omega \mathrm{Cr}\{f(\xi(\omega)) \leq x-\varepsilon_n\} \Pr(\mathrm{d}\,\omega) \\
&\quad = \int_\Omega \lim_{n\to\infty} \mathrm{Cr}\{f(\xi(\omega)) \leq x-\varepsilon_n\} \Pr(\mathrm{d}\,\omega) \\
&\quad = s\int_\Omega \mathrm{Cr}\{f(\xi(\omega)) \leq x\} \Pr(\mathrm{d}\,\omega) = \mathrm{Ch}\{f(\xi) \leq x\}.
\end{aligned}$$

The left continuity of $\mathrm{Ch}\{f(\xi) \leq x\}$ is proved. □

Theorem 2.5 ([147]). *Assume that f is a continuous and nonincreasing function, and ξ is a fuzzy random variable such that for almost every $\omega \in \Omega$, fuzzy variable $\xi(\omega)$ is right continuous and lower semicontinuous. Then distribution function $G_{f(\xi)_L}(x)$ is left continuous.*

Proof. For almost every $\omega \in \Omega$, since fuzzy variable $\xi(\omega)$ is right continuous and lower semicontinuous, by Theorem B.2 in Appendix B, we get credibility function $\mathrm{Cr}\{\xi(\omega) \geq x\}$ is right continuous. Using this result, we can obtain the desired result by the same method as that of Theorem 2.4. □

In the following, we discuss on the semicontinuity conditions for distribution function $G_{f(\xi)_U}(x)$.

Theorem 2.6 ([147]). *Let ξ be a fuzzy random variable, and f a nondecreasing and continuous function. If for almost every $\omega \in \Omega$, fuzzy variable ξ satisfies condition C2, then distribution function $G_{f(\xi)_U}(x)$ is left continuous.*

Proof. From the proof of Theorem 2.3, we know for almost every $\omega \in \Omega$, credibility function $\mathrm{Cr}\{\xi(\omega) \geq x\}$ is a left continuous function under the assumptions of the theorem. Since f is a nondecreasing and continuous function, the inverse function f^{-1} is also nondecreasing and continuous. Therefore, composite function $\mathrm{Cr}\{\xi(\omega) \geq f^{-1}(x)\}$ is a left continuous function with respect to x. Thus,

$$\lim_{n\to\infty} \mathrm{Cr}\{f(\xi(\omega)) \geq x - \varepsilon_n\} = \mathrm{Cr}\{f(\xi(\omega)) \geq x\}$$

for any sequence $\{\xi_n\}$ with $\xi_n \downarrow 0$ and almost every $\omega \in \Omega$.

Furthermore, noting that

$$\mathrm{Cr}\{f(\xi(\omega)) \geq x - \varepsilon_n\} \leq 1$$

for $n = 1, 2, \cdots$, it follows from the dominated convergence theorem that

$$\lim_{n\to\infty} \mathrm{Ch}\{f(\xi) \geq x - \varepsilon_n\} = \mathrm{Ch}\{f(\xi) \geq x\}.$$

This implies the left continuity of $G_{f(\xi)_U}(x)$. The proof of the theorem is completed. □

Corollary 2.3 ([147]). *If f is a nondecreasing and continuous function, and ξ is a fuzzy random variable such that for almost every $\omega \in \Omega$, fuzzy variable $\xi(\omega)$ is left continuous or upper semicontinuous, then distribution function $G_{f(\xi)_U}(x)$ is left continuous.*

Proof. The proof is similar to that of Corollary 2.2. □

Theorem 2.7 ([147]). *Let ξ be a fuzzy random variable, and f a nonincreasing and continuous function. If for almost every $\omega \in \Omega$, fuzzy variable $\xi(\omega)$ satisfies C1, then distribution function $G_{f(\xi)_U}(x)$ is left continuous.*

Proof. Under the assumptions of the theorem, from the proof of Theorem 2.2, we know for almost every $\omega \in \Omega$, credibility function $\mathrm{Cr}\{\xi(\omega) \leq x\}$ is a right continuous function. We note that f is a nonincreasing and continuous function; the inverse function f^{-1} is also nonincreasing and continuous. Therefore, the composite function $\mathrm{Cr}\{\xi(\omega) \leq f^{-1}(x)\}$ is a left continuous function w.r.t. x. Thus, by the same reasoning as that of Theorem 2.6, we can obtain the required result. □

Corollary 2.4 ([147]). *If f is a nonincreasing and continuous function, and ξ is a fuzzy random variable such that for almost every $\omega \in \Omega$, fuzzy variable $\xi(\omega)$ is right continuous or upper semicontinuous, then distribution function $G_{f(\xi)_U}(x)$ is left continuous.*

Proof. The proof is similar to that of Corollary 2.1. □

Theorem 2.8 ([147]). *Assume that f is a nondecreasing and continuous function, and ξ is a fuzzy random variable such that for almost every $\omega \in \Omega$, fuzzy variable $\xi(\omega)$ is right continuous and lower semicontinuous. Then distribution function $G_{f(\xi)_U}(x)$ is right continuous.*

Proof. Under assumptions of the theorem and by the proof of Theorem 2.5, we know for almost every $\omega \in \Omega$, credibility function $\mathrm{Cr}\{\xi(\omega) \geq x\}$ is right continuous. Furthermore, since inverse function f^{-1} is continuous and nondecreasing, the composite function $\mathrm{Cr}\{\xi(\omega) \geq f^{-1}(x)\}$ is a right continuous function w.r.t. x. Therefore, we have

$$\lim_{n\to\infty} \mathrm{Cr}\{f(\xi(\omega)) \geq x+\varepsilon_n\} = \mathrm{Cr}\{f(\xi(\omega)) \geq x\}$$

for any sequence $\{\varepsilon_n\}$ with $\varepsilon_n \downarrow 0$ and almost every $\omega \in \Omega$.

Furthermore, noting that $\mathrm{Cr}\{\xi(\omega) \geq x+\varepsilon_n\} \leq 1$ for $n = 1,2,\cdots$, by the dominated convergence theorem, we obtain

$$\lim_{n\to\infty} \mathrm{Ch}\{f(\xi) \geq x+\varepsilon_n\} = \mathrm{Ch}\{f(\xi) \geq x\},$$

which completes the proof of the theorem. □

Theorem 2.9 ([147]). *Assume that f is a nonincreasing and continuous function, and ξ is a fuzzy random variable such that for almost every $\omega \in \Omega$, fuzzy variable $\xi(\omega)$ is left continuous and lower semicontinuous. Then distribution function $G_{f(\xi)_U}(x)$ is right continuous.*

Proof. By the assumptions and the proof of Theorem 2.4, we know credibility function $\mathrm{Cr}\{\xi(\omega) \leq x\}$ is left continuous for almost every $\omega \in \Omega$. Since inverse function f^{-1} is continuous and nonincreasing, composite function $\mathrm{Cr}\{\xi(\omega) \leq f^{-1}(x)\}$ is a right continuous function w.r.t. x. By the same reasoning as that of Theorem 2.8, we can obtain the required result. □

2.2.2 Continuity Conditions

Based on the obtained semicontinuity theorems, in this subsection, we shall establish a continuity theorem for the distribution functions of the fuzzy random variables. First of all, we give the following useful result.

Theorem 2.10 ([147]). *Let ξ be a fuzzy random variable defined on a probability space $(\Omega, \Sigma, \mathrm{Pr})$, and f a real-valued function.*

(i) *If f is strictly monotone and ω is a discrete random variable, then we have*

$$\mathrm{Ch}\{f(\xi) \leq x\} = \mathrm{Ch}\{f(\xi) < x\} \tag{2.14}$$

and

$$\mathrm{Ch}\{f(\xi)\geq x\}=\mathrm{Ch}\{f(\xi)>x\} \tag{2.15}$$

hold except on an at most countable subset of $\Re$.

(**ii**) *If f is monotone, and fuzzy variable $\xi(\omega)$ is lower semicontinuous for almost every $\omega\in\Omega$, then (2.14) and (2.15) hold for every $x\in\Re$.*

Proof. Here we only prove both assertions (i) and (ii) only for (2.14), since (2.15) can be obtained by the same reasoning.

We now prove assertion (i) of the theorem. Without losing any generality, we suppose f is an increasing function. Since ω is a discrete random variable, it takes on at most countable values, say, $\omega_1,\omega_2,\cdots\omega_n,\cdots$. For each $\omega_i, i=1,2,\cdots$, by Theorem B.3 in Appendix B, we have

$$\mathrm{Cr}\{\xi(\omega_i)\leq x\}=\mathrm{Cr}\{\xi(\omega_i)<x\} \tag{2.16}$$

hold except on an at most countable subset E_i of $\Re$. Noting that the function f is increasing, the inverse function f^{-1} is also increasing. Therefore, for every $f^{-1}(x)\in\Re\setminus E_i$, on the one hand,

$$\mathrm{Cr}\{f(\xi(\omega_i))\leq x\}=\mathrm{Cr}\{\xi(\omega_i)\leq f^{-1}(x)\}.$$

On the other hand,

$$\mathrm{Cr}\{f(\xi(\omega_i))<x\}=\mathrm{Cr}\{\xi(\omega_i)<f^{-1}(x)\}.$$

Here, $x\in f(\Re\setminus E_i)$. Since f is a one-to-one function, we have $f(\Re\setminus E_i)=\Re\setminus f(E_i)$, and $f(E_i)$ is an at most countable subset of $\Re$. By (2.16), we have

$$\mathrm{Cr}\{f(\xi(\omega_i))\leq x\}=\mathrm{Cr}\{f(\xi(\omega_i))<x\}$$

for every $x\in\Re\setminus f(E_i)$.

Furthermore, for every $x\in\Re\setminus\cup_{i=1}^{\infty}f(E_i)$, we have

$$\mathrm{Cr}\{f(\xi(\omega_i))\leq x\}=\mathrm{Cr}\{f(\xi(\omega_i))<x\}$$

holds for each $i=1,2,\cdots$. As a consequence,

$$\begin{aligned}
&\mathrm{Ch}\{f(\xi)\leq x\}\\
&=\int_{\Omega}\mathrm{Cr}\{f(\xi(\omega))\leq x\}\,\mathrm{Pr}(\mathrm{d}\,\omega)\\
&=\int_{\Omega}\mathrm{Cr}\{f(\xi(\omega))<x\}\,\mathrm{Pr}(\mathrm{d}\,\omega)\\
&=\mathrm{Ch}\{f(\xi)<x\}
\end{aligned}$$

holds except on an at most countable subset $\cup_{i=1}^{\infty}f(E_i)$ of $\Re$. The assertion (i) is proved.

Now, we prove assertion (ii). Suppose that f is nondecreasing. Since for almost every $\omega \in \Omega$, fuzzy variable $\xi(\omega)$ is lower semicontinuous, by Theorem B.3 in Appendix B, we can deduce that for almost every $\omega \in \Omega$,

$$\mathrm{Cr}\{\xi(\omega) \le x\} = \mathrm{Cr}\{\xi(\omega) < x\} \tag{2.17}$$

holds for every $x \in \Re$.

For every $x \in \Re$ and almost every $\omega \in \Omega$, on the one hand,

$$\mathrm{Cr}\{f(\xi(\omega)) \le x\} = \mathrm{Cr}\{\xi(\omega) \le f^{-1}(x)\}.$$

On the other hand, we have

$$\mathrm{Cr}\{f(\xi(\omega)) < x\} = \mathrm{Cr}\{\xi(\omega) \le f^{-1}(x)\},$$

or

$$\mathrm{Cr}\{f(\xi(\omega)) < x\} = \mathrm{Cr}\{\xi(\omega) < f^{-1}(x)\}$$

for the case that f is strictly increasing.

Therefore, combining with (2.17), we get

$$\mathrm{Cr}\{f(\xi(\omega)) \le x\} = \mathrm{Cr}\{f(\xi(\omega)) < x\}$$

for any $x \in \Re$ and almost every $\omega \in \Omega$. Furthermore, we have for any $x \in \Re$,

$$\begin{aligned}
&\mathrm{Ch}\{f(\xi) \le x\} \\
&= \int_\Omega \mathrm{Cr}\{f(\xi(\omega)) \le x\} \Pr(\mathrm{d}\,\omega) \\
&= \int_\Omega \mathrm{Cr}\{f(\xi(\omega)) < x\} \Pr(\mathrm{d}\,\omega) \\
&= \mathrm{Ch}\{f(\xi) < x\}
\end{aligned}$$

as required. The theorem is proved. □

Theorem 2.11 ([147]). *Suppose ξ is a fuzzy random variable, f is a monotone and continuous function. If for almost every $\omega \in \Omega$, $\xi(\omega)$ is a lower semicontinuous fuzzy variable, and membership function $\mu_{\xi(\omega)}$ satisfies conditions C1 and C2, then both distribution functions $G_{f(\xi)_L}(x)$ and $G_{f(\xi)_U}(x)$ are continuous functions.*

Proof. Here we only prove the continuity of distribution function $G_{f(\xi)_L}(x)$. The continuity of $G_{f(\xi)_U}(x)$ can be proved in a similar way.

Since fuzzy variable $\xi(\omega)$ is lower semicontinuous for almost every $\omega \in \Omega$, by Theorem 2.10, we know

$$\mathrm{Ch}\{f(\xi) \ge x\} = \mathrm{Ch}\{f(\xi) > x\}. \tag{2.18}$$

Without losing any generality, we assume that f is nonincreasing. From condition C2 and Theorem 2.3, we know distribution function $G_{f(\xi)_L}(x)$ is right continuous.

On the other hand, by condition C1, Theorem 2.7 implies

$$G_{f(\xi)_U}(x) = \mathrm{Ch}\{f(\xi) \geq x\}$$

is left continuous. Hence, combining with (2.18), by the self-duality of Ch, we have

$$G_{f(\xi)_L}(x) = \mathrm{Ch}\{f(\xi) \leq x\} = 1 - \mathrm{Ch}\{f(\xi) \geq x\}$$

is also a left continuous function w.r.t. x.

As a consequence, the distribution function $G_{f(\xi)_L}(x)$ is a continuous function w.r.t. x. The proof is completed. □

2.3 $\top$-Independence Condition for Fuzzy Random Vector

It is known that in classic cases, the operations of fuzzy variables or fuzzy numbers (realizations of fuzzy random variables) base the *extension principle* with the minimum-operator (i.g., minimum triangular norm). Nevertheless, studies in the past two decades already showed that the classic extension principle is not always the optimal way to combine fuzzy variables, the operations associated with different kinds of triangular norms may be required in different specific applications and situations (e.g., [9, 15, 19, 26, 49, 56, 75, 83, 110, 138, 140]). A generalized extension principle makes use of the general t-norm operators to yield different operators for fuzzy variables, in accordance with different types of t-norms, which further yield a generalized form of independence for the fuzzy variables and more general identification criteria for fuzzy random vectors. The generalized extension principle will be spelt out in Sect. 2.4.

2.3.1 $\top$-*Independent Fuzzy Variables*

A triangular norm (t-norm for short) is a function $\top : [0,1]^2 \to [0,1]$ such that for any $x, y, z \in [0,1]$ the following four axioms are satisfied (see [66]):

(T1) Commutativity: $\top(x,y) = \top(y,x)$.
(T2) Associativity: $\top(x,\top(y,z)) = \top(\top(x,y),z)$.
(T3) Monotonicity: $\top(x,y) \leq \top(x,z)$ whenever $y \leq z$.
(T4) Boundary condition: $\top(x,1) = x$.

The associativity (T2) allows us to extend each t-norm $\top$ in a unique way to an n-ary operation in the usual way by induction, defining for each n-tuple $(x_1, x_2, \cdots, x_n) \in [0,1]^n$

$$\top_{k=1}^{n} x_k = \top(\top_{k=1}^{n-1} x_k, x_n) = \top(x_1, x_2, \cdots, x_n).$$

Furthermore, we say function $\perp$ is the t-conorm of $\top$, if $\perp(x,y) = 1 - \top(1-x, 1-y)$ for any $x, y \in [0,1]$.

Definition 2.8 ([25]). Let $\top$ be a t-norm. A family of fuzzy variables $\{X_i, i \in I\}$ is called $\top$-independent iff for any subset $\{i_1, i_2, \cdots, i_n\} \subset I$, where $n \geq 2$,

$$\mathrm{Pos}\{X_{i_k} \in B_k, k = 1,2,\cdots,n\} = \top_{k=1}^{n} \mathrm{Pos}\{X_{i_k} \in B_k\} \tag{2.19}$$

for any sets $B_1, B_2, \cdots, B_n$ of $\Re$.

Particularly, fuzzy variables $X_1, \cdots, X_n$ are $\top$-independent if

$$\mathrm{Pos}\{X_k \in B_k, k = 1,2,\cdots,n\} = \top_{k=1}^{n} \mathrm{Pos}\{X_k \in B_k\} \tag{2.20}$$

for any sets $B_1, B_2, \cdots, B_n$ of $\Re$.

The following proposition gives a basic property of the $\top$-independent fuzzy variables.

Proposition 2.1 ([157]). *Let $\top$ be a t-norm, $X_1, X_2, \cdots, X_n$ fuzzy variables. If they are $\top$-independent, then for any $m \geq 2$ and $1 \leq i_1 < \cdots < i_m \leq n$, the fuzzy variables $X_{i_k}, k = 1,2,\cdots,m$ are also $\top$-independent.*

Proof. Let $m \geq 2$ and $1 \leq i_1 < \cdots < i_m \leq n$. Noting that $X_1, X_2, \cdots, X_n$ are $\top$-independent fuzzy variables, for any $B_{i_k} \subset \Re, k = 1,2,\cdots,m$, we have

$$\begin{aligned}
&\mathrm{Pos}\left\{X_{i_1} \in B_{i_1}, \cdots, X_{i_n} \in B_{i_m}\right\} \\
&= \mathrm{Pos}\left\{X_{i_1} \in B_{i_1}, \cdots, X_{i_m} \in B_{i_m}; X_{j_1} \in \Re, \cdots, X_{j_{n-m}} \in \Re\right\} \\
&= \top\left(\mathrm{Pos}\{X_{i_1} \in B_{i_1}\}, \cdots, \mathrm{Pos}\{X_{i_m} \in B_{i_m}\}; \mathrm{Pos}\{X_{j_1} \in \Re\}, \cdots, \mathrm{Pos}\{X_{j_{n-m}} \in \Re\}\right) \\
&= \top\left(\mathrm{Pos}\{X_{i_1} \in B_{i_1}\}, \cdots, \mathrm{Pos}\{X_{i_m} \in B_{i_m}\}; 1, \cdots, 1\right) \\
&= \top_{k=1}^{m} \mathrm{Pos}\{X_{i_k} \in B_{i_k}\},
\end{aligned}$$

where $\{j_1, \cdots, j_{n-m}\} = \{1, \cdots, n\} \setminus \{i_1, \cdots, i_m\}$. As a consequence, fuzzy variables $X_{i_k}, k = 1,2,\cdots,m$ are $\top$-independent. □

Theorem 2.12 ([157]). *Let $\top$ be a t-norm, and $\perp$ the t-conorm of $\top$. Then fuzzy variables $X_1, X_2, \cdots, X_n$ are $\top$-independent if and only if*

$$\mathrm{Nec}\left(\bigcup_{k=1}^{n} \{X_k \in B_k\}\right) = \perp_{k=1}^{n} \mathrm{Nec}\{X_k \in B_k\} \tag{2.21}$$

for any subsets $B_1, B_2, \cdots, B_n$ of $\Re$.

Proof. Recall the properties of t-conorm $\perp$; for any subsets $B_1, B_2, \cdots, B_n$ of $\Re$, one has

$$\mathrm{Nec}\left(\bigcup_{k=1}^{n}\{X_k \in B_k\}\right) = \perp_{k=1}^{n}\mathrm{Nec}\{X_k \in B_k\}$$

$$\Longleftrightarrow 1-\mathrm{Nec}\left(\bigcup_{k=1}^{n}\{X_k \in B_k\}\right) = 1-\perp_{k=1}^{n}\mathrm{Nec}\{X_k \in B_k\}$$

$$\Longleftrightarrow \mathrm{Pos}\left(\bigcap_{k=1}^{n}\{X_k \in B_k^c\}\right) = \top_{k=1}^{n}\Big(1-\mathrm{Nec}\{X_k \in B_k\}\Big)$$

$$\Longleftrightarrow \mathrm{Pos}\left(\bigcap_{k=1}^{n}\{X_k \in B_k^c\}\right) = \top_{k=1}^{n}\mathrm{Pos}\{X_k \in B_k^c\}.$$

From (2.20) and the arbitrary of $B_k, k = 1, 2, \cdots, n$, we can deduce that (2.21) is equivalent to the $\top$-independence of fuzzy variables. □

Remark 2.2. If the t-norm $\top$ in Theorem 2.12 is taken as the minimum t-norm, whose t-conorm is the maximum t-norm, then (2.20) degenerates to

$$\mathrm{Nec}\left(\bigcup_{k=1}^{n}\{X_k \in B_k\}\right) = \bigvee_{k=1}^{n}\mathrm{Nec}\{X_k \in B_k\},$$

for any subsets $B_1, B_2, \cdots, B_n$ of $\Re$. That is the classical independence of fuzzy variables [95].

Theorem 2.13 ([157]). *Let $\top$ be a t-norm. Then fuzzy variables $X_1, X_2, \cdots, X_n$ are $\top$-independent if and only if*

$$2\mathrm{Cr}\left(\bigcap_{k=1}^{n}\{X_k \in B_k\}\right) \wedge 1 = \top_{k=1}^{n}\Big(2\mathrm{Cr}\{X_k \in B_k\} \wedge 1\Big) \tag{2.22}$$

for any subsets $B_1, B_2, \cdots, B_n$ of $\Re$.

Proof. We note that

$$\mathrm{Pos}\{\cdot\} = 2\mathrm{Cr}\{\cdot\} \wedge 1;$$

then replacing each $\mathrm{Pos}\{\cdot\}$ in (2.20) with $2\mathrm{Cr}\{\cdot\} \wedge 1$ proves the theorem. □

Remark 2.3. If the t-norm $\top$ in Theorem 2.13 is the minimum t-norm, then (2.22) degenerates to

$$\mathrm{Cr}\left(\bigcap_{k=1}^{n}\{X_k \in B_k\}\right) = \bigwedge_{k=1}^{n}\mathrm{Cr}\{X_k \in B_k\}$$

for any subsets $B_1, B_2, \cdots, B_n$ of $\Re$. That is just the classical independence of fuzzy variables [87, 95].

Theorem 2.14 ([157]). *Let $\top$ be a t-norm, and $\perp$ the t-conorm of $\top$. Then fuzzy variables $X_1, X_2, \cdots, X_n$ are $\top$-independent if and only if*

$$\left[2\mathrm{Cr}\left(\bigcup_{k=1}^{n}\{X_k \in B_k\}\right) - 1\right] \vee 0 = \perp_{k=1}^{n}\left[2\left(\mathrm{Cr}\{X_k \in B_k\} - 1\right) \vee 0\right] \tag{2.23}$$

for any subsets $B_1, B_2, \cdots, B_n$ of $\Re$.

Proof. By Theorem 2.13, the fuzzy variables $X_1, X_2, \cdots, X_n$ are $\top$-independent if and only if

$$2\mathrm{Cr}\{X_1 \in B_1^c, X_2 \in B_2^c, \cdots, X_n \in B_n^c\} \wedge 1 = \top_{k=1}^{n}\left(2\mathrm{Cr}\{X_k \in B_k^c\} \wedge 1\right)$$

for any subsets $B_1, B_2, \cdots, B_n$ of $\Re$.

By the self-duality of credibility measure, we have

$$1 - \mathrm{Cr}\left(\bigcup_{k=1}^{n}\{X_k \in B_k\}\right) = \mathrm{Cr}\{X_1 \in B_1^c, X_2 \in B_2^c, \cdots, X_n \in B_n^c\}.$$

Therefore,

$$2\left[1 - \mathrm{Cr}\left(\bigcup_{k=1}^{n}\{X_k \in B_k\}\right)\right] \wedge 1 = \top_{k=1}^{n}\left[2\left(1 - \mathrm{Cr}\{X_k \in B_k^c\}\right) \wedge 1\right].$$

Furthermore, it follows from the properties of t-conorm $\perp$ that

$$2\left[1 - \mathrm{Cr}\left(\bigcup_{k=1}^{n}\{X_k \in B_k\}\right)\right] \wedge 1 = \top_{k=1}^{n}\left[2\left(1 - \mathrm{Cr}\{X_k \in B_k^c\}\right) \wedge 1\right]$$

$$\Longleftrightarrow 2\left[1 - \mathrm{Cr}\left(\bigcup_{k=1}^{n}\{X_k \in B_k\}\right)\right] \wedge 1$$

$$= 1 - \perp_{k=1}^{n}\left[1 - 2\left(1 - \mathrm{Cr}\{X_k \in B_k^c\}\right) \wedge 1\right]$$

$$\Longleftrightarrow 1 - 2\left[1 - \mathrm{Cr}\left(\bigcup_{k=1}^{n}\{X_k \in B_k\}\right)\right] \wedge 1$$

$$= \perp_{k=1}^{n}\left[1 - 2\left(1 - \mathrm{Cr}\{X_k \in B_k^c\}\right) \wedge 1\right]$$

$$\Longleftrightarrow \left[2\mathrm{Cr}\left(\bigcup_{k=1}^{n}\{X_k \in B_k\}\right)-1\right] \vee 0 = \perp_{k=1}^{n}\left[2\left(\mathrm{Cr}\{X_k \in B_k\}-1\right)\vee 0\right].$$

The proof of the theorem is completed. □

Remark 2.4. If the t-norm $\top$ in Theorem 2.14 is taken as the minimum t-norm, then (2.23) degenerates to

$$\mathrm{Cr}\left(\bigcup_{k=1}^{n}\{X_k \in B_k\}\right) = \bigvee_{k=1}^{n}\mathrm{Cr}\{X_k \in B_k\},$$

for any subsets $B_1, B_2, \cdots, B_n$ of $\Re$. That is also the classical independence of fuzzy variables [95].

For the $\top$-independence of functions of fuzzy variables, we have the following results.

Theorem 2.15 ([157]). *Let $\top$ be a t-norm, and g_k for $k = 1,2,\cdots,n$ be real-valued functions on $\Re$. If X_k for $k = 1,2,\cdots,n$ are $\top$-independent fuzzy variables, then $g_k(X_k)$ for $k = 1,2,\cdots,n$ are $\top$-independent fuzzy variables.*

Proof. Let $\zeta_k = g_k(X_k)$ for $k = 1,2,\cdots,n$. Noting that for any subset B_k of $\Re$, $g_k^{-1}(B_k)$ is a subset of $\Re$, by the $\top$-independence of fuzzy variables $X_k, k = 1,2,\cdots,n$, we have

$$\begin{aligned}
&\mathrm{Pos}\left\{\zeta_1 \in B_1, \zeta_2 \in B_2, \cdots, \zeta_n \in B_n\right\}\\
&= \mathrm{Pos}\left\{g_1(X_1) \in B_1, g_2(X_2) \in B_2, \cdots, g_n(X_n) \in B_n\right\}\\
&= \mathrm{Pos}\left\{X_1 \in g_1^{-1}(B_1), X_2 \in g_2^{-1}(B_2), \cdots, X_n \in g_n^{-1}(B_n)\right\}\\
&= \top_{k=1}^{n}\mathrm{Pos}\left\{X_k \in g_k^{-1}(B_k)\right\} = \top_{k=1}^{n}\mathrm{Pos}\left\{g_k(X_k) \in B_k\right\}\\
&= \top_{k=1}^{n}\mathrm{Pos}\{\zeta_k \in B_k\}.
\end{aligned}$$

It follows from Definition 2.8 that $\zeta_k, k = 1,2,\cdots,n$ are $\top$-independent fuzzy variables. The proof of the theorem is completed. □

Theorem 2.16 ([157]). *Let $\top$ be a t-norm, and X_k for $k = 1,2,\cdots,n$ be $\top$-independent fuzzy variables. For any partition of $\{X_k, k = 1,2,\cdots,n\}$, i.e., $\{X_{i_1}, X_{i_2}, \cdots, X_{i_{m_i}}\}$ for $i = 1,2,\cdots,p$, where $p \geq 2$ and $\sum_{i=1}^{p}\sum_{j=1}^{m_i} i_j = n$, if g_i for $i = 1,2,\cdots,p$ are functions from $\Re^{m_i}$ to $\Re$, then we have $g_i(X_{i_1}, X_{i_2}, \cdots, X_{i_{m_i}})$ for $i = 1,2,\cdots,p$ are $\top$-independent fuzzy variables.*

Proof. Let $\eta_i = g_i(X_{i_1}, X_{i_2}, \cdots, X_{m_i}), i = 1,2,\cdots,p$. Note that for any subset B_i of $\Re$, $g_i^{-1}(B_i) = (A_{i_1}, A_{i_2}, \cdots, A_{i_{m_i}})$ is a subset of $\Re^{m_i}$, where A_{i_j} is a subset of $\Re$ for $j = 1,2,\cdots,m_i$. It follows from the $\top$-independence of $X_k, k = 1,2,\cdots,n$, that

$$\begin{aligned}
&\mathrm{Pos}\Big\{\eta_1 \in B_1, \eta_2 \in B_2, \cdots, \eta_p \in B_p\Big\}\\
&= \mathrm{Pos}\Big\{g_i(X_{i_1}, X_{i_2}, \cdots, X_{i_{m_i}}) \in B_i, i = 1, 2, \cdots, p\Big\}\\
&= \mathrm{Pos}\Big\{\Big(X_{i_1}, X_{i_2}, \cdots, X_{i_{m_i}}\Big) \in g_i^{-1}(B_i), i = 1, 2, \cdots, p\Big\}\\
&= \mathrm{Pos}\left(\bigcap_{i=1}^{p}\bigcap_{j=1}^{m_i}\{X_{i_j} \in A_{i_j}\}\right) = \top_{i=1}^{p}\top_{j=1}^{m_i}\mathrm{Pos}\{X_{i_j} \in A_{i_j}\}\\
&= \top_{i=1}^{p}\mathrm{Pos}\Big\{\Big(X_{i_1}, X_{i_2}, \cdots, X_{i_{m_i}}\Big) \in g_i^{-1}(B_i)\Big\} \quad \text{(Proposition2.1)}\\
&= \top_{i=1}^{p}\mathrm{Pos}\{\eta_i \in B_i\}.
\end{aligned}$$

By Definition 2.8, $\eta_1, \eta_2, \cdots, \eta_p$ are ⊤-independent fuzzy variables. The theorem is proved. □

Example 2.9. Suppose $X_1, X_2, \cdots, X_8$ are ⊤-independent fuzzy variables. Then, by Theorem 2.16, we know the following new fuzzy variables

$$Y_1 = (X_1 - X_2)\cdot X_3, \quad Y_2 = \sqrt{X_4^6 + X_5^4 + X_6^2}, \quad \text{and} \quad Y_3 = X_7^3/X_8$$

are also ⊤-independent.

In various of applications of ⊤-independence, the t-norm ⊤ should be carefully selected. As a matter of fact, it is not true that every t-norm can be used as an independence operator. For example, if events A, B are ⊤-independent with a t-norm ⊤ such that $\mathrm{Pos}\{A, B\} = \top(\mathrm{Pos}\{A\}, \mathrm{Pos}\{B\}) = 0$, where A, B are assigned with positive possibilities, then A, B are inconsistent events which contradicts the independence between A and B. In such a case, the t-norm ⊤ is independence-inconsistent, and is absolutely unsuitable to be applied to the ⊤-independence-based problems. The independence-consistency of a t-norm is critical in practical applications. It should be built as a criterion in the selection of t-norm, such as, a t-norm ⊤ is said to be *independence-consistent*, if $\top(a, b) > 0$, for any $a, b > 0$. Such condition ensures the rationality of the ⊤-independence induced by the selected t-norm ⊤. There is still much room for this open issue.

2.3.2 ⊤-Independence Condition for Fuzzy Random Vector

In this section, we discuss the ⊤-independence condition for fuzzy random vector, where ⊤ can be any continuous t-norm. Let us begin with the measurability criteria for fuzzy random vectors that were established in [35] in which it has been concluded that under min-independence condition, i.e., $\xi_k(\omega), k = 1, 2, \cdots, n$ are min-independent fuzzy variables for any $\omega \in \Omega$, if $\xi_k, k = 1, 2, \cdots, n$ are upper semicontinuous fuzzy random variables, then $\xi = (\xi_1, \xi_2, \cdots, \xi_n)$ is a fuzzy random vector. The major task of this section is to extend such result to a more general case of $\top$-independence situation.

We say a fuzzy vector X is upper semicontinuous (abbreviated by usc) if its membership function $\mu_X(x)$ is usc at every $t \in \Re^n$. Furthermore, a fuzzy random vector ξ is said to be usc if for each $\omega \in \Omega$, fuzzy vector $\xi(\omega)$ is usc. Here, the measurability criteria for fuzzy random vectors are listed as a Proposition without proof for which any interested reader may refer to [35].

Proposition 2.2 ([35]). *Let* $(\Omega, \Sigma, \Pr)$ *be a complete probability space, and* ξ *a map from* Ω *to* usc-$\mathscr{F}_v^n$. *Then the following six statements are equivalent:*

(i) ξ *is a fuzzy random vector[0.2cm],*
(ii) *For every open subset* $G \subset \Re^n$, $\text{Pos}\{\xi(\omega) \in G\}$ *is* Σ*-measurable.*
(iii) *For every open ball* $B(t;r)$ $(t \in \Re^n, r > 0)$, $\text{Pos}\{\xi(\omega) \in B(t;r)\}$ *is* Σ*-measurable.*
(iv) *For every compact set* $K \subset \Re^n$, $\text{Pos}\{\xi(\omega) \in K\}$ *is* Σ*-measurable.*
(v) *For each* $\alpha \in (0,1]$, ξ^α *is a random set from* Ω *to* $\Re^n$.
(vi) *For every Borel subset* $B \subset \Re^n$, $\text{Pos}\{\gamma \mid \xi(\omega, \gamma) \in B\}$ *is* Σ*-measurable.*

Before the discussion on $\top$-independence condition, we first derive the following two additional measurability criteria for fuzzy random vector.

Proposition 2.3 ([146]). *Let* $(\Omega, \Sigma, \Pr)$ *be a complete probability space, and* ξ *a map from* Ω *to* usc-$\mathscr{F}_v^n$. *Then* ξ *is a fuzzy random vector if and only if for every open-closed interval* $I \subset \Re^n$, $\text{Pos}\{\xi(\omega) \in I\}$ *is* Σ*- measurable.*

Proof. From assertion (vi) in Proposition 2.2, the *Necessity* is obviously valid since every interval $I = (c,d] \subset \Re^n$ is a Borel subset of $\Re^n$.

Sufficiency: Since ξ is a map from Ω to usc-$\mathscr{F}_v^n$, $\xi(\omega) = (\xi_1(\omega), \cdots, \xi_n(\omega))$ is an n-ary fuzzy vector, for any $\omega \in \Omega$. Note that every open subset $G \subset \Re^n$ can be expressed as the union of at most countable many disjoint open-closed intervals $\{I_k\}$, $G = \cup_{k=1}^{\infty} I_k$, where

$$I_k = \prod_{j=1}^{n} \left(c_j^k, d_j^k\right], \quad \left(c_j^k, d_j^k\right] \subset \Re.$$

Therefore,

$$\text{Pos}\{\xi(\omega) \in G\} = \text{Pos}\left\{\xi(\omega) \in \bigcup_{n=1}^{\infty} I_n\right\} = \sup_{n \geq 1} \text{Pos}\{\xi(\omega) \in I_n\}.$$

Since $\text{Pos}\{\xi(\omega) \in I\}$ is Σ- measurable, we have $\text{Pos}\{\xi(\omega) \in G\}$ is Σ- measurable. Furthermore, by assertion (ii) in Proposition 2.2, ξ is a fuzzy random vector. The proof of the theorem is completed. □

Proposition 2.4 ([146]). *Let* $(\Omega, \Sigma, \Pr)$ *be a complete probability space, and* ξ *a map from* Ω *to* usc-$\mathscr{F}_v^n$. *Then* ξ *is a fuzzy random vector if and only if for every open-closed interval* $I \subset \Re^n$, $\mathrm{Cr}\{\xi(\omega) \in I\}$ *is* Σ*- measurable.*

Proof. Necessity: Suppose that ξ is a fuzzy random vector. For any open-closed interval $I \subset \Re^n$, $\mathrm{Cr}\{\xi(\omega) \in I\}$ can be expressed by

$$\mathrm{Cr}\{\xi(\omega) \in I\} = \frac{1}{2}\Big(1 + \mathrm{Pos}\{\xi(\omega) \in I\} - \mathrm{Pos}\{\xi(\omega) \in I^c\}\Big).$$

Noting that I and I^c both are Borel subset of $\Re^n$, by assertion (vi) in Proposition 2.2, $\mathrm{Cr}\{\xi(\omega) \in I\}$ is a Σ- measurable function.

Sufficiency: We note that for any open-closed interval $I \subset \Re^n$, $\mathrm{Pos}\{\xi(\omega) \in I\}$ can be written as

$$\mathrm{Pos}\{\xi(\omega) \in I\} - 2\mathrm{Cr}\{\xi(\omega) \in I\} \wedge 1.$$

Therefore, the Σ-measurability of $\mathrm{Cr}\{\xi(\omega) \in I\}$ implies that $\mathrm{Pos}\{\xi(\omega) \in I\}$ is also Σ- measurable. Furthermore, by Proposition 2.3, ξ is a fuzzy random vector. □

Example 2.10. Assume that Ω is a complete probability space, and C and W are random variables on Ω. Try to testify ξ is a fuzzy random variable, where

$$\mu_{\xi(\omega)}(x) = EXP\left(-\left(\frac{x - C(\omega)}{W(\omega)}\right)^2\right), \quad x \in \Re.$$

We use Proposition 2.3 to verify ξ is a fuzzy random variable. For any open-closed interval $(a,b] \subset \Re$, to testify $\mathrm{Pos}\{\xi(\omega) \in (a,b]\}$ is Σ-measurable, it suffices to show the equation

$$\Big\{\omega \in \Omega \mid \mathrm{Pos}\{a < \xi(\omega) \leq b\} \geq t\Big\} \in \Sigma$$

holds for any $t \in (0,1]$. Since

$$\begin{aligned}
&\Big\{\omega \in \Omega \mid \mathrm{Pos}\{a < \xi(\omega) \leq b\} \geq t\Big\} \\
&= \Big\{\omega \in \Omega \mid \Big[C(\omega) - W(\omega)\sqrt{-\ln t}, C(\omega) + W(\omega)\sqrt{-\ln t}\Big] \cap (a,b] \neq \emptyset\Big\} \\
&= \Big\{\omega \in \Omega \mid C(\omega) - W(\omega)\sqrt{-\ln t} \leq b\Big\} \\
&\quad \bigcap \Big\{\omega \in \Omega \mid C(\omega) - W(\omega)\sqrt{-\ln t} > a\Big\},
\end{aligned}$$

noting that

$$\Big\{\omega \in \Omega \mid C(\omega) - W(\omega)\sqrt{-\ln t} \leq b\} \in \Sigma,$$

and

$$\left\{\omega \in \Omega \mid C(\omega) - W(\omega)\sqrt{-\ln t} > a\right\} \in \Sigma,$$

we have $\text{Pos}\{\xi(\omega) \in (a,b]\}$ is Σ-measurable. Furthermore, since $\mu_{\xi(\omega)}(x)$ is continuous, by Proposition 2.3, ξ is a fuzzy random variable.

Theorem 2.17 ([146]). *Let $(\Omega, \Sigma, \text{Pr})$ be a complete probability space, and $\xi = (\xi_1, \xi_2, \cdots, \xi_n)$ a map from Ω to usc-$\mathscr{F}_v^n$. If $\xi_k, k = 1, 2, \cdots, n$ are fuzzy random variables defined on $(\Omega, \Sigma, \text{Pr})$, and $\xi_k(\omega), k = 1, 2, \cdots, n$ are $\top$-independent fuzzy variables for any $\omega \in \Omega$, where $\top$ is a continuous t-norm, then ξ is a fuzzy random vector.*

Proof. From Proposition 2.3, it suffices to prove $\text{Pos}\{\xi(\omega) \in I\}$ is Σ- measurable for any open-closed interval $I \subset \Re^n$. Denoting

$$I = \prod_{k=1}^{n} J_k = \prod_{k=1}^{n} (c_k, d_k],$$

we have

$$\begin{aligned}
\text{Pos}\{\xi(\omega) \in I\} &= \text{Pos}\left\{(\xi_1(\omega), \xi_2(\omega), \cdots, \xi_n(\omega)) \in \prod_{k=1}^{n} J_k\right\} \\
&= \text{Pos}\left\{\bigcap_{k=1}^{n} \{\xi_k(\omega) \in J_k\}\right\}.
\end{aligned}$$

Note that $\xi_k(\omega), k = 1, 2, \cdots, n$ are $\top$-independent fuzzy variables, we have

$$\begin{aligned}
\text{Pos}\{\xi(\omega) \in I\} &= \text{Pos}\left\{\bigcap_{k=1}^{n} \{\xi_k(\omega) \in J_k\}\right\} \\
&= \top_{k=1}^{n} \text{Pos}\{\xi_k(\omega) \in J_k\}.
\end{aligned}$$

By Proposition 2.3, we know $\text{Pos}\{\xi_k(\omega) \in J_k\}$ is Σ-measurable; this fact together with that every continuous t-norm $\top$ is a Borel measurable function deduces that $\text{Pos}\{\xi(\omega) \in I\}$ is a Σ-measurable function. This completes the proof of the theorem. □

2.4 Some Limit Theorems

In this section, the limit results of the sum of fuzzy random variables are based on an important class of t-norms: Archimedean t-norms. A t-norm $\top$ is said to be Archimedean if $\top(x, x) < x$ for all $x \in (0, 1)$. It is easy to check that the minimum t-norm is not Archimedean.

Moreover, every continuous Archimedean t-norm $\top$ can be represented by a continuous and strictly decreasing function (see [129])

$$f : [0,1] \to [0,\infty] \quad \text{with} \quad f(1) = 0$$

and

$$\top(x_1,\ldots,x_n) = f^{[-1]}(f(x_1)+\cdots+f(x_n)) \tag{2.24}$$

for all $x_i \in (0,1), 1 \le i \le n$, where $f^{[-1]}$ is the pseudo-inverse of f, defined by

$$f^{[-1]}(y) = \begin{cases} f^{-1}(y), & \text{if } y \in [0, f(0)] \\ 0, & \text{if } y \in (f(0),\infty). \end{cases}$$

The function f is called the additive generator of $\top$.

Example 2.11. Some continuous Archimedean t-norms with additive generators are listed as follows:

(1) Yager t-norm ($\lambda \in (0,\infty)$):

$$\top^Y_\lambda(x,y) = \max\left\{1 - \sqrt[\lambda]{(1-x)^\lambda + (1-y)^\lambda}, 0\right\}$$

with additive generator $f^Y_\lambda(x) = (1-x)^\lambda$.

(2) Dombi t-norm ($\lambda \in (0,\infty)$):

$$\top^D_\lambda(x,y) = \frac{1}{1 + \sqrt[\lambda]{(\frac{1-x}{x})^\lambda + (\frac{1-y}{y})^\lambda}}$$

with additive generator $f^D_\lambda(x) = ((1-x)/x)^\lambda$.

(3) Product t-norm: $\top^P(x,y) = xy$ with additive generator $f^P = -log$.

Let g be a function from $\Re^m$ to $\Re$. For fuzzy variables $Y_k, 1 \le k \le m$ with membership functions $\mu_k, 1 \le k \le m$, the membership function of $g(Y_1,Y_2,\cdots,Y_m)$ is determined by $\mu_1,\mu_2,\cdots,\mu_m$ via the following generalized extension principle:

$$\mu_{g(Y_1,Y_2,\cdots,Y_m)}(x) = \sup_{x_1,x_2,\cdots x_m \in \Re} \left\{ \top^m_{k=1}\mu_k(x_k) \mid x = g(x_1,x_2,\cdots,x_m) \right\}, \tag{2.25}$$

where $\top$ can be any general t-norm. By extension principle (2.25), the membership function of arithmetic mean $(Y_1+\cdots+Y_n)/n$ is

$$\mu_{\frac{1}{n}(Y_1+\cdots+Y_n)}(z) = \sup_{x_1+\cdots+x_n = nz} \top\left(\mu_{Y_1}(x_1),\cdots,\mu_{Y_n}(x_n)\right). \tag{2.26}$$

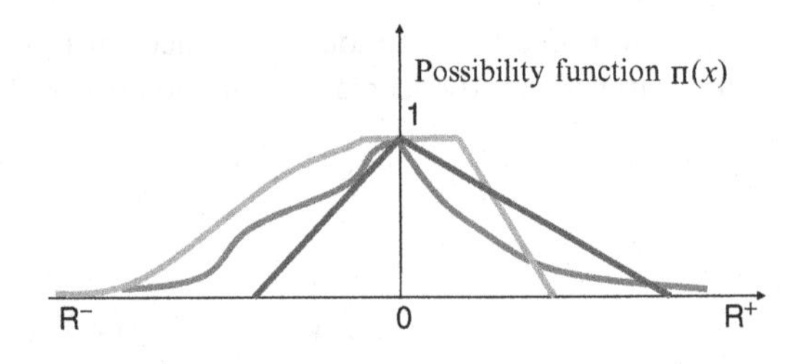

Fig. 2.8 Several types of function Π

Furthermore, for a continuous Archimedean t-norm with additive generator f, from (2.24) and (2.26), the membership function of $(Y_1 + \cdots + Y_n)/n$ can be determined by

$$\mu_{\frac{1}{n}(Y_1+\cdots+Y_n)}(r) = f^{[-1]}\left(\inf_{x_1+\cdots+x_n=nr} \sum_{k=1}^{n} f(\mu_{Y_k}(x_k))\right). \quad (2.27)$$

2.4.1 Limit Theorems of the Sum of Fuzzy Random Variables

The discussion in this section is limited to the following assumptions:

A1. The operations of fuzzy variables are determined by the generalized extension principle (2.25), and $\top$ denotes any continuous Archimedean t-norm with an additive generator f.

A2. Π is a nonnegative real-valued function with $\Pi(0) = 1$, and Π is nonincreasing on $\Re^+$ and nondecreasing on $\Re^-$ (Fig. 2.8).

With condition A1, the generalized extension principle provides any continuous Archimedean t-norm operator for fuzzy variables. As to condition A2, the function Π is called a possibility function, through which we can construct convex membership functions such as triangular and norm distributions of fuzzy variables, where such convexity, as to be showed in the following theorems, is critical to our desired results.

Denote Ξ as the support of the possibility function Π, i.e., the closure of subset $\{t \in \Re \mid \Pi(t) > 0\}$ of $\Re$. For function Π and Archimedean t-norm $\top$ with additive generator f, since $f : [0,1] \to [0,\infty]$ is continuous and strictly decreasing, we know that the composition function $f \circ \Pi : \Re \to [0,\infty]$ is nonincreasing on $\Re^-$ and nondecreasing on $\Re^+$ with $f \circ \Pi(0) = 0$, and $f \circ \Pi(x) = f(0)$ for any $x \notin \Xi$.

Now, we consider the convex hull of the composition function $f \circ \Pi$ on Ξ, denoted $\mathrm{co}(f \circ \Pi)_\Xi$, which is defined as

$$\mathrm{co}(f \circ \Pi)_\Xi(z) = \inf\left\{\sum_{k=1}^{n} \lambda_k (f \circ \Pi)(x_k)\right\} \quad (z \in \Xi), \quad (2.28)$$

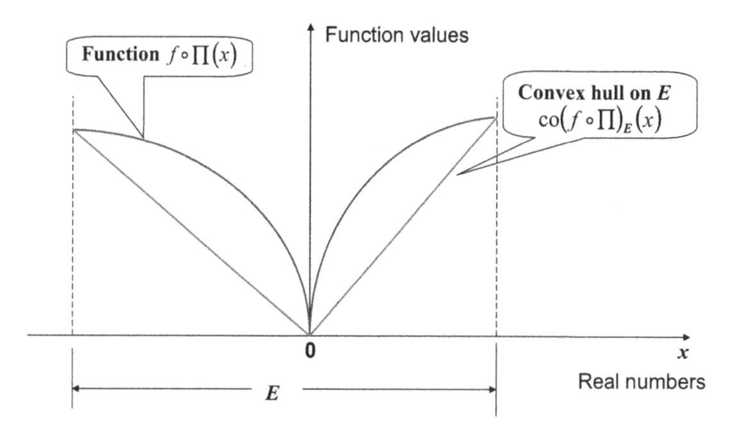

Fig. 2.9 Convex hull $\text{co}(f \circ \Pi)_E$ of composition function $f \circ \Pi$ on E

where the infimum is taken over all representations of z as a (finite) convex combination $\sum_{k=1}^{n} \lambda_k x_k$ of points of Ξ. From the knowledge of convex analysis (refer to Tiel [139]), we know $\text{co}(f \circ \Pi)_\Xi$ is the largest convex function $h(x)$ such that $h(x) \leq f \circ \Pi(x)$, for $x \in \Xi$ (see Fig. 2.9).

Example 2.12. Let Π be the membership function of triangular fuzzy variable $(-1,0,3)$ and $\top_\lambda^Y(x,y)$ is a Yager t-norm with additive generator $f_\lambda^Y(x) = (1-x)^\lambda$, where $\lambda \geq 1$. Find the convex hull of $f_\lambda^Y \circ \Pi$ on $\Xi = [-1,3]$.

Since the membership function Π is

$$\Pi(x) = \begin{cases} x+1, & \text{if } x \in [-1,0] \\ 1-\frac{x}{3}, & \text{if } x \in [0,3] \\ 0, & \text{otherwise,} \end{cases}$$

we have

$$f_\lambda^Y \circ \Pi(x) = \begin{cases} (-x)^\lambda, & \text{if } x \in [-3,0] \\ (\frac{x}{3})^\lambda, & \text{if } x \in [0,6] \\ 1, & \text{otherwise,} \end{cases}$$

which is a convex function on $[-1,3]$. Thus, we have $\text{co}(f_\lambda^Y \circ \Pi)_{[-1,3]} = f_\lambda^Y \circ \Pi$.

Before the discussion of the limit theorems for fuzzy random variables, it is better that we introduce the following basic result of the sum of fuzzy variables.

Lemma 2.1 ([148]). *Let $Y_k, k = 1,2,\cdots$ be a sequence of fuzzy variables with identical membership function Π, and $W_n = Y_1 + \cdots + Y_n$. If $\text{co}(f \circ \Pi)_\Xi(x) > 0$ for any nonzero $x \in \Xi$, then*

$$\lim_{n \to \infty} \mu_{\frac{1}{n}W_n}(z) = \begin{cases} 1, & \text{if } z = 0 \\ 0, & \text{otherwise.} \end{cases}$$

Proof. The proof can be divided into the following three cases.

Case 1. $z = 0$. We have

$$\mu_{\frac{1}{n}W_n}(0) = \sup_{x_1+\cdots+x_n=0} \top\Big(\mu_{Y_1}(x_1), \cdots, \mu_{Y_n}(x_n)\Big)$$
$$\geq \top(\Pi(0), \cdots, \Pi(0)) = 1.$$

Case 2. $z \neq 0, z \notin \Xi$. For any $\{x_1, x_2, \cdots, x_n\}$ with $x_1 + \cdots + x_n = nz$, there must be two points x_i and x_j, $1 \leq i, j \leq n$ such that $x_i \leq z$ and $x_j \geq z$, which implies $\top(\Pi(x_i), \Pi(x_j)) \leq \Pi(z)$. Therefore $\top(\Pi(x_1), \cdots, \Pi(x_n)) \leq \Pi(z)$. It follows that

$$\mu_{\frac{1}{n}W_n}(z) = \sup_{x_1+\cdots+x_n=nz} \top(\Pi(x_1), \cdots, \Pi(x_n)) \leq \Pi(z) = 0.$$

Case 3. $z \neq 0, z \in \Xi$. In this case, we note that if $x_k \notin \Xi$ for some k, then

$$f^{[-1]}\left(\sum_{k=1}^{n} f \circ \Pi(x_k)\right) = 0.$$

Therefore, by (2.24) and (2.26), we have

$$\mu_{\frac{1}{n}W_n}(z) = f^{[-1]}\left(\inf_{x_1+\cdots+x_n=nz} \sum_{k=1}^{n} f(\mu_{Y_k}(x_k))\right)$$
$$= f^{[-1]}\left(\inf_{\substack{x_1+\cdots+x_n=nz \\ x_k\in\Xi, 1\leq k\leq n}} \sum_{k=1}^{n} f \circ \Pi(x_k)\right).$$

By (2.28), we obtain

$$\inf_{\substack{x_1+\cdots+x_n=nz \\ x_k\in\Xi, 1\leq k\leq n}} \frac{1}{n}\sum_{k=1}^{n} f \circ \Pi(x_k) \geq \mathrm{co}(f \circ \Pi)_{\Xi}(z),$$

or, equivalently,

$$\inf_{\substack{x_1+\cdots+x_n=nz \\ x_k\in\Xi, 1\leq k\leq n}} \sum_{k=1}^{n} f \circ \Pi(x_k) \geq n \cdot \mathrm{co}(f \circ \Pi)_{\Xi}(z).$$

Since $f^{[-1]}$ is nonincreasing, we can deduce

$$\mu_{\frac{1}{n}W_n}(z) \leq f^{[-1]}\Big(n \cdot \mathrm{co}(f \circ \Pi)_{\Xi}(z)\Big).$$

Noting that $\mathrm{co}(f \circ \Pi)(r) > 0$ for any nonzero $r \in \Xi$, we have

$$\mu_{\frac{1}{n}W_n}(z) \leq f^{[-1]}\Big(n \cdot \mathrm{co}(f \circ \Pi)_{\Xi}(z)\Big) \to 0 \quad (n \to \infty).$$

Combining the above three cases proves the Lemma. □

In what follows, we discuss the convergence properties of sum of fuzzy random variables $\xi_k, k = 1, 2, \cdots$ which are of "similar" shapes. More precisely, the fuzzy random variables are formed by random lateral moves on possibility function Π, i.e., $\Pi(x - U_k(\omega))$, where $U_k, k = 1, 2, \cdots$, are random variables.

Letting $l = \inf\{x \mid \Pi(x) = 1\}$ and $q = \sup\{x \mid \Pi(x) = 1\}$, we have the following result.

Theorem 2.18 ([153]). *Assume $\{\xi_n\}$ is a sequence of fuzzy random variables with $\mu_{\xi_k(\omega)}(x) = \Pi(x - U_k(\omega))$ for almost every $\omega \in \Omega$, where $U_k, k = 1, 2, \cdots$, are random variables, if $\mathrm{co}(f \circ \Pi)(r) > 0$ for any $r \in \Xi$ with $\Pi(r) < 1$, then for every $\varepsilon > 0$, we have*

$$\lim_{n\to\infty} \mathrm{Ch}\left\{ l - \varepsilon < \frac{1}{n}\sum_{k=1}^{n}(\xi_k - U_k) < q + \varepsilon \right\} = 1. \tag{2.29}$$

Furthermore, if $U_k, k = 1, 2, \cdots$, are independent and identically distributed (i.i.d.) random variables with finite expected value, then for every $\varepsilon > 0$, we have

$$\lim_{n\to\infty} \mathrm{Ch}\left\{ l - \varepsilon < \frac{1}{n}\sum_{k=1}^{n}\xi_k - E[U_1] < q + \varepsilon \right\} = 1. \tag{2.30}$$

Proof. Denoting $\zeta_k = \xi_k - U_k$ and $S_n = \zeta_1 + \cdots + \zeta_n$, we know that $\zeta_k(\omega) = \xi_k(\omega) - U_k(\omega)$ and $S_n(\omega) = \zeta_1(\omega) + \zeta_k(\omega)$ are fuzzy variables for every $\omega \in \Omega$, and $\mu_{\zeta_k(\omega)}(x) = \Pi(x), k = 1, 2, \cdots$ for almost every $\omega \in \Omega$. By the self-duality of Ch, we can deduce for every n that

$$\begin{aligned}
&\mathrm{Ch}\left\{ l - \varepsilon < \frac{1}{n}\sum_{k=1}^{n}(\xi_k - U_k) < q + \varepsilon \right\} \\
&= \mathrm{Ch}\left\{ l - \varepsilon < \frac{S_n}{n} < q + \varepsilon \right\} \\
&= 1 - \mathrm{Ch}\left(\left\{ \frac{S_n}{n} \leq l - \varepsilon \right\} \bigcup \left\{ \frac{S_n}{n} \geq q + \varepsilon \right\} \right),
\end{aligned}$$

to achieve (2.32), it suffices to prove

$$\lim_{n\to\infty} \mathrm{Ch}\left(\left\{ \frac{S_n}{n} \leq l - \varepsilon \right\} \bigcup \left\{ \frac{S_n}{n} \geq q + \varepsilon \right\} \right) = 0. \tag{2.31}$$

We note that $\Pi(r) < 1$ for any $r > q$ and $r < l$, by Lemma 2.1, given $\omega \in \Omega$, we have for any $n \geq 1$ and $\varepsilon > 0$ that

$$\begin{aligned}
&\mathrm{Cr}\left(\left\{\frac{S_n(\omega)}{n} \leq l-\varepsilon\right\} \bigcup \left\{\frac{S_n(\omega)}{n} \geq q+\varepsilon\right\}\right) \\
&\leq \mathrm{Pos}\left\{\frac{S_n(\omega)}{n} \leq l-\varepsilon\right\} \bigvee \mathrm{Pos}\left\{\frac{S_n(\omega)}{n} \geq q+\varepsilon\right\} \\
&= \sup_{r \leq l-\varepsilon} \mu_{\frac{S_n(\omega)}{n}}(r) \bigvee \sup_{r \geq q+\varepsilon} \mu_{\frac{S_n(\omega)}{n}}(r) \\
&= \mu_{\frac{S_n(\omega)}{n}}(l-\varepsilon) \bigvee \mu_{\frac{S_n(\omega)}{n}}(q+\varepsilon) \to 0 \ (n \to \infty).
\end{aligned}$$

Using Lebesgue's dominated convergence theorem, we have

$$\begin{aligned}
&\lim_{n\to\infty} \mathrm{Ch}\left(\left\{\frac{S_n}{n} \leq l-\varepsilon\right\} \bigcup \left\{\frac{S_n}{n} \geq q+\varepsilon\right\}\right) \\
&= \lim_{n\to\infty} \int_{\Omega} \mathrm{Cr}\left(\left\{\frac{S_n(\omega)}{n} \leq l-\varepsilon\right\} \bigcup \left\{\frac{S_n(\omega)}{n} \geq q+\varepsilon\right\}\right) \mathrm{Pr}(\mathrm{d}\omega) \\
&= \int_{\Omega} \lim_{n\to\infty} \mathrm{Cr}\left(\left\{\frac{S_n(\omega)}{n} \leq l-\varepsilon\right\} \bigcup \left\{\frac{S_n(\omega)}{n} \geq q+\varepsilon\right\}\right) \mathrm{Pr}(\mathrm{d}\omega) = 0,
\end{aligned}$$

which proves (2.31).

Furthermore, since $U_k, k = 1, 2, \cdots$ are i.i.d. random variables, by the strong law of large numbers for random variables, we get

$$\mathrm{Pr}\left\{\omega \in \Omega \mid \frac{1}{n}\sum_{k=1}^{n} U_k(w) \to E[U_1]\right\} = 1.$$

Since for almost every $\omega \in \Omega$,

$$\begin{aligned}
&\lim_{n\to\infty} \mathrm{Cr}\left\{l-\varepsilon < \frac{S_n(\omega)}{n} < q+\varepsilon\right\} \\
&= 1 - \lim_{n\to\infty} \mathrm{Cr}\left(\left\{\frac{S_n(\omega)}{n} \leq l-\varepsilon\right\} \bigcup \left\{\frac{S_n(\omega)}{n} \geq q+\varepsilon\right\}\right) = 1,
\end{aligned}$$

we can obtain

$$\begin{aligned}
&\lim_{n\to\infty} \mathrm{Cr}\left\{l-\varepsilon < \frac{1}{n}\sum_{k=1}^{n} \xi_k(\omega) - U_1(\omega) < q+\varepsilon\right\} \\
&= \lim_{n\to\infty} \mathrm{Cr}\left\{l-\varepsilon < \frac{1}{n}S_n(\omega) + \frac{1}{n}\sum_{k=1}^{n} U_k(\omega) - U_1(\omega) < q+\varepsilon\right\} = 1
\end{aligned}$$

for almost every $\omega \in \Omega$.

It follows from the dominated convergence theorem that

$$\lim_{n\to\infty} \mathrm{Ch}\left\{l-\varepsilon<\frac{1}{n}\sum_{k=1}^{n}\xi_k-U_1<q+\varepsilon\right\}$$

$$=\lim_{n\to\infty}\int_{\Omega}\mathrm{Cr}\left\{l-\varepsilon<\frac{1}{n}\sum_{k=1}^{n}\xi_k(\omega)-U_1(\omega)<q+\varepsilon\right\}\mathrm{Pr}(\mathrm{d}\omega)=1.$$

The proof of the theorem is completed. □

Corollary 2.5 ([153]). *Assume $\{\xi_n\}$ is a sequence of fuzzy random variables with $\mu_{\xi_k(\omega)}(x)=\Pi(x-U_k(\omega))$ for almost every $\omega\in\Omega$, where Π is unimodal, and $U_k, k=1,2,\cdots$, are random variables, if $\mathrm{co}(f\circ\Pi)(r)>0$ for any nonzero $r\in\Xi$, then we have*

$$\frac{1}{n}\sum_{k=1}^{n}(\xi_k-U_k)\xrightarrow{\mathrm{Ch}}0, \tag{2.32}$$

as $n\to\infty$. Furthermore, if $U_k, k=1,2,\cdots$, are i.i.d random variables with finite expected value, then we have

$$\frac{1}{n}\sum_{k=1}^{n}\xi_k\xrightarrow{\mathrm{Ch}}E[U_1], \tag{2.33}$$

as $n\to\infty$.

Proof. Since possibility function Π is unimodal, we have $l=q=0$. By Theorem 2.18, one can deduce

$$\lim_{n\to\infty}\mathrm{Ch}\left\{-\varepsilon<\frac{1}{n}\sum_{k=1}^{n}(\xi_k-U_k)<\varepsilon\right\}=1$$

or, equivalently,

$$\lim_{n\to\infty}\mathrm{Ch}\left\{\left|\frac{1}{n}\sum_{k=1}^{n}(\xi_k-U_k)\right|\geq\varepsilon\right\}=0.$$

Similarly, we can obtain

$$\lim_{n\to\infty}\mathrm{Ch}\left\{\left|\frac{1}{n}\sum_{k=1}^{n}\xi_k-E[U_1]\right|\geq\varepsilon\right\}=0.$$

The proof of the theorem is completed. □

Theorem 2.19 ([153]). *Assume $\{\xi_n\}$ is a sequence of fuzzy random variables with $\mu_{\xi_k(\omega)}(x) = \Pi(x - U_k(\omega))$ for almost every $\omega \in \Omega$, where Π is unimodal, $U_k, k = 1, 2, \cdots$, are random variables, if $\mathrm{co}(f \circ \Pi)(r) > 0$ for any nonzero $r \in \Xi$, and $E[\xi_1] < \infty$, $E[U_1] < \infty$, then we have*

$$E\left[\frac{1}{n}\sum_{k=1}^{n}\xi_k - \frac{1}{n}\sum_{k=1}^{n}U_k\right] \to 0 \quad (n \to \infty). \tag{2.34}$$

Proof. Denoting $\zeta_k = \xi_k - U_k$, $S_n = \zeta_1 + \cdots + \zeta_k$, we have (2.38) is equivalent to

$$E\left[\frac{S_n}{n}\right] \to 0 \quad (n \to \infty).$$

Noting that

$$E\left[\frac{S_n}{n}\right] = \int_0^{\infty} \mathrm{Ch}\left\{\frac{S_n}{n} \geq r\right\} \mathrm{d}r - \int_{-\infty}^{0} \mathrm{Ch}\left\{\frac{S_n}{n} \leq r\right\} \mathrm{d}r,$$

to prove the theorem, it suffices to prove

$$\int_0^{\infty} \mathrm{Ch}\left\{\frac{S_n}{n} \geq r\right\} \mathrm{d}r \to 0 \tag{2.35}$$

and

$$\int_{-\infty}^{0} \mathrm{Ch}\left\{\frac{S_n}{n} \leq r\right\} \mathrm{d}r \to 0, \tag{2.36}$$

respectively, as $n \to \infty$.

On one hand, since Π is unimodal and $\mathrm{co}(f \circ \Pi)(r) > 0$ for any nonzero $r \in \Xi$, by Lemma 2.1, for any $r > 0$ and almost every $\omega \in \Omega$, we have

$$\mathrm{Cr}\left\{\frac{S_n(\omega)}{n} \geq r\right\} \leq \sup_{t \geq r} \mu_{\frac{S_n(\omega)}{n}}(t) = \mu_{\frac{S_n(\omega)}{n}}(r) \to 0$$

as $n \to \infty$.

By the dominated convergence theorem, we obtain

$$\lim_{n\to\infty} \mathrm{Ch}\left\{\frac{S_n}{n} \geq r\right\} = \int_{\Omega} \lim_{n\to\infty} \mathrm{Cr}\left\{\frac{S_n(\omega)}{n} \geq r\right\} \Pr(\mathrm{d}\omega) = 0.$$

Noting that for almost every $\omega \in \Omega$, $\zeta_k(\omega), k = 1,2,\cdots$ has identical possibility distribution Π, one has

$$\begin{aligned}
\mathrm{Cr}\left\{\frac{S_n(\omega)}{n} \geq r\right\} &\leq \mu_{\frac{S_n(\omega)}{n}}(r) \\
&= \sup_{(x_1+x_2+\cdots+x_n)/n=r} \top_{k=1}^{n} \Pi(x_k) \\
&\leq \Pi(r) = \mu_{\zeta_1(\omega)}(r)
\end{aligned} \tag{2.37}$$

for any $r > 0$ and $n = 1,2,\cdots$.

Since $\mu_{\zeta_1(\omega)}(r) = \mathrm{Pos}\{\zeta_1(\omega) \geq r\} < 1$, we obtain

$$\mu_{\zeta_1(\omega)}(r) = 2\mathrm{Cr}\{\zeta_1(\omega) \geq r\}$$

for any $r > 0$. Hence, the fact of (2.37) implies

$$\mathrm{Cr}\left\{\frac{S_n(\omega)}{n} \geq r\right\} \leq 2\mathrm{Cr}\{\zeta_1(\omega) \geq r\}, \quad n = 1,2,\cdots.$$

Integrating with respect to ω on the above inequality, we obtain

$$\mathrm{Ch}\left\{\frac{S_n}{n} \geq r\right\} \leq 2\mathrm{Ch}\{\zeta_1 \geq r\}, \quad n = 1,2,\cdots.$$

Since

$$E[\zeta_1] = E[\xi_1 - U_1] = E[\xi_1] - E[U_1] < \infty,$$

applying Lebesgue's dominated convergence theorem, we have

$$\lim_{n\to\infty} \int_0^\infty \mathrm{Ch}\left\{\frac{S_n}{n} \geq r\right\} \mathrm{d}r = \int_0^\infty \lim_{n\to\infty} \mathrm{Ch}\left\{\frac{S_n}{n} \geq r\right\} \mathrm{d}r = 0,$$

which proves (2.35).

By the same reasoning, it can be proved that (2.36) is valid. The proof of the theorem is completed. □

Corollary 2.6 ([153]). *Assume $\{\xi_n\}$ is a sequence of fuzzy random variables with $\mu_{\xi_k(\omega)}(x) = \Pi(x - U_k(\omega))$ for almost every $\omega \in \Omega$, where Π is unimodal, $U_k, k = 1,2,\cdots$, are i.i.d. random variables with finite expected value, if $\mathrm{co}(f \circ \Pi)(r) > 0$ for any nonzero $r \in \Xi$, and $E[\xi_1] < \infty$, then we have*

$$\lim_{n\to\infty} E\left[\frac{1}{n}\sum_{k=1}^{n} \xi_k\right] \to E[U_1]. \tag{2.38}$$

Proof. Since $U_k, k = 1, 2, \cdots$, are i.i.d. random variables with finite expected value, by Theorem 2.19, we obtain

$$\begin{aligned}
&E\left[\frac{1}{n}\sum_{k=1}^{n}\xi_k\right] - E[U_1] \\
&= E\left[\frac{1}{n}\sum_{k=1}^{n}\xi_k\right] - E\left[\frac{1}{n}\sum_{k=1}^{n}U_k\right] \\
&= E\left[\frac{1}{n}\sum_{k=1}^{n}\xi_k - \frac{1}{n}\sum_{k=1}^{n}U_k\right] \to 0
\end{aligned}$$

as $n \to \infty$. The desired result follows. □

Remark 2.5. In Lemma 2.1 and Theorems 2.18 and 2.19, we note that a key condition is

$$\mathrm{co}(f \circ \Pi)_{\Xi}(x) > 0 \text{ for any nonzero } x \in \Xi,$$

which will prove indispensable to the main results of the next section. This convexity condition is determined completely by the composition of the possibility function Π and the additive function of the chosen t-norm. The following example gives a situation that it is invalid.

Example 2.13. Consider the following possibility function Π,

$$\Pi(x) = \begin{cases} x+1, & \text{if } x \in [-1,0] \\ e^{-x}, & \text{if } x \geq 0 \\ 0, & \text{if } x < -1, \end{cases}$$

and Yager t-norm $T_1^Y(x,y)$ with additive generator $f_1^Y(x) = 1 - x$. Let us find and analyze the convex hull $\mathrm{co}(f_1^Y \circ \Pi)_{\Xi}(x)$.

Since

$$f_1^Y \circ \Pi(x) = \begin{cases} -x, & \text{if } x \in [-1,0] \\ 1 - e^{-x}, & \text{if } x \in [0,\infty) \\ 1, & \text{otherwise,} \end{cases}$$

we can find that the convex hull of $f_1^Y \circ \Pi$ on $\Xi = [-1,\infty)$ is

$$\mathrm{co}(f_1^Y \circ \Pi)_{\Xi}(x) = \begin{cases} -x, & \text{if } x \in [-1,0] \\ 0, & \text{if } x \in [0,\infty). \end{cases}$$

Hence, in the case of the above, the condition that $\mathrm{co}(f_1^Y \circ \Pi)_{\Xi}(x) > 0$ for any nonzero $x \in \Xi$ does not hold (see Fig. 2.10).

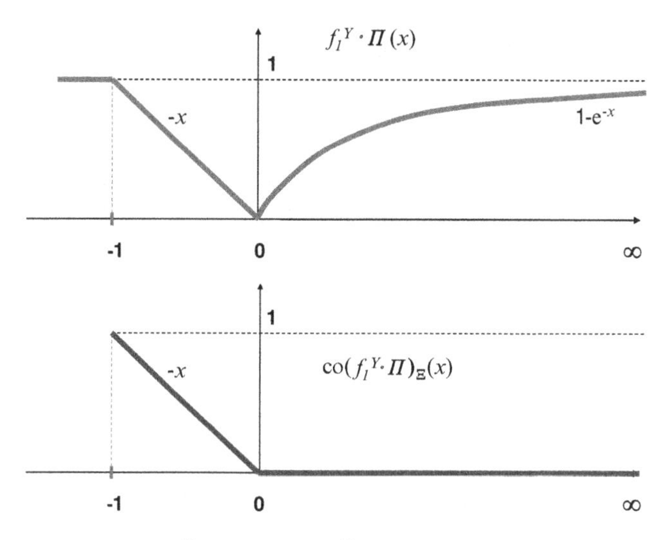

Fig. 2.10 Function curves of $f_1^Y \circ \Pi(x)$ and $\mathrm{co}(f_1^Y \circ \Pi)_\Xi(x)$ in Example 2.13

2.4.2 *Laws of Large Numbers*

On the basis of the preceding convergence properties derived in Sect. 2.4.1, we shall prove two laws of large numbers for fuzzy random variables in this section.

Theorem 2.20 ([153]). *Assume $\{\xi_n\}$ is a sequence of fuzzy random variables with $\mu_{\xi_k(\omega)}(x) = \Pi(x - U_k(\omega))$ for almost every $\omega \in \Omega$, where Π is unimodal, $U_k, k = 1, 2, \cdots$, are i.i.d. random variables with finite expected value, if $\mathrm{co}(f \circ \Pi)(r) > 0$ for any nonzero $r \in \Xi$, and $E[\xi_1] < \infty$, then we have the following results*

$$\frac{1}{n}\sum_{k=1}^{n}\xi_k - E\left[\frac{1}{n}\sum_{k=1}^{n}\xi_k\right] \xrightarrow{\mathrm{Ch}} 0, \tag{2.39}$$

as $n \to \infty$.

Proof. By the subadditivity of credibility measure Cr, for almost every $\omega \in \Omega$, we have

$$\begin{aligned}
&\mathrm{Cr}\left\{\left|\frac{1}{n}\sum_{k=1}^{n}\xi_k(\omega) - E\left[\frac{1}{n}\sum_{k=1}^{n}\xi_k\right]\right| \geq \varepsilon\right\}\\
&= \mathrm{Cr}\left\{\left|\frac{1}{n}\sum_{k=1}^{n}\xi_k(\omega) - E[U_1] + E[U_1] - E\left[\frac{1}{n}\sum_{k=1}^{n}\xi_k\right]\right| \geq \varepsilon\right\}\\
&\leq \mathrm{Cr}\left\{\left|\frac{1}{n}\sum_{k=1}^{n}\xi_k(\omega) - E[U_1]\right| \geq \frac{\varepsilon}{2}\right\} + \mathrm{Cr}\left\{\left|E\left[\frac{1}{n}\sum_{k=1}^{n}\xi_k\right] - E[U_1]\right| \geq \frac{\varepsilon}{2}\right\}.
\end{aligned}$$

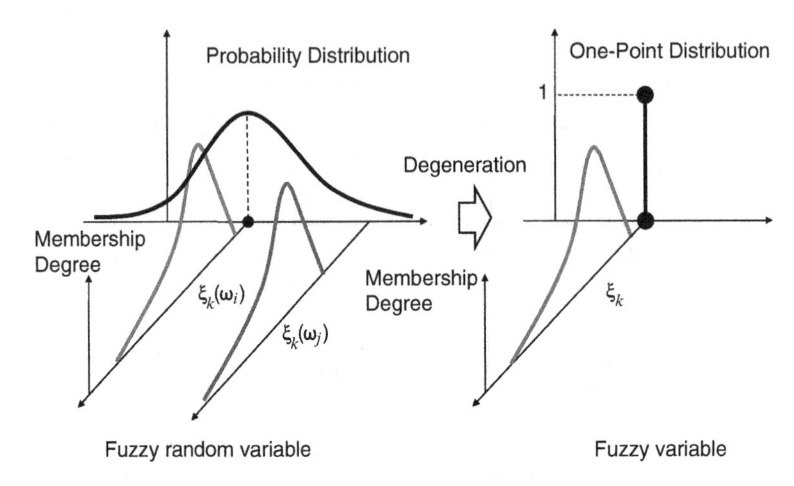

Fig. 2.11 Degeneration from fuzzy random variable to fuzzy variable

Integrating with respect to ω on the above inequality, we get

$$\begin{aligned}&\mathrm{Ch}\left\{\left|\frac{1}{n}\sum_{k=1}^{n}\xi_k - E\left[\frac{1}{n}\sum_{k=1}^{n}\xi_k\right]\right| \geq \varepsilon\right\}\\&\leq \mathrm{Ch}\left\{\left|\frac{1}{n}\sum_{k=1}^{n}\xi_k - E[U_1]\right| \geq \frac{\varepsilon}{2}\right\} + \mathrm{Ch}\left\{\left|E\left[\frac{1}{n}\sum_{k=1}^{n}\xi_k\right] - E[U_1]\right| \geq \frac{\varepsilon}{2}\right\}.\end{aligned}$$

It follows from Corollaries 2.5 and 2.6 that for every $\varepsilon > 0$,

$$\lim_{n\to\infty}\mathrm{Ch}\left\{\left|\frac{1}{n}\sum_{k=1}^{n}\xi_k - E\left[\frac{1}{n}\sum_{k=1}^{n}\xi_k\right]\right| \geq \varepsilon\right\} = 0,$$

which proves (2.39). The proof of the theorem is completed. □

Remark 2.6. If $\{\xi_n\}$ of Theorem 2.20 degenerates to a sequence of fuzzy variables (Fig. 2.11), then for all $\omega \in \Omega$, $\xi_k(\omega) \equiv \xi_k, k = 1, 2, \cdots$. In such a situation, the original probability distribution degenerates to a one-point distribution, or we can say, the randomness vanishes.

It is not difficult to get that (2.39) degenerates to

$$\lim_{n\to\infty}\mathrm{Cr}\left\{\left|\frac{1}{n}\sum_{k=1}^{n}\xi_k - E\left[\frac{1}{n}\sum_{k=1}^{n}\xi_k\right]\right| \geq \varepsilon\right\} = 0$$

for any $\varepsilon > 0$. That is exactly the strong law of large numbers for fuzzy variables (see [18, 157]).

Theorem 2.21 ([153]). *Under the conditions of Theorem 2.20, we have there exist* $E \in \Sigma, F \in \mathscr{A}$ *with* $\Pr(E) = \mathrm{Cr}(F) = 0$ *such that*

$$\lim_{n\to\infty}\left(\frac{1}{n}\sum_{k=1}^{n}\xi_k(\omega,\gamma) - E\left[\frac{1}{n}\sum_{k=1}^{n}\xi_k\right]\right) = 0 \tag{2.40}$$

for every $(\omega,\gamma) \in \Omega\backslash E \times \Gamma\backslash F$.

Proof. Denoting $\zeta_k = \xi_k - U_k$ and $S_n = \zeta_1 + \cdots + \zeta_n$, we have

$$\begin{aligned}
&\mathrm{Cr}\left\{\left|\frac{1}{n}\sum_{k=1}^{n}\xi_k(\omega) - E\left[\frac{1}{n}\sum_{k=1}^{n}\xi_k\right]\right| \ge \varepsilon\right\} \\
&= \mathrm{Cr}\left\{\left|\frac{S_n(\omega)}{n} + \frac{1}{n}\sum_{k=1}^{n}U_k(\omega) - E[U_1] + E[U_1] - E\left[\frac{1}{n}\sum_{k=1}^{n}\xi_k\right]\right| \ge \varepsilon\right\} \\
&\le \mathrm{Cr}\left\{\left|\frac{S_n(\omega)}{n}\right| \ge \frac{\varepsilon}{3}\right\} + \mathrm{Cr}\left\{\left|\frac{1}{n}\sum_{k=1}^{n}U_k(\omega) - E[U_1]\right| \ge \frac{\varepsilon}{3}\right\} \\
&\quad + \mathrm{Cr}\left\{\left|E\left[\frac{1}{n}\sum_{k=1}^{n}\xi_k\right] - E[U_1]\right| \ge \frac{\varepsilon}{3}\right\}
\end{aligned}$$

for almost every $\omega \in \Omega$.

On the one hand, noting that for almost every $\omega \in \Omega$, $\zeta_k(\omega), k = 1,2,\cdots$ has identical possibility distribution $\mu_{\zeta_k(\omega)}(x) = \Pi(x)$ which is unimodal, we have

$$\mathrm{Cr}\left\{\left|\frac{S_n(\omega)}{n}\right| \ge \frac{\varepsilon}{3}\right\} \le \mu_{\frac{1}{n}S_n(\omega)}\left(-\frac{\varepsilon}{3}\right) \bigvee \mu_{\frac{1}{n}S_n(\omega)}\left(\frac{\varepsilon}{3}\right) \to 0 \tag{2.41}$$

as $n \to \infty$.

On the other hand, since $U_k, k = 1,2,\cdots$, are i.i.d. random variables with finite expected value, it follows from the strong law of large numbers for fuzzy random variable sequences that

$$\frac{1}{n}\sum_{k=1}^{n}U_k(\omega) \to E[U_1] \quad (n \to \infty) \tag{2.42}$$

holds almost surely with respect to $\omega \in \Omega$, that is, there exists a set $E \in \Sigma$ with $\Pr(E) = 0$ such that the result (2.42) holds for $\omega \in \Omega \setminus E$. In addition, by Corollary 2.6, it is easy to get

$$E\left[\frac{1}{n}\sum_{k=1}^{n}\xi_k\right] \to E[U_1] \quad (n \to \infty). \tag{2.43}$$

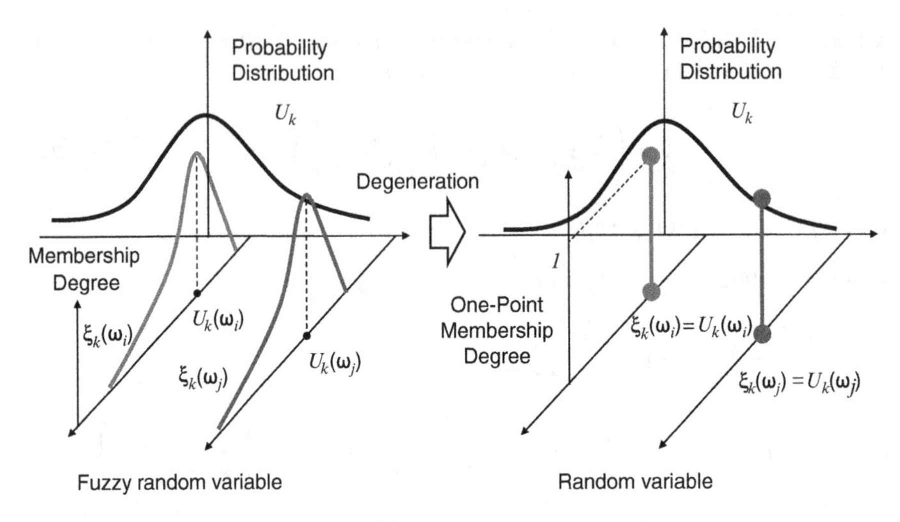

Fig. 2.12 Degeneration from fuzzy random variable to random variable

Combining (2.41), (2.42), and (2.43) deduces

$$\lim_{n\to\infty} \mathrm{Cr}\left\{\left|\frac{1}{n}\sum_{k=1}^{n}\xi_k(\omega)-E\left[\frac{1}{n}\sum_{k=1}^{n}\xi_k\right]\right|\geq\varepsilon\right\}=0$$

holds for $\omega\in\Omega\setminus E$.

Since convergence in Cr implies convergence almost surely with respect to $\gamma\in\Gamma$ (see [87]), there exists a set $F\in\Gamma$ such that

$$\frac{1}{n}\sum_{k=1}^{n}\xi_k(\omega)(\gamma)-E\left[\frac{1}{n}\sum_{k=1}^{n}\xi_k\right]\to 0 \quad (n\to\infty)$$

holds with respect to $(\omega,\gamma)\in\Omega\setminus E\times\Gamma\setminus F$. The required result holds, and the proof of the theorem is completed. □

Remark 2.7. If $\{\xi_n\}$ in Theorem 2.21 degenerates to a sequence of random variables (Fig. 2.12), then for any given $\omega\in\Omega$, $\xi_k(\omega)(\gamma)\equiv\xi_k(\omega)$ for all $\gamma\in\Gamma, k=1,2,\cdots$. Hence the possibility distribution of $\xi_k(\omega)$ degenerates to

$$\mu_{\xi_k(\omega)}(r)=\Pi(r-U_k(\omega))=\begin{cases}1, & \text{if } r=U_k(\omega)\\ 0, & \text{otherwise,}\end{cases}$$

for any $\omega\in\Omega$, which implies $\xi_k=U_k, k=1,2,\cdots$. Therefore, $\{\xi_n\}$ is a sequence of i.i.d. random variables with finite expected value $E[U_1]$, and

$$E\left[\frac{1}{n}\sum_{k=1}^{n}\xi_k\right]=E[\xi_1],$$

for $n=1,2,\cdots$. Thus, result (2.40) in Theorem 2.21 degenerates to

$$\Pr\left\{\omega\in\Omega \mid \frac{1}{n}\sum_{k=1}^{n}\xi_k(\omega)\to E[\xi_1]\right\}=1.$$

That is the strong law of large numbers for random variables.

Example 2.14. Considering fuzzy random variables $\xi_k, k=1,2,\cdots$ with

$$\xi_k(\omega)=\Big(U_k(\omega)-1,U_k(\omega),5+U_k(\omega)\Big)$$

for $k=1,2,\cdots$, where $\top$ is product t-norm, $U_k, k=1,2,\cdots$ are i.i.d. random variables, and $U_k\sim\mathscr{N}(0,1), k=1,2,\cdots$.

The possibility distribution of $\xi_k(\omega)$ can be given as

$$\mu_{\xi_k(\omega)}(x)=\Pi(x-U_k(\omega))=\begin{cases} x+1-U_k(\omega), & x\in[U_k(\omega)-1,U_k(\omega)]\\ \dfrac{5+U_k(\omega)-x}{5}, & x\in[U_k(\omega),5+U_k(\omega)]\\ 0, & \text{otherwise}\end{cases}$$

$k=1,2,\cdots$, and here the possibility function Π is

$$\Pi(x)=\begin{cases} x+1, & x\in[-1,0]\\ 1-\dfrac{x}{5}, & x\in[0,5]\\ 0, & \text{otherwise,}\end{cases}$$

where the support Ξ of Π is $[-1,5]$.

From Example 2.11 we know the product t-norm $\top(x,y)=xy$ is a continuous Archimedean t-norm with additive generator $f^P=-log$. Hence, $f^P\circ\Pi$ can be presented as

$$f^P\circ\Pi(x)=\begin{cases} -log(x+1), & x\in[-1,0]\\ -log(\frac{5-x}{5}), & x\in[0,5]\\ \infty, & \text{otherwise,}\end{cases}$$

which is a convex function on $[-1,5]$, and $\mathrm{co}(f^P \circ \Pi)(x) = f^P \circ \Pi(x) > 0$ for any nonzero $x \in [-1,5]$. Therefore, from Theorems 2.20 and 2.21, we have

$$\frac{1}{n}\sum_{k=1}^{n}\xi_k - E\left[\frac{1}{n}\sum_{k=1}^{n}\xi_k\right] \xrightarrow{\mathrm{Ch}} 0$$

as $n \to \infty$, and

$$\lim_{n\to\infty}\left(\frac{1}{n}\sum_{k=1}^{n}\xi_k(\omega)(\gamma) - E\left[\frac{1}{n}\sum_{k=1}^{n}\xi_k\right]\right) = 0$$

holds for all $(\omega,\gamma) \in \Omega \setminus E \times \Gamma \setminus F$ with $\Pr(E) = \mathrm{Cr}(F) = 0$.

Chapter 3
Fuzzy Stochastic Renewal Processes

In practical applications, chances are pretty good that randomness and fuzziness often coexist in a single process on which a stochastic process considering solely the randomness is established completely in the context of probability. When dealing with those different sources of uncertainty at a time we cannot treat them separately. Fuzzy random variable carries an integrality of such twofold uncertainty information that goes beyond the stochastic information contained in the random variable.

In this chapter, we discuss a class of renewal processes modeled by fuzzy random variables that is more general than the stochastic renewal processes, named fuzzy stochastic renewal processes. The content of this chapter can be regarded as an extension of the limit theorems introduced in Chap. 2. For more on this subject, see Hwang[53], Popova and Wu [121], and Zhao and Tang [180].

In Sect. 3.1, we model a fuzzy stochastic renewal process in which two long-term states of both fuzzy stochastic renewal time and renewal rate are determined in chance measure, and a fuzzy random elementary renewal theorem is proved that gives the limit for the long-term expected renewal rate.

In Sect. 3.2, we discuss a fuzzy stochastic renewal reward process. The long-term states of both fuzzy random average reward and reward rate are derived in chance measure, and a fuzzy stochastic renewal reward theorem is proved that presents the limit for the long-term expected reward rate.

In Sect. 3.3, we present two explanatory examples on a multiservice system and a replacement problem that serve as simple applications of the fuzzy stochastic renewal processes discussed in the preceding sections.

3.1 Fuzzy Stochastic Renewal Process

3.1.1 Problem Settings

Let $\xi_n, n = 1, 2, \cdots$ be a sequence of positive fuzzy random variables defined on the probability space $(\Omega, \Sigma, \Pr)$. Here a positive fuzzy random variable ξ means

S. Wang and J. Watada, *Fuzzy Stochastic Optimization: Theory, Models and Applications*, DOI 10.1007/978-1-4419-9560-5_3, © Springer Science+Business Media New York 2012

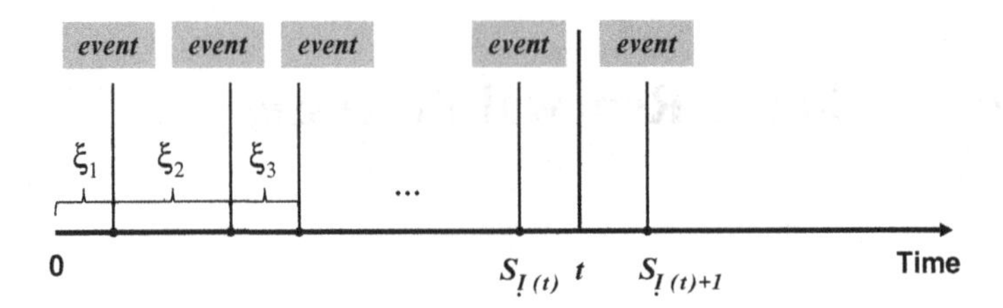

Fig. 3.1 $S_{N(t)}$ and $S_{N(t)+1}$ in a renewal process

that for almost every $\omega \in \Omega$, fuzzy realization $\xi(\omega)$ is positive almost surely, i.e., $\mathrm{Pos}\{\xi(\omega) \leq 0\} = 0$. For each n, we denote ξ_n the interarrival time between the $(n-1)$th and nth event (renewal). Define

$$S_0 = 0, \ S_n = \sum_{k=1}^{n} \xi_k, \ n \geq 1.$$

It is clear that S_n is the time when the nth renewal occurs. Let $N(t)$ denote the total number of the events that have occurred by time t. Then we have

$$N(t) = \max\{n \mid S_n \leq t\}.$$

For any $\omega \in \Omega$, $N(t)(\omega) = \max\{n \mid S_n(\omega) \leq t\}$ is a nonnegative integer-valued fuzzy variable on the possibility space $(\Gamma, \mathscr{A}, \mathrm{Pos})$, and furthermore, $N(t)(\omega)(\gamma)$ is a nonnegative real integer for any $\gamma \in \Gamma$. We call $N(t)$ a fuzzy random renewal variable, and the process $\{N(t), t > 0\}$ a fuzzy stochastic renewal process.

What's more, for any given $\omega \in \Omega$ and integer n, $S_n(\omega) = \xi_1(\omega) + \cdots + \xi_n(\omega)$ is also a fuzzy variable, and for any $t > 0$, we have the following equivalent events:

$$N(t)(\omega) < n \Longleftrightarrow S_n(\omega) > t,$$

$$N(t)(\omega) \geq n \Longleftrightarrow S_n(\omega) \leq t,$$

$$N(t)(\omega) = n \Longleftrightarrow S_n(\omega) \leq t < S_{n+1}(\omega).$$

In addition, $S_{N(t)}$ represents the time of the last renewal prior to or at time t, while $S_{N(t)+1}$ represents the time of the first renewal after time t (see Fig. 3.1).

Here, let us recall the conditions A1 and A2 in Chap. 2:

A1. The operations of fuzzy variables are determined by the generalized extension principle (2.25), and $\top$ denotes any continuous Archimedean t-norm with an additive generator f.

A2. Π is a nonnegative real-valued function with $\Pi(0) = 1$, and Π is nonincreasing on $\Re^+$ and nondecreasing on $\Re^-$.

In this chapter, we assume that the interarrival times and rewards are under the conditions A1 and A2, and furthermore, we assume that the t-norm $\top$ with additive generator f and possibility function Π satisfy the following condition.

A3. $\mathrm{co}(f \circ \Pi)_{\Xi}(x) > 0$ for any nonzero $x \in \Xi$, and there exists a positive real number a such that $\Pi(-a) = 0$.

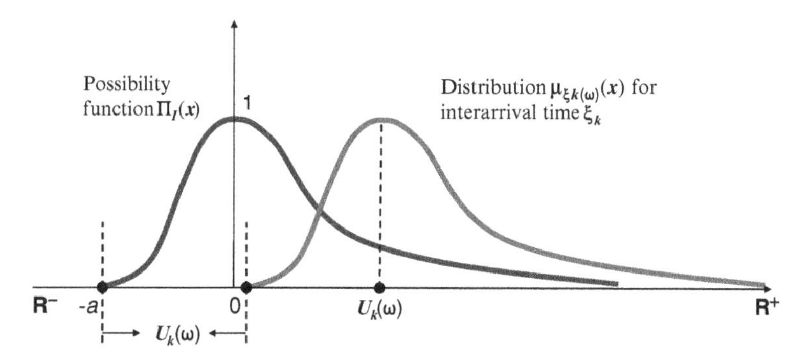

Fig. 3.2 Constructing distributions for interarrival times

In A3, the requirement on the positive real number a such that $\Pi(-a) = 0$ enables the function Π to construct the positive distributions of the interarrival times and rewards (the renewal reward process is to be discussed in the next section). To avoid any confusion, we use the Π_I and Π_R to denote the possibility functions of the interarrival times and rewards, respectively. Here in this section, the support Ξ for Π_I is always assumed to be $[-a, \infty)$ unless otherwise mentioned.

In fact, we note that in fuzzy stochastic renewal processes, the interarrival times and rewards between two events are positive fuzzy random variables; hence, making use of the function Π with $\Pi(-a) = 0$ in condition A2, we can characterize the interarrival times by fuzzy random variables $\xi_k, k = 1, 2, \cdots$ with distributions

$$\mu_{\xi_k(\omega)}(x) = \Pi_I(x - U_k(\omega)), \quad U_k \geq a \quad \text{almost sure},$$

where $U_k, k = 1, 2, \cdots$ are random variables, and Π_I satisfies the condition that $\Pi_I(-a) = 0, a > 0$ (Fig. 3.2).

In renewal models, a critical task is determining the long-term expected renewal rate as time t goes to infinity, i.e., $\lim_{t \to \infty} E\left[\frac{N(t)}{t}\right]$. In order to derive this limit, let us investigate and determine the long-term states for average renewal time $\frac{S_{N(t)}}{N(t)}$ and renewal rate $\frac{N(t)}{t}$.

3.1.2 Long-Term States of Average Renewal Time and Renewal Rate

Theorem 3.1 ([148]). *Assume $\{\xi_n\}$ is a sequence of fuzzy random interarrival times with $\mu_{\xi_k(\omega)}(x) = \Pi_I(x - U_k(\omega))$ for almost every $\omega \in \Omega$, where random variables $U_k \geq a$ almost surely, and $N(t)$ is the fuzzy stochastic renewal variable. Then we have*

$$\frac{1}{N(t)} \xrightarrow{\text{Ch}} 0.$$

Proof. Denoting $X_k = \xi_k - U_k$, $S_n^* = X_1 + \cdots + X_n$, we have

$$\begin{aligned}\Pi_I(n) &= \top\big(\Pi_I(0), \cdots, \Pi_I(0), \Pi_I(n)\big)\\ &\le \sup_{x_1+x_2+\cdots+x_n=n} \top_{k=1}^{n} \mu_{X_k(\omega)}(x_k)\\ &= \mu_{S_n^*(\omega)}(n) = \mu_{\frac{1}{n}S_n^*(\omega)}(1).\end{aligned}$$

By Lemma 2.1, we have

$$\lim_{t\to\infty} \Pi_I(t) = 0.$$

Moreover, for any $\varepsilon > 0$, we let M be the smallest integer such that $M > 1/\varepsilon$. Since for any $\omega \in \Omega$ and $t > 0$,

$$\frac{1}{N(t)(\omega)} \ge \varepsilon \Longleftrightarrow N(t)(\omega) < M,$$

we have

$$\begin{aligned}N(t)(\omega) < M &\Longleftrightarrow S_M(\omega) > t\\ &\Longleftrightarrow \frac{S_M(\omega)}{M} - \frac{1}{M}\sum_{k=1}^{M} U_k(\omega) > \frac{t}{M} - \frac{1}{M}\sum_{k=1}^{M} U_k(\omega)\\ &\Longleftrightarrow \frac{S_M^*(\omega)}{M} > \frac{t}{M} - \frac{1}{M}\sum_{k=1}^{M} U_k(\omega).\end{aligned}$$

Without losing any generality, we let $t > \sum_{k=1}^{M} U_k(\omega)$. Therefore,

$$\begin{aligned}\mathrm{Ch}\left\{\frac{1}{N(t)} \ge \varepsilon\right\} &= \int_\Omega \mathrm{Cr}\left\{\frac{1}{N(t)(\omega)} \ge \varepsilon\right\} \Pr(\mathrm{d}\omega)\\ &= \int_\Omega \mathrm{Cr}\left\{\frac{S_M^*(\omega)}{M} > \frac{t}{M} - \frac{1}{M}\sum_{k=1}^{M} U_k(\omega)\right\} \Pr(\mathrm{d}\omega)\\ &\le \int_\Omega \mathrm{Pos}\left\{\frac{S_M^*(\omega)}{M} > \frac{t}{M} - \frac{1}{M}\sum_{k=1}^{M} U_k(\omega)\right\} \Pr(\mathrm{d}\omega)\\ &\le \int_\Omega \Pi_I\left(\frac{1}{M}\cdot\left(t - \sum_{k=1}^{M} U_k(\omega)\right)\right) \Pr(\mathrm{d}\omega).\end{aligned}$$

Noting that $\lim_{t\to\infty} \Pi_I(t) = 0$, it follows from dominated convergence theorem that

$$\lim_{t\to\infty} \mathrm{Ch}\left\{\frac{1}{N(t)} \ge \varepsilon\right\} = \int_\Omega \lim_{t\to\infty} \mathrm{Cr}\left\{\frac{1}{N(t)(\omega)} \ge \varepsilon\right\} \Pr(\mathrm{d}\omega) = 0.$$

The proof is completed. □

Remark 3.1. Theorem 3.1 is critical, since it implies the rationality of the fuzzy stochastic renewal process modeled in this chapter. From the result of Theorem 3.1, we can see that the total number of renewals that occur in the fuzzy random process is infinite, which implies each interarrival time should be finite.

Theorem 3.2 ([148]). *Suppose that $\{\xi_k\}$ is a sequence of fuzzy random interarrival times with $\mu_{\xi_k(\omega)}(x) = \Pi_I(x - U_k(\omega))$ for almost every $\omega \in \Omega$, where $U_k, k \geq 1$ are i.i.d. random variables with finite expected values such that $U_k \geq a$ almost surely, and $N(t)$ is the fuzzy random renewal variable. Then we have*

$$\frac{S_{N(t)}}{N(t)} \xrightarrow{\text{Ch}} E[U_1].$$

Proof. To obtain the required result, we need to claim that given almost every $\omega \in \Omega$, the following limit

$$\lim_{t\to\infty} \text{Cr}\left\{ \left| \frac{S_{N(t)(\omega)}(\omega)}{N(t)(\omega)} - E[U_1] \right| \geq \varepsilon \right\} = 0 \tag{3.1}$$

for any $\varepsilon > 0$. To obtain the above (3.1), it suffices to prove that for any $\delta > 0$, there exists an $E_\delta \in \mathscr{A}$ (ample field) with $\text{Cr}(E_\delta) < \delta$ such that $\{S_{N(t)(\omega)}(\omega)/N(t)(\omega)\}$ converges to $E[U_1]$ uniformly on $\Gamma \setminus E_\delta$, that is,

$$\frac{S_{N(t)(\omega,\gamma)}(\omega,\gamma)}{N(t)(\omega,\gamma)} \to E[U_1] \quad (t \to \infty) \tag{3.2}$$

holds for all $\gamma \in \Gamma \setminus E_\delta$.

The proof for (3.2) breaks into the following three steps:

Step 1. We note that $U_k, k \geq 1$ are i.i.d. random variables with finite expected values. Then, by the proof of Theorem 2.18, we get that for almost every $\omega \in \Omega$,

$$\lim_{n\to\infty} \text{Cr}\left\{ \left| \frac{S_n(\omega)}{n} - E[U_1] \right| \geq \varepsilon \right\} = 0$$

for any $\varepsilon > 0$, which is equivalent to the fact that for any $\varepsilon > 0$ and $\delta > 0$, there exist a set $A_\delta \in \mathscr{A}$ with $\text{Cr}(A_\delta) < \delta/2$ and a positive integer M such that for every $\gamma \in \Gamma \setminus A_\delta$,

$$\left| \frac{S_n(\omega,\gamma)}{n} - E[U_1] \right| < \varepsilon$$

whenever $n \geq M$.

Step 2. From the proof of Theorem 3.1, we get that given almost every $\omega \in \Omega$, for any $\varepsilon > 0$, we have

$$\lim_{t\to\infty} \text{Cr}\left\{ \frac{1}{N(t)(\omega)} \geq \varepsilon \right\} = 0.$$

Therefore, for any $\delta > 0$, there exists $B_\delta \in \mathscr{A}$ with $\mathrm{Cr}(B_\delta) < \delta/2$, and for the above positive integer M, there exists a positive real number t_M, such that for every $\gamma \in \Gamma \setminus B_\delta$,

$$\frac{1}{N(t)(\omega,\gamma)} \leq \frac{1}{M}$$

whenever $t \geq t_M$.

Step 3. Now combining the above results in Steps 1 and 2, we have that for any $\varepsilon > 0, \delta > 0$, there exist $E_\delta = A_\delta \cup B_\delta \in \mathscr{A}$ with

$$\mathrm{Cr}\{E_\delta\} = \mathrm{Cr}\{A_\delta \cup B_\delta\} \leq \mathrm{Cr}\{A_\delta\} + \mathrm{Cr}\{B_\delta\} < \delta,$$

and a positive real number t_M, such that for all

$$\gamma \in \Gamma \setminus E_\delta = \Gamma \setminus (A_\delta \cup B_\delta),$$

we have

$$N(t)(\omega,\gamma) \geq M,$$

and therefore,

$$\left| \frac{S_{N(t)(\omega,\gamma)}(\omega,\gamma)}{N(t)(\omega,\gamma)} - E[U_1] \right| < \varepsilon$$

whenever $t \geq t_M$. That is,

$$\frac{S_{N(t)(\omega,\gamma)}(\omega,\gamma)}{N(t)(\omega,\gamma)} \to E[U_1] \quad (t \to \infty)$$

for all $\gamma \in \Gamma \setminus E_\delta$.

Thus, for almost every $\omega \in \Omega$, we have

$$\lim_{t\to\infty} \mathrm{Cr}\left\{ \left| \frac{S_{N(t)(\omega)}(\omega)}{N(t)(\omega)} - E[U_1] \right| \geq \varepsilon \right\} = 0$$

for any $\varepsilon > 0$.

Finally, applying dominated convergence theorem, we have

$$\begin{aligned}
&\lim_{t\to\infty} \mathrm{Ch}\left\{ \left| \frac{S_{N(t)}}{N(t)} - E[U_1] \right| \geq \varepsilon \right\} \\
&= \lim_{t\to\infty} \int_\Omega \mathrm{Cr}\left\{ \left| \frac{S_{N(t)(\omega)}(\omega)}{N(t)(\omega)} - E[U_1] \right| \geq \varepsilon \right\} \Pr(\mathrm{d}\omega) \\
&= \int_\Omega \lim_{t\to\infty} \mathrm{Cr}\left\{ \left| \frac{S_{N(t)(\omega)}(\omega)}{N(t)(\omega)} - E[U_1] \right| \geq \varepsilon \right\} \Pr(\mathrm{d}\omega) = 0.
\end{aligned}$$

The proof is completed. □

Theorem 3.3 ([148]). *Under the same assumptions in Theorem 3.2, we have*

$$\frac{N(t)}{t} \xrightarrow{\text{Ch}} \frac{1}{E[U_1]}.$$

Proof. The whole proof goes into the following four steps.

Step 1. Recalling the result of Theorem 3.2, we have that for any $\varepsilon > 0$,

$$\lim_{t\to\infty} \text{Ch}\left\{\left|\frac{S_{N(t)}}{N(t)} - E[U_1]\right| \geq \varepsilon\right\} = 0. \tag{3.3}$$

Step 2. We now prove

$$\lim_{t\to\infty} \text{Ch}\left\{\left|\frac{S_{N(t)+1}}{N(t)} - E[U_1]\right| \geq \varepsilon\right\} = 0. \tag{3.4}$$

In fact, for any $\omega \in \Omega$, we have

$$\begin{aligned}
&\left|\frac{S_{N(t)(\omega)+1}(\omega)}{N(t)(\omega)} - E[U_1]\right| \\
&= \left|\frac{S_{N(t)(\omega)+1}(\omega)}{N(t)(\omega)+1} \cdot \frac{N(t)(\omega)+1}{N(t)(\omega)} - E[U_1]\right| \\
&= \left|\frac{S_{N(t)(\omega)+1}(\omega)}{N(t)(\omega)+1} \cdot \frac{N(t)(\omega)+1}{N(t)(\omega)} - \Theta(\omega) + \Theta(\omega) - E[U_1]\right| \\
&\leq \frac{1}{N(t)(\omega)}\left|\frac{S_{N(t)(\omega)+1}(\omega)}{N(t)(\omega)+1} - E[U_1]\right| + \left|\frac{S_{N(t)(\omega)+1}(\omega)}{N(t)(\omega)+1} - E[U_1]\right| + \frac{E[U_1]}{N(t)(\omega)},
\end{aligned}$$

where

$$\Theta(\omega) = \frac{N(t)(\omega)+1}{N(t)(\omega)} \cdot E[U_1].$$

Without any loss of generality, letting $0 < \varepsilon < 1$, we can obtain

$$\begin{aligned}
&\text{Ch}\left\{\left|\frac{S_{N(t)+1}}{N(t)} - E[U_1]\right| \geq \varepsilon\right\} \\
&= \int_\Omega \text{Cr}\left\{\left|\frac{S_{N(t)(\omega)+1}(\omega)}{N(t)(\omega)} - E[U_1]\right| \geq \varepsilon\right\} \Pr(\mathrm{d}\omega) \\
&\leq \int_\Omega \text{Cr}\left\{\frac{\Lambda(\omega)}{N(t)(\omega)} + \Lambda(\omega) + \frac{E[U_1]}{N(t)(\omega)} \geq \varepsilon\right\} \Pr(\mathrm{d}\omega)
\end{aligned}$$

$$\leq \mathrm{Ch}\left\{\frac{1}{N(t)} \geq \frac{\varepsilon}{3}\right\} + 2\mathrm{Ch}\left\{\left|\frac{S_{N(t)+1}}{N(t)+1} - E[U_1]\right| \geq \frac{\varepsilon}{3}\right\}$$
$$+\mathrm{Ch}\left\{\frac{1}{N(t)} \geq \frac{\varepsilon}{3E[U_1]}\right\},$$

where

$$\Lambda(\omega) = \left|\frac{S_{N(t)(\omega)+1}(\omega)}{N(t)(\omega)+1} - E[U_1]\right|.$$

From (3.3) and Theorem 3.1, we can obtain (3.4).

Step 3. Next, for any $\varepsilon > 0$, we claim

$$\lim_{t\to\infty} \mathrm{Ch}\left\{\left|\frac{N(t)}{S_{N(t)}} - \frac{1}{E[U_1]}\right| \geq \varepsilon\right\} = 0, \tag{3.5}$$

and

$$\lim_{t\to\infty} \mathrm{Ch}\left\{\left|\frac{N(t)}{S_{N(t)+1}} - \frac{1}{E[U_1]}\right| \geq \varepsilon\right\} = 0. \tag{3.6}$$

We take $0 < \varepsilon < 1/E[U_1]$ without any loss of generality; for any $\omega \in \Omega$, the following inequality

$$\left|\frac{N(t)(\omega)}{S_{N(t)(\omega)}(\omega)} - \frac{1}{E[U_1]}\right| \geq \varepsilon$$

implies

$$\left|\frac{S_{N(t)(\omega)}(\omega)}{N(t)(\omega)} - E[U_1]\right| \geq \frac{E^2[U_1]\varepsilon}{1+E[U_1]\varepsilon}.$$

By (3.3) and the relation between the above two inequalities, we have that

$$\lim_{t\to\infty} \mathrm{Ch}\left\{\left|\frac{N(t)}{S_{N(t)}} - \frac{1}{E[U_1]}\right| \geq \varepsilon\right\}$$
$$= \lim_{t\to\infty} \int_\Omega \mathrm{Cr}\left\{\left|\frac{N(t)(\omega)}{S_{N(t)(\omega)}(\omega)} - \frac{1}{E[U_1]}\right| \geq \varepsilon\right\} \Pr(\mathrm{d}\omega)$$
$$\leq \lim_{t\to\infty} \int_\Omega \mathrm{Cr}\left\{\left|\frac{S_{N(t)(\omega)}(\omega)}{N(t)(\omega)} - E[U_1]\right| \geq \frac{E^2[U_1]\varepsilon}{1+E[U_1]\varepsilon}\right\} \Pr(\mathrm{d}\omega)$$
$$= \lim_{t\to\infty} \mathrm{Ch}\left\{\left|\frac{S_{N(t)}}{N(t)} - E[U_1]\right| \geq \frac{E^2[U_1]\varepsilon}{1+E[U_1]\varepsilon}\right\} = 0.$$

Similarly, by (3.4) we can prove

$$\lim_{t\to\infty} \mathrm{Ch}\left\{\left|\frac{N(t)}{S_{N(t)+1}} - \frac{1}{E[U_1]}\right| \geq \varepsilon\right\} = 0.$$

Step 4. By the definitions of $N(t)$, $S_{N(t)}$ and $S_{N(t)+1}$, we know that for any $\omega \in \Omega$,

$$S_{N(t)(\omega)}(\omega) \leq t < S_{N(t)(\omega)+1}(\omega),$$

which implies

$$\frac{S_{N(t)(\omega)}(\omega)}{N(t)(\omega)} \leq \frac{t}{N(t)(\omega)} < \frac{S_{N(t)(\omega)+1}(\omega)}{N(t)(\omega)}, \tag{3.7}$$

or equivalently,

$$\frac{N(t)(\omega)}{S_{N(t)(\omega)+1}(\omega)} < \frac{N(t)(\omega)}{t} \leq \frac{N(t)(\omega)}{S_{N(t)(\omega)}(\omega)}.$$

Therefore, for any $\varepsilon > 0$, one has

$$\begin{aligned}
&\mathrm{Ch}\left\{\left|\frac{N(t)}{t} - \frac{1}{E[U_1]}\right| \geq \varepsilon\right\} \\
&= \int_\Omega \mathrm{Cr}\left(\left\{\frac{N(t)(\omega)}{t} \geq \frac{1}{E[U_1]} + \varepsilon\right\} \bigcup \left\{\frac{N(t)(\omega)}{t} \leq \frac{1}{E[U_1]} - \varepsilon\right\}\right) \mathrm{Pr}(\mathrm{d}\omega) \\
&\leq \int_\Omega \left[\mathrm{Cr}\left\{\frac{N(t)(\omega)}{t} \geq \frac{1}{E[U_1]} + \varepsilon\right\} + \mathrm{Cr}\left\{\frac{N(t)(\omega)}{t} \leq \frac{1}{E[U_1]} - \varepsilon\right\}\right] \mathrm{Pr}(\mathrm{d}\omega) \\
&\leq \mathrm{Ch}\left\{\frac{N(t)(\omega)}{S_{N(t)(\omega)}(\omega)} \geq \frac{1}{E[U_1]} + \varepsilon\right\} + \mathrm{Ch}\left\{\frac{N(t)(\omega)}{S_{N(t)(\omega)+1}(\omega)} \leq \frac{1}{E[U_1]} - \varepsilon\right\}.
\end{aligned}$$

By (3.5) and (3.6), the above inequality implies

$$\lim_{t\to\infty} \mathrm{Ch}\left\{\left|\frac{N(t)}{t} - \frac{1}{E[U_1]}\right| \geq \varepsilon\right\} = 0.$$

The proof of the theorem is completed. □

3.1.3 Long-Term Expected Renewal Rate

Lemma 3.1 ([148]). *Suppose $\xi_k, k = 1, 2, \cdots$ are fuzzy random interarrival times with $\mu_{\xi_k(\omega)} = \Pi_I(x - U_k(\omega))$ for almost every $\omega \in \Omega$, where U_k are random* *variables, and $N(t)$ is the fuzzy random renewal variable. If $U_k \geq a$ almost surely, then for any real number $t, r > 0$, we have*

$$\mathrm{Ch}\left\{\frac{N(t)}{t} \geq r\right\} \leq \int_{\Omega} \mathrm{Pos}\left\{\frac{1}{\xi_1(\omega) - (U_1(\omega) - a)} \geq r\right\} \mathrm{Pr}(\mathrm{d}\omega).$$

Proof. We denote $X_k = \xi_k - U_k, S_n^* = X_1 + \cdots + X_n$. Since for every $\omega \in \Omega$, $X_k(\omega)$ are identically distributed fuzzy variables, we have for any given real number $t, r > 0$ and almost every $\omega \in \Omega$,

$$\begin{aligned}
&\mathrm{Pos}\left\{\frac{N(t)(\omega)}{t} \geq r\right\} = \mathrm{Pos}\left\{N(t)(\omega) \geq rt\right\} \\
&= \mathrm{Pos}\left\{N(t)(\omega) \geq M\right\} = \mathrm{Pos}\left\{S_M(\omega) \leq t\right\} \\
&= \mathrm{Pos}\left\{S_M(\omega) - \sum_{k=1}^{M} U_k(\omega) \leq t - \sum_{k=1}^{M} U_k(\omega)\right\} \\
&= \mathrm{Pos}\left\{S_M^*(\omega) \leq t - \sum_{k=1}^{M} U_k(\omega)\right\} \\
&= \sup_{\sum_{k=1}^{M} X_k(\omega) \leq t - \sum_{k=1}^{M} U_k(\omega)} \top_{k=1}^{M} \mu_{X_k(\omega)}(x_k) \\
&\leq \mathrm{Pos}\left\{X_1(\omega) \leq \frac{t}{M} - \sum_{k=1}^{M} \frac{U_k(\omega)}{M}\right\},
\end{aligned}$$

where M is the smallest integer such that $M \geq rt$. Furthermore, since $U_k(\omega) \geq a$ almost surely for any k, the following inequalities

$$\begin{aligned}
&\mathrm{Pos}\left\{\frac{N(t)(\omega)}{t} \geq r\right\} \\
&\leq \mathrm{Pos}\left\{X_1(\omega) \leq \frac{1}{r} - a\right\} \\
&\leq \mathrm{Pos}\left\{\xi_1(\omega) - (U_1(\omega) - a) \leq \frac{1}{r}\right\} \\
&= \mathrm{Pos}\left\{\frac{1}{\xi_1(\omega) - (U_1(\omega) - a)} \geq r\right\}
\end{aligned}$$

hold for almost every $\omega \in \Omega$. Thus,

$$\mathrm{Cr}\left\{\frac{N(t)(\omega)}{t} \geq r\right\} \leq \mathrm{Pos}\left\{\frac{1}{\xi_1(\omega) - (U_1(\omega) - a)} \geq r\right\}.$$

Integrating with respect to ω on the above inequality, we have

$$\mathrm{Ch}\left\{\frac{N(t)}{t} \geq r\right\} \leq \int_{\Omega} \mathrm{Pos}\left\{\frac{1}{\xi_1(\omega)-(U_1(\omega)-a)} \geq r\right\} \mathrm{Pr}(\mathrm{d}\omega).$$

The proof of the lemma is completed. □

Theorem 3.4 (Fuzzy Random Elementary Renewal Theorem [148]). *Assume $\{\xi_k\}$ is a sequence of fuzzy random interarrival times with $\mu_{\xi_k(\omega)}(x) = \Pi_I(x - U_k(\omega))$ for almost every $\omega \in \Omega$, where $U_k, k \geq 1$ are i.i.d. random variables with finite expected values such that $U_k \geq a+h$ with $h > 0$ almost surely, and $N(t)$ is the fuzzy random renewal variable. Then we have*

$$\lim_{t\to\infty} \frac{E[N(t)]}{t} = \frac{1}{E[U_1]}. \tag{3.8}$$

Furthermore, if $E[\xi_1]$ is finite, then we have

$$\lim_{t\to\infty} \frac{E[N(t)]}{t} = \frac{1}{\lim_n E[S_n/n]}. \tag{3.9}$$

Proof. First of all, we are to prove

$$\int_0^{\infty} \left[\int_{\Omega} \mathrm{Pos}\left\{\frac{1}{\xi_1(\omega)-(U_1(\omega)-a-h)} \geq r\right\} \mathrm{Pr}(\mathrm{d}\omega)\right] \mathrm{d}r < \infty. \tag{3.10}$$

Since $U_k(\omega) \geq a+h$ holds almost surely for each k, we have for almost every $\omega \in \Omega$,

$$\mathrm{Pos}\{\xi_1(\omega)-(U_1(\omega)-a-h) \leq h\} = 0$$

which is equivalent to

$$\mathrm{Pos}\left\{\frac{1}{\xi_1(\omega)-(U_1(\omega)-a-h)} \geq \frac{1}{h}\right\} = 0.$$

Therefore, the following inequality

$$\int_0^{\infty} \mathrm{Pos}\left\{\frac{1}{\xi_1(\omega)-(U_1(\omega)-a-h)} \geq r\right\} \mathrm{d}r$$

$$= \int_0^{1/h} \mathrm{Pos}\left\{\frac{1}{\xi_1(\omega)-(U_1(\omega)-a-h)} \geq r\right\} \mathrm{d}r \leq \frac{1}{h}$$

holds for almost $\omega \in \Omega$. Furthermore, by Fubini's theorem (see [43]), we can obtain

$$\int_0^\infty \left[\int_\Omega \text{Pos}\left\{ \frac{1}{\xi_1(\omega) - (U_1(\omega) - a - h)} \ge r \right\} \Pr(\mathrm{d}\omega) \right] \mathrm{d}r$$
$$= \int_\Omega \left[\int_0^\infty \text{Pos}\left\{ \frac{1}{\xi_1(\omega) - (U_1(\omega) - a - h)} \ge r \right\} \mathrm{d}r \right] \Pr(\mathrm{d}\omega) \le \frac{1}{h}.$$

This proves (3.10).

Since $N(t)(\omega)$ are nonnegative fuzzy variables for any $\omega \in \Omega$, we have

$$\frac{E[N(t)]}{t} = \int_\Omega \left[\int_0^\infty \text{Cr}\left\{ \frac{N(t)(\omega)}{t} \ge r \right\} \mathrm{d}r \right] \Pr(\mathrm{d}\omega)$$
$$= \int_0^\infty \text{Ch}\left\{ \frac{N(t)}{t} \ge r \right\} \mathrm{d}r.$$

According to Theorem 3.3, we get

$$\frac{N(t)}{t} \xrightarrow{\text{Ch}} \frac{1}{E[U_1]},$$

from which it is not difficult to obtain

$$\lim_{t\to\infty} \text{Ch}\left\{ \frac{N(t)}{t} \ge r \right\} = \text{Ch}\left\{ \frac{1}{E[U_1]} \ge r \right\} \tag{3.11}$$

for almost every $r \in \Re$. Furthermore, by Lemma 3.1, for any $t, r > 0$, one has

$$\text{Ch}\left\{ \frac{N(t)}{t} \ge r \right\} \le \int_\Omega \text{Pos}\left\{ \frac{1}{\xi_1(\omega) - (U_1(\omega) - a - h)} \ge r \right\} \Pr(\mathrm{d}\omega).$$

It follows from (3.11) and Lebesgue's dominated convergence theorem that

$$\lim_{t\to\infty} \frac{E[N(t)]}{t} = \int_0^\infty \lim_{t\to\infty} \text{Ch}\left\{ \frac{N(t)}{t} \ge r \right\} \mathrm{d}r$$
$$= \int_0^\infty \text{Ch}\left\{ \frac{1}{E[U_1]} \ge r \right\} \mathrm{d}r$$
$$= \frac{1}{E[U_1]}.$$

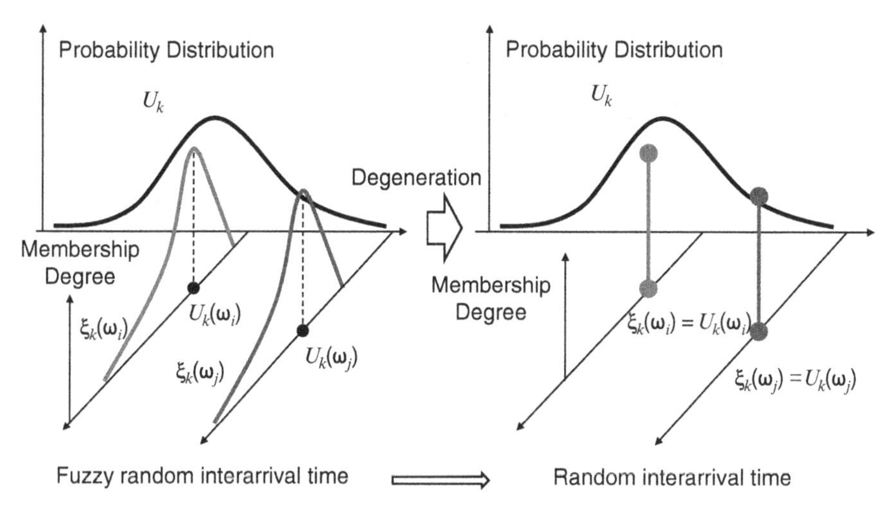

Fig. 3.3 Fuzzy random interarrival times reducing to random interarrival times

Moreover, if $E[\xi]$ is finite, we can deduce directly from Theorem 2.19 that

$$\lim_{t\to\infty}\frac{E[N(t)]}{t}=\frac{1}{E[U_1]}=\frac{1}{\lim_n E\left[S_n/n\right]}.$$

The proof of the theorem is completed. □

Remark 3.2. If $\{\xi_k\}$ in Theorem 3.4 degenerates to a sequence of random variables (Fig. 3.3), then both of results (3.8) and (3.9) can degenerate to that of the stochastic elementary renewal theorem.

As we know, for any given $\omega\in\Omega$, fuzzy variables $\xi_k(\omega), k=1,2,\cdots$ degenerate to crisp numbers. Hence, the membership function of $\xi_k(\omega)$ degenerates to

$$\mu_{\xi_k(\omega)}(r)=\Pi_I(r-U_k(\omega))=\begin{cases}1, & \text{if } r=U_k(\omega)\\ 0, & \text{otherwise,}\end{cases}$$

which implies $\xi_k=U_k$ for $k=1,2,\cdots$.

Therefore, on one hand, $\{\xi_k\}$ is a sequence of i.i.d. random variables with the same finite expected value $E[\xi_1]=E[U_1]$, and the result (3.8) of Theorem 3.4 degenerates to

$$\lim_{t\to\infty}\frac{E[N(t)]}{t}=\frac{1}{E[\xi_1]},$$

which is exactly the result of stochastic elementary renewal theorem (see [127]). On the other hand, it is clear that

$$\frac{1}{\lim_n E\left[S_n/n\right]}=\frac{1}{\lim_n E\left[(\xi_1+\xi_2+\cdots+\xi_n)/n\right]}$$
$$=\frac{1}{E[\xi_1]}$$

for i.i.d. random variables $\xi_1, \xi_2, \cdots$. Therefore, (3.9) also degenerates to the result of stochastic elementary renewal theorem.

Furthermore, if fuzzy random interarrival times are symmetric, then we can obtain the following interesting result.

Corollary 3.1 ([148]). *Under the same condition as in Theorem 3.4, if Π_I is symmetric with respect to zero, then we have*

$$\lim_{t\to\infty} \frac{E[N(t)]}{t} = \frac{1}{E[\xi_1]}.$$

Proof. Since Π_I is symmetric with respect to zero, we know for every $\omega \in \Omega$, $\mu_{\xi_1(\omega)}$ is symmetric with respect to $U_1(w)$, and we can calculate that

$$\int_0^\infty \mathrm{Cr}\{\xi_1(\omega) \geq r\}\mathrm{d}r = U_1(\omega).$$

Integrating with respect to ω on the above equality, we have

$$E[\xi_1] = \int_\Omega \left[\int_0^\infty \mathrm{Cr}\{\xi_1(\omega) \geq r\}\mathrm{d}r \right] \Pr(\mathrm{d}\omega) = E[U_1].$$

It follows from Theorem 3.4 that

$$\lim_{t\to\infty} \frac{E[N(t)]}{t} = \frac{1}{E[\xi_1]}.$$

The proof of the lemma is completed. □

Remark 3.3. Corollary 3.1 has the same result with stochastic elementary renewal theorem (see [127]).

3.2 Fuzzy Stochastic Renewal Reward Process

3.2.1 Problem Settings

On the basis of the preceding renewal process $\{N(t), t > 0\}$ with fuzzy random interarrival times ξ_n, $n \geq 1$ in the Sect. 3.1, now we suppose that each time when a renewal occurs, a reward is received which is a positive fuzzy random variable. We denote η_n the reward earned at each time of the nth renewal (see Fig. 3.4). Let $C(t)$ represent the total reward earned by time t; then we have

$$C(t) = \sum_{k=1}^{N(t)} \eta_k, \tag{3.12}$$

where $N(t)$ is the fuzzy random renewal variable.

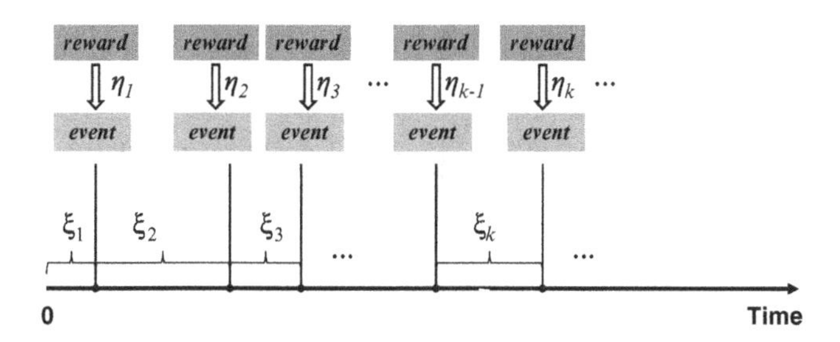

Fig. 3.4 Renewal reward process

Similarly as the distributions of fuzzy random interarrival times which has been discussed in the Sect. 3.1, under the conditions A2 and A3, we characterize the rewards by positive fuzzy random variables $\eta_k, k = 1, 2, \cdots$ with distributions $\mu_{\eta_k(\omega)}(x) = \Pi_R(x - V_k(\omega))$, $V_k \geq b$ almost surely, where $V_k, k = 1, 2, \cdots$ are random variables, and Π_R is a possibility function with $\Pi_R(-b) = 0, b > 0$. Just as in Sect. 3.1, the support Ξ for Π_R in this section is always assumed to be $[-b, \infty)$ unless other additional specifications are added.

In the renewal reward process, a key issue is to determine the long-term expected reward rate (reward obtained per unit time) as time t goes to infinity, i.e., $\lim_{t\to\infty} E\left[\frac{C(t)}{t}\right]$. To obtain this, we need to examine the long-term states of average reward $\frac{C(t)}{N(t)}$ and reward rate $\frac{C(t)}{t}$ first.

3.2.2 Long-Term States of Average Reward and Reward Rate

Lemma 3.2 ([150]). *Assume $\{\eta_k\}$ is a sequence of fuzzy random rewards with $\mu_{\eta_k(\omega)}(x) = \Pi_R(x - V_k(\omega))$ for almost every $\omega \in \Omega$, where $V_k, k = 1, 2, \cdots$ are i.i.d. random variables with finite expected values. Then we have*

$$\frac{\eta_1 + \cdots + \eta_n}{n} \xrightarrow{\text{Ch}} E[V_1]. \tag{3.13}$$

Proof. We denote $X_k = \eta_k - V_k$. Hence, for any $\omega \in \Omega$, fuzzy variables $X_k(\omega) = \eta_k(\omega) - V_k(\omega)$ have the same membership functions $\mu_{X_k(\omega)}(x) = \Pi_R(x)$ for $k = 1, 2, \cdots$. We first claim that

$$\frac{X_1 + \cdots + X_n}{n} \xrightarrow{\text{Ch}} 0. \tag{3.14}$$

From the assumption A2 on Π_R, for any $\omega \in \Omega$, we know the membership function $\mu_{X_1(\omega)+\cdots+X_n(\omega)}$ is nonincreasing on $\Re^+$, and nondecreasing on $\Re^-$. Therefore, for any $\omega \in \Omega$ and $\varepsilon > 0$, we have

$$\begin{aligned}
&\mathrm{Cr}\left\{\left|\frac{X_1(\omega)+\cdots+X_n(\omega)}{n}\right|\geq\varepsilon\right\}\\
&\quad\leq \mathrm{Pos}\left\{\frac{X_1(\omega)+\cdots+X_n(\omega)}{n}\geq\varepsilon\right\}\bigvee \mathrm{Pos}\left\{\frac{X_1(\omega)+\cdots+X_n(\omega)}{n}\leq-\varepsilon\right\}\\
&\quad= \mathrm{Pos}\left\{\frac{X_1(\omega)+\cdots+X_n(\omega)}{n}=\varepsilon\right\}\bigvee \mathrm{Pos}\left\{\frac{X_1(\omega)+\cdots+X_n(\omega)}{n}=-\varepsilon\right\}.
\end{aligned}$$

Therefore, Lemma 2.1 implies

$$\lim_{n\to\infty}\mathrm{Cr}\left\{\left|\frac{X_1(\omega)+\cdots+X_n(\omega)}{n}\right|\geq\varepsilon\right\}=0.$$

Using the dominated convergence theorem, we have

$$\begin{aligned}
&\lim_{n\to\infty}\mathrm{Ch}\left\{\left|\frac{X_1+\cdots+X_n}{n}\right|\geq\varepsilon\right\}\\
&\quad=\int_{\Omega}\lim_{n\to\infty}\mathrm{Cr}\left\{\left|\frac{X_1(\omega)+\cdots+X_n(\omega)}{n}\right|\geq\varepsilon\right\}\Pr(\mathrm{d}\omega)=0.
\end{aligned}$$

This proves our claim.

On the other hand, since $V_k, k=1,2,\cdots$ are i.i.d. random variables with finite expected values, by the strong law of large numbers for random variables, we get

$$\frac{V_1+\cdots+V_n}{n}\xrightarrow{a.s.}E[V_1]. \tag{3.15}$$

Noting that for any $\varepsilon>0$,

$$\begin{aligned}
&\mathrm{Ch}\left\{\left|\frac{\eta_1+\cdots+\eta_n}{n}-E[V_1]\right|\geq\varepsilon\right\}\\
&\quad\leq\mathrm{Ch}\left\{\left|\frac{X_1+\cdots+X_n}{n}\right|\geq\frac{\varepsilon}{2}\right\}+\mathrm{Ch}\left\{\left|\frac{V_1+\cdots+V_n}{n}-E[V_1]\right|\geq\frac{\varepsilon}{2}\right\}.
\end{aligned}$$

Combining (3.14) with (3.15) implies

$$\frac{\eta_1+\cdots+\eta_n}{n}\xrightarrow{\mathrm{Ch}}E[V_1].$$

The lemma is proved. □

Theorem 3.5 ([150]). *Suppose $(\xi_1,\eta_1),(\xi_2,\eta_2),\cdots$ is a sequence of pairs of fuzzy random interarrival times and rewards, where $\mu_{\xi_k(\omega)}(x)=\Pi_I(x-U_k(\omega))$, $U_k(\omega)\geq a$, and $\mu_{\eta_k(\omega)}(x)=\Pi_R(x-V_k(\omega))$, $V_k(\omega)\geq b$, for almost every $\omega\in\Omega$, $k=1,2,\cdots$,*

$N(t)$ is the fuzzy random renewal variable, and $C(t)$ is the total reward. If $\{U_k\}$ and $\{V_k\}$ are i.i.d. random variable sequences with finite expected values, respectively, then

$$\frac{C(t)}{N(t)} \xrightarrow{\text{Ch}} E[V_1].$$

Proof. The proof breaks into four steps as below.

Step 1. For any $\omega \in \Omega$, from the definition we know

$$C(t)(\omega) = \eta_1(\omega) + \cdots + \eta_{N(t)(\omega)}(\omega).$$

Now, for almost every $\omega \in \Omega$, let us prove

$$\lim_{t\to\infty} \text{Cr}\left\{\left|\frac{C(t)(\omega)}{N(t)(\omega)} - E[V_1]\right| \geq \varepsilon\right\} = 0$$

for any $\varepsilon > 0$. Equivalently, it suffices to obtain the fact that for any $\delta > 0$, there is a set $F_\delta \in \mathscr{A}$ with $\text{Cr}(F_\delta) < \delta$ such that

$$\frac{C(t)(\omega)}{N(t)(\omega)} \to E[V_1] \quad (t \to \infty)$$

uniformly w.r.t. $\gamma \in \Gamma \setminus F_\delta$.

Step 2. From the proof of Lemma 3.2, we have

$$\lim_{t\to\infty} \text{Cr}\left\{\left|\frac{\eta_1(\omega) + \cdots + \eta_n(\omega)}{n} - E[V_1]\right| \geq \varepsilon\right\} = 0$$

for any $\varepsilon > 0$, which is equivalent to the fact that, for any $\varepsilon > 0$ and $\delta > 0$, there exist $A_\delta \in \mathscr{A}$ with $\text{Cr}(A_\delta) < \delta/2$ and a positive integer M such that for all $\gamma \in \Gamma \setminus A\delta$,

$$\left|\frac{\eta_1(\omega,\gamma) + \cdots + \eta_n(\omega,\gamma)}{n} - E[V_1]\right| < \varepsilon$$

whenever $n \geq M$.

Step 3. Recalling the proof of Theorem 3.1, we have that for almost every $\omega \in \Omega$, given any $\varepsilon > 0$,

$$\lim_{t\to\infty} \text{Cr}\left\{\frac{1}{N(t)(\omega)} \geq \varepsilon\right\} = 0.$$

Hence, for the above positive integer M, there exist $B_\delta \in \mathscr{A}$ with $\text{Cr}(B_\delta) < \delta/2$ and a positive real number t_M such that for all $\gamma \in \Gamma \setminus B_\delta$,

$$\frac{1}{N(t)(\omega,\gamma)} \leq \frac{1}{M}$$

or equivalently,

$$N(t)(\omega,\gamma) \geq M,$$

whenever $t \geq t_M$.

Step 4. Combining the above results in Steps 2 and 3, for any $\varepsilon > 0$ and $\delta > 0$, there exist

$$F_\delta = A_\delta \cup B_\delta \in \mathscr{A}$$

and a positive real number t_M such that for all $\gamma \in \Gamma \setminus F_\delta$, where $\mathrm{Cr}(F_\delta) < \delta$, we have

$$N(t)(\omega,\gamma) \geq M,$$

and

$$\begin{aligned}
&\left|\frac{\eta_1(\omega,\gamma)+\cdots+\eta_{N(t)(\omega,\gamma)}(\omega,\gamma)}{N(t)(\omega,\gamma)} - E[V_1]\right| \\
&= \left|\frac{C(t)(\omega,\gamma)}{N(t)(\omega,\gamma)} - E[V_1]\right| < \varepsilon
\end{aligned}$$

provided $t \geq t_M$. As a consequence, for any $\varepsilon > 0$,

$$\lim_{t\to\infty} \mathrm{Cr}\left\{\left|\frac{C(t)(\omega)}{N(t)(\omega)} - E[V_1]\right| \geq \varepsilon\right\} = 0.$$

This result holds for almost every $\omega \in \Omega$.

Finally, applying the dominated convergence theorem, we have

$$\begin{aligned}
&\lim_{t\to\infty} \mathrm{Ch}\left\{\left|\frac{C(t)}{N(t)} - E[V_1]\right| \geq \varepsilon\right\} \\
&= \lim_{t\to\infty} \int_\Omega \mathrm{Cr}\left\{\left|\frac{C(t)(\omega)}{N(t)(\omega)} - E[V_1]\right| \geq \varepsilon\right\} \mathrm{Pr}(\mathrm{d}\omega) \\
&= \int_\Omega \lim_{t\to\infty} \mathrm{Cr}\left\{\left|\frac{C(t)(\omega)}{N(t)(\omega)} - E[V_1]\right| \geq \varepsilon\right\} \mathrm{Pr}(\mathrm{d}\omega) = 0,
\end{aligned}$$

which proves the required result. □

Theorem 3.6 ([150]). *Under the same conditions as in Theorem 3.5, we have*

$$\frac{C(t)}{t} \xrightarrow{\mathrm{Ch}} \frac{E[V_1]}{E[U_1]}.$$

Proof. By Theorem 3.5, we have the result

$$\frac{C(t)}{N(t)} \xrightarrow{\mathrm{Ch}} E[V_1]. \tag{3.16}$$

In addition, from Theorem 3.3, we have

$$\frac{N(t)}{t} \xrightarrow{\text{Ch}} \frac{1}{E[U_1]}. \tag{3.17}$$

Furthermore, we can get the following equivalent events:

$$\begin{aligned}
\frac{C(t)}{t} - \frac{E[V_1]}{E[U_1]} &\Longleftrightarrow \frac{C(t)}{N(t)} \cdot \frac{N(t)}{t} - \frac{E[V_1]}{E[U_1]} \\
&\Longleftrightarrow \frac{C(t)}{N(t)} \cdot \frac{N(t)}{t} - \frac{N(t)}{t} \cdot E[V_1] + \frac{N(t)}{t} \cdot E[V_1] - \frac{E[V_1]}{E[U_1]} \\
&\Longleftrightarrow \left(\frac{N(t)}{t} - \frac{1}{E[U_1]}\right)\left(\frac{C(t)}{N(t)} - E[V_1]\right) + \frac{1}{E[U_1]}\left(\frac{C(t)}{N(t)} - E[V_1]\right) \\
&\quad + E[V_1]\left(\frac{N(t)}{t} - \frac{1}{E[U_1]}\right).
\end{aligned}$$

Without losing any generality, let $0 < \varepsilon < 1$. We can obtain that

$$\begin{aligned}
&\text{Ch}\left\{\left|\frac{C(t)}{t} - \frac{E[V_1]}{E[U_1]}\right| \geq \varepsilon\right\} \\
&\leq \text{Ch}\left\{\left|\frac{N(t)}{t} - \frac{1}{E[U_1]}\right| \cdot \left|\frac{C(t)}{N(t)} - E[V_1]\right| \geq \frac{\varepsilon}{3}\right\} \\
&\quad + \text{Ch}\left\{\left|\frac{C(t)}{N(t)} - E[V_1]\right| \geq \frac{E[U_1]\varepsilon}{3}\right\} + \text{Ch}\left\{\left|\frac{N(t)}{t} - \frac{1}{E[U_1]}\right| \geq \frac{\varepsilon}{3E[V_1]}\right\} \\
&\leq \text{Ch}\left\{\left|\frac{N(t)}{t} - \frac{1}{E[U_1]}\right| \geq \frac{\varepsilon}{3}\right\} + \text{Ch}\left\{\left|\frac{C(t)}{N(t)} - E[V_1]\right| \geq \frac{\varepsilon}{3}\right\} \\
&\quad + \text{Ch}\left\{\left|\frac{C(t)}{N(t)} - E[V_1]\right| \geq \frac{E[U_1]\varepsilon}{3}\right\} + \text{Ch}\left\{\left|\frac{N(t)}{t} - \frac{1}{E[U_1]}\right| \geq \frac{\varepsilon}{3E[V_1]}\right\}.
\end{aligned}$$

On one hand, by (3.16), we have

$$\text{Ch}\left\{\left|\frac{C(t)}{N(t)} - E[V_1]\right| \geq \frac{\varepsilon}{3}\right\} \to 0 \quad (t \to \infty),$$

and

$$\text{Ch}\left\{\left|\frac{C(t)}{N(t)} - E[V_1]\right| \geq \frac{E[U_1]\varepsilon}{3}\right\} \to 0 \quad (t \to \infty).$$

On the other hand, from (3.17), we have

$$\text{Ch}\left\{\left|\frac{N(t)}{t} - \frac{1}{E[U_1]}\right| \geq \frac{\varepsilon}{3}\right\} \to 0 \quad (t \to \infty),$$

and

$$\mathrm{Ch}\left\{\left|\frac{N(t)}{t}-\frac{1}{E[U_1]}\right|\geq\frac{\varepsilon}{3E[V_1]}\right\}\to 0 \quad (t\to\infty).$$

As a consequence, we obtain

$$\lim_{t\to\infty}\mathrm{Ch}\left\{\left|\frac{C(t)}{t}-\frac{E[V_1]}{E[U_1]}\right|\geq\varepsilon\right\}=0.$$

The theorem is proved. □

3.2.3 Long-Term Expected Reward Rate

In what follows, a fuzzy stochastic renewal reward theorem will be presented. Here we require the fuzzy random variable sequences in question to be uniformly and essentially bounded. A sequence of fuzzy random variables $\{\xi_n\}$ is said to be uniformly and essentially bounded if there are two real numbers b_L and b_U such that for each $k=1,2,\cdots$, we have $\mathrm{Ch}\{b_L\leq\xi_n\leq b_U\}=1$.

Theorem 3.7 (Fuzzy Random Renewal Reward Theorem [150]). *Suppose that $(\xi_1,\eta_1),(\xi_2,\eta_2),\cdots$ is a sequence of pairs of fuzzy random interarrival times and rewards, where $\{\eta_k\}$ is uniformly essentially bounded, $\mu_{\xi_k(\omega)}(x)=\Pi_I(x-U_k(\omega))$ with $U_k\geq a+h$ almost surely, and $\mu_{\eta_k(\omega)}(x)=\Pi_R(x-V_k(\omega))$, for almost every $\omega\in\Omega$, $k=1,2,\cdots$, $N(t)$ is the fuzzy random renewal variable, and $C(t)$ is the total reward. If $\{U_k\}$ and $\{V_k\}$ are i.i.d. random variable sequences with finite expected values, respectively, and $\{\xi_k(\omega)\}$ and $\{\eta_k(\omega)\}$ are mutually $\top$-independent for almost every $\omega\in\Omega$, then*

$$\lim_{t\to\infty}\frac{E[C(t)]}{t}=\frac{E[V_1]}{E[U_1]}. \tag{3.18}$$

Proof. From the supposition, rewards $\{\eta_k\}$ are a sequence of uniformly and essentially bounded fuzzy random variables, i.e., there exist b_L and b_U such that for each $k=1,2,\cdots$, we have $\mathrm{Ch}\{b_L\leq\eta_k\leq b_U\}=1$. It follows from the self-duality of Ch that

$$\mathrm{Ch}\{\eta_k\notin[b_L,b_U]\}=0.$$

Therefore, by the definition of the mean chance, the following equalities

$$\mathrm{Cr}\{\eta_k(\omega)\notin[b_L,b_U]\}=0,\ k=1,2,\cdots$$

which is equivalent to

$$\mathrm{Cr}\{b_L\leq\eta_k(\omega)\leq b_U\}=1,\ k=1,2,\cdots \tag{3.19}$$

hold for almost every $\omega \in \Omega$. On the other hand, since $U_k \geq a+h$ almost surely, we have

$$\mathrm{Cr}\{\xi_k(\omega) \geq h\} = 1,\ k = 1,2,\cdots \tag{3.20}$$

for almost every $\omega \in \Omega$.

In the following, we first claim

$$\mathrm{Ch}\left\{\frac{C(t)}{t} \geq r\right\} \leq I_{\left\{\frac{b_U}{h} \geq r\right\}}(r) \tag{3.21}$$

for any $t, r > 0$, where $I_{\{\cdot\}}$ is the indicator function of event $\{\cdot\}$. Since for almost every $\omega \in \Omega$, the sequences $\{\xi_k(\omega)\}$ and $\{\eta_k(\omega)\}$ are mutually $\top$-independent, we have

$$\begin{aligned}
&\mathrm{Cr}\left\{\frac{C(t)(\omega)}{t} \geq r\right\} \\
&\leq \mathrm{Pos}\left\{\sum_{k=1}^{N(t)(\omega)} \eta_k(\omega) \geq rt\right\} \\
&= \sup_n \mathrm{Pos}\left(\left\{\sum_{k=1}^{n} \eta_k(\omega) \geq rt\right\} \cap \{N(t)(\omega) = n\}\right) \\
&= \sup_n \top\left(\mathrm{Pos}\left\{\sum_{k=1}^{n} \eta_k(\omega) \geq rt\right\}, \mathrm{Pos}\{N(t)(\omega) = n\}\right).
\end{aligned}$$

for any $t, r > 0$. Furthermore, from (3.19), we can deduce

$$\begin{aligned}
\mathrm{Cr}\left\{\frac{C(t)(\omega)}{t} \geq r\right\} &\leq \sup_n \top\Big(\mathrm{Pos}\{n \cdot b_U \geq rt\}, \mathrm{Pos}\{N(t)(\omega) = n\}\Big) \\
&= \mathrm{Pos}\{N(t)(\omega) \cdot b_U \geq rt\} \\
&= \mathrm{Pos}\{N(t)(\omega) \geq [rt/b_U]\} \\
&= \mathrm{Pos}\left\{S_{[rt/b_U]}(\omega) \leq t\right\},
\end{aligned}$$

where $[x]$ is the smallest integer larger than or equal to $x \in \Re$. Recalling (3.20), we have

$$\begin{aligned}
\mathrm{Pos}\left\{S_{[rt/b_U]}(\omega) \leq t\right\} &\leq \mathrm{Pos}\left\{h \leq \frac{t}{[rt/b_U]}\right\} \\
&\leq \mathrm{Pos}\left\{r \leq \frac{b_U}{h}\right\} = I_{\left\{\frac{b_U}{h} \geq r\right\}}(r).
\end{aligned}$$

As a consequence, we have for any $t, r > 0$, the following inequality

$$\mathrm{Cr}\left\{\frac{C(t)(\omega)}{t} \geq r\right\} \leq I_{\left\{\frac{b_U}{h} \geq r\right\}}(r)$$

holds almost surely w.r.t. $\omega \in \Omega$. Integrating w.r.t. ω on the above inequality, we obtain (3.21).

Since $C(t)/t$ is a positive fuzzy random variable for any $t > 0$, from the definition (2.6) we have

$$\frac{E[C(t)]}{t} = \int_0^\infty \mathrm{Ch}\left\{\frac{C(t)}{t} \geq r\right\} \mathrm{d}r.$$

According to Theorem 3.6, we know

$$\frac{C(t)}{t} \xrightarrow{\mathrm{Ch}} \frac{E[V_1]}{E[U_1]},$$

it follows that

$$\lim_{t\to\infty} \mathrm{Ch}\left\{\frac{C(t)}{t} \geq r\right\} = \mathrm{Ch}\left\{\frac{E[V_1]}{E[U_1]} \geq r\right\} \tag{3.22}$$

for almost every $r \in \Re$. Noting that

$$\int_0^\infty I_{\left\{\frac{b_U}{h} \geq r\right\}}(r)\mathrm{d}r = \frac{b_U}{h} < \infty,$$

and combining (3.21) with (3.22), Lebesgue's dominated convergence theorem implies

$$\lim_{t\to\infty} \frac{E[C(t)]}{t} = \int_0^\infty \lim_{t\to\infty} \mathrm{Ch}\left\{\frac{C(t)}{t} \geq r\right\} \mathrm{d}r = \frac{E[V_1]}{E[U_1]}.$$

The proof of the theorem is completed. □

Remark 3.4. In Theorem 3.7, if $\{(\xi_k, \eta_k)\}$ becomes a sequence of random variable pairs, then the result (3.18) degenerates to that of the stochastic renewal reward theorem.

In fact, since each pair of ξ_k and η_k becomes random variables, for any given $\omega \in \Omega$, both fuzzy variables $\xi_k(\omega)$ and $\eta_k(\omega)$ degenerate to crisp numbers. Hence, the membership functions of $\xi_k(\omega)$ and $\eta_k(\omega)$ degenerate to

$$\mu_{\xi_k(\omega)}(r) = \Pi_I\Big(r - U_k(\omega)\Big) = \begin{cases} 1, & \text{if } r = U_k(\omega) \\ 0, & \text{otherwise,} \end{cases}$$

and

$$\mu_{\eta_k(\omega)}(r) = \Pi_R\Big(r - V_k(\omega)\Big) = \begin{cases} 1, & \text{if } r = V_k(\omega) \\ 0, & \text{otherwise,} \end{cases}$$

respectively, which implies $\xi_k = U_k$ for $k = 1, 2, \cdots$. Therefore, the result (3.18) of Theorem 3.7 becomes

$$\lim_{t\to\infty} \frac{E[C(t)]}{t} = \frac{E[\eta_1]}{E[\xi_1]},$$

which is the result of stochastic renewal reward theorem (see [127]).

Furthermore, if both of the fuzzy random interarrival times and rewards have symmetric distributions, then we can obtain the following result which is exactly the same as that of the stochastic renewal reward theorem.

Corollary 3.2 ([150]). *Under the condition of Theorem 3.7, furthermore, if for the distributions of interarrival times and rewards, the Π_I and Π_R are symmetric with respect to zero. Then we have*

$$\lim_{t\to\infty} \frac{E[C(t)]}{t} = \frac{E[\eta_1]}{E[\xi_1]}.$$

Proof. Since Π_I and Π_R are symmetric with respect to zero, respectively, we know that for every $\omega \in \Omega$, $\mu_{\xi_1(\omega)}$ and $\mu_{\eta_1(\omega)}$ are symmetric with respect to $U_1(\omega)$ and $V_1(\omega)$, respectively, and we can calculate that

$$\int_0^\infty \mathrm{Cr}\{\xi_1(\omega) \geq r\}\mathrm{d}r = U_1(\omega),$$

and

$$\int_0^\infty \mathrm{Cr}\{\eta_1(\omega) \geq r\}\mathrm{d}r = V_1(\omega).$$

Integrating with respect to ω on the above equalities, we obtain

$$E[\xi_1] = \int_\Omega \left[\int_0^\infty \mathrm{Cr}\{\xi_1(\omega) \geq r\}\mathrm{d}r\right] \Pr(\mathrm{d}\omega) = E[U_1], \tag{3.23}$$

and

$$E[\eta_1] = \int_\Omega \left[\int_0^\infty \mathrm{Cr}\{\eta_1(\omega) \geq r\}\mathrm{d}r\right] \Pr(\mathrm{d}\omega) = E[V_1]. \tag{3.24}$$

It follows from Theorem 3.7 and (3.23)–(3.24) that

$$\lim_{t\to\infty} \frac{E[C(t)]}{t} = \frac{E[\eta_1]}{E[\xi_1]}.$$

The proof of the corollary is completed. □

3.3 Explanatory Examples

To explain the utility of the fuzzy random elementary renewal theorem (Theorem 3.4), and fuzzy random renewal reward theorem (Theorem 3.7), we provide in this section two application examples of a multiservice system and a replacement problem.

3.3.1 An Application to Multiservice Systems

Consider a multi-service station (see Fig. 3.5). Assume that there are 6 kinds of services provided by the system and that customers come for the service i at probability p_i, where $p_i = 1/6, i = 1,2,\cdots,6$. The customers are served independently of each other, and service time T_i (minute) provided by service i and the associated cost C_i (\$) of service i in the station are assumed to be positive triangular fuzzy variables, as shown in Table 3.1.

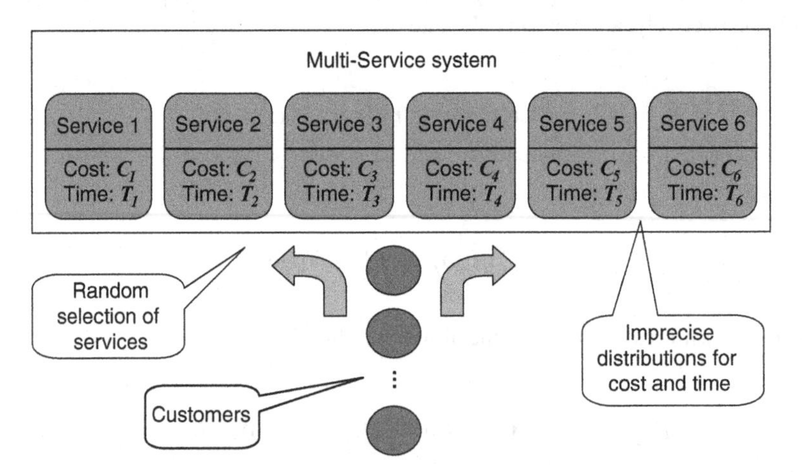

Fig. 3.5 Fuzzy random multi-service system

Table 3.1 Distributions of the service time and cost

Service i	Service time T_i (M)	Cost C_i (\$)
1	$T_1 = (2,3,5)$	$C_1 = (10,50,60)$
2	$T_2 = (3,4,6)$	$C_2 = (30,70,80)$
3	$T_3 = (4,5,7)$	$C_3 = (20,60,70)$
4	$T_4 = (5,6,8)$	$C_4 = (40,80,90)$
5	$T_5 = (6,7,9)$	$C_5 = (60,100,110)$
6	$T_6 = (7,8,10)$	$C_6 = (50,90,100)$

Table 3.2 Distributions of the random parameters U_k and V_k

Service i	Realization $U_k(\omega_i)$	Realization $V_k(\omega_i)$	Probability
1	$U_k(\omega_1) = 3$	$V_k(\omega_1) = 50$	1/6
2	$U_k(\omega_2) = 4$	$V_k(\omega_2) = 70$	1/6
3	$U_k(\omega_3) = 5$	$V_k(\omega_3) = 60$	1/6
4	$U_k(\omega_4) = 6$	$V_k(\omega_4) = 80$	1/6
5	$U_k(\omega_5) = 7$	$V_k(\omega_5) = 100$	1/6
6	$U_k(\omega_6) = 8$	$V_k(\omega_6) = 90$	1/6

Here, for each service i, the service time T_i and the cost C_i are assumed to be $\top$-independent. Taking the t-norm $\top$ as a Yager t-norm $\top^Y_\lambda, \lambda > 1$, we shall calculate the long-term expected number of customers served and expected rewards earned by this multiservice station, respectively.

Let ξ_k be the interarrival time between the $(k-1)$th and kth customers requesting services, and η_k be the reward gained from the kth customer, $k = 1, 2, \cdots$. Under the above assumptions, we know the service requested by the kth customer and the associated reward are stochastic, while the service time and the cost of each service in the station are fuzzy, and are characterized by triangular fuzzy variable T_i and C_i, respectively, for $i = 1, 2, \ldots, 6$. Taking all the scenarios into account, the interarrival times $\{\xi_k\}$ and rewards $\{\eta_k\}$ can be considered as two sequences of fuzzy random variables.

The distributions of the interarrival times ξ_k and rewards η_k, for $k = 1, 2, \cdots$ can be presented as follows:

$$\xi_k \sim \begin{pmatrix} T_1 & T_2 & T_3 & T_4 & T_5 & T_6 \\ p_1 & p_2 & p_3 & p_4 & p_5 & p_6 \end{pmatrix},$$

and

$$\eta_k \sim \begin{pmatrix} C_1 & C_2 & C_3 & C_4 & C_5 & C_6 \\ p_1 & p_2 & p_3 & p_4 & p_5 & p_6 \end{pmatrix},$$

respectively.

The total service time S_n for the first n customers is calculated by $S_n = \xi_1 + \cdots + \xi_n$, and the total number $N(t)$ of customers who have been served and the total reward $C(t)$ by time t, are given by

$$N(t) = \max\{n > 0 \mid 0 < S_n \le t\},$$

and

$$C(t) = \eta_1 + \eta_2 + \cdots + \eta_{N(t)},$$

respectively. Given the above distributions of fuzzy random interarrival times $\{\xi_k\}$ and rewards $\{\eta_k\}$, without losing any generality, we assign values of two i.i.d. random variable sequences $\{U_k\}$ and $\{V_k\}$ on probability space Ω, as shown in Table 3.2. Hence, the distributions of each pair of ξ_k and η_k for $k = 1, 2, \cdots$ can be rewritten as

$$\xi_k(\omega) = \Big(U_k(\omega) - 1, U_k(\omega), U_k(\omega) + 2\Big) \text{ (minute)},$$

and

$$\eta_k(\omega) = \Big(V_k(\omega) - 40, V_k(\omega), V_k(\omega) + 10\Big) \text{ (\$)},$$

for any $\omega \in \Omega$.

Therefore, we can find the possibility function $\Pi_I = (-1, 0, 2)$. Furthermore, since

$$\Pi_I(x) = \begin{cases} x+1, & \text{if } x \in [-1,0] \\ \dfrac{2-x}{2}, & \text{if } x \in [0,2] \\ 0, & \text{otherwise,} \end{cases}$$

we have

$$f_\lambda^Y \circ \Pi_I(x) = \begin{cases} (-x)^\lambda, & \text{if } x \in [-1,0] \\ \left(\dfrac{x}{2}\right)^\lambda, & \text{if } x \in [0,2] \\ 1, & \text{otherwise,} \end{cases}$$

which is a convex function on $[-1,2]$. Thus, we have

$$\text{co}(f_\lambda^Y \circ \Pi_I)(x) = f_\lambda^Y \circ \Pi_I(x) > 0$$

for any nonzero $x \in [-1,2]$. Similarly, we can get $\text{co}(f_\lambda^Y \circ \Pi_R)(x) > 0$ for any nonzero $x \in [-40, 10]$.

As a consequence, by Theorems 3.4 and 3.7, we can calculate that the long-term expected number of customers served per minute and the expected reward earned per minute by this service station in the long run are

$$\lim_{t \to \infty} \frac{E[N(t)]}{t} = \frac{1}{\sum_{i=1}^{6} p_i U_1(\omega_i)} \doteq 0.03,$$

and

$$\lim_{t \to \infty} \frac{E[C(t)]}{t} = \frac{\sum_{i=1}^{6} p_i V_1(\omega_i)}{\sum_{i=1}^{6} p_i U_1(\omega_i)} \doteq 13.6(\$),$$

respectively.

3.3.2 A Replacement Application

We now address the calculation for the long-term expected number of the elements replaced as well as the cost of the replacement per unit time in a machine. Suppose that the lifetimes ξ_k of some kind of elements are positive triangular fuzzy random variables with distributions as follows:

$$\xi_k = \left(U_k - 3, U_k, U_k + 3\right) \text{ (month)}, \quad k = 1, 2, \cdots \tag{3.25}$$

where U_k are i.i.d. uniformly distributed random variables with $U_k \sim \mathscr{U}(4,6)$. Each time when we replace a broken element with a new one, we have to pay for it. The costs $\eta_k, k = 1, 2, \cdots$ of the element are also assumed to be positive triangular fuzzy random variables that have the distributions given below:

$$\eta_k = \left(V_k - 20, V_k, V_k + 40\right) \text{ (\$)}, \quad k = 1, 2, \cdots \tag{3.26}$$

where V_k are i.i.d. random variables with $V_k \sim \mathscr{U}(30,90)$. The fuzzy realizations of the fuzzy random lifetimes and costs, i.e., $\{\xi_k(\omega)\}$ and $\{\eta_k(\omega)\}$, are two sequences of $\top$-independent fuzzy variables, respectively. Here, we take t-norm $\top$ as the product t-norm with the additive generator $f^P = -log$. Moreover, we assume that the costs and the lifetimes of the elements are mutually independent. Similar to the previous example, the total number of elements that have been replaced by time t, denoted $N(t)$, can be expressed by

$$N(t) = \max\left\{n > 0 \mid 0 < \xi_1 + \xi_2 + \cdots + \xi_n \leq t\right\},$$

and the total cost of the replacement by time t, denoted $C(t)$, can be given by

$$C(t) = \eta_1 + \eta_2 + \cdots + \eta_{N(t)}.$$

Now we utilize Theorems 3.4 and 3.7 to compute the long-run expected number of elements replaced and expected average cost per month in this replacement process. We note that the lifetimes and the costs are positive triangular fuzzy random variables, and according to (3.25) and (3.26), for any $\omega \in \Omega$,

$$\Pi_I = \xi_k(\omega) - U_k(\omega) = (-3, 0, 3)$$

and

$$\Pi_R = \eta_k(\omega) - V_k(\omega) = (-20, 0, 40)$$

for $k = 1, 2, \cdots$ are the distributions of triangular fuzzy variables which are bounded and unimodal. Furthermore, we can calculate that

$$\Pi_I(x) = \begin{cases} 1 + \frac{x}{3}, & \text{if } x \in [-3, 0] \\ 1 - \frac{x}{3}, & \text{if } x \in [0, 3] \\ 0, & \text{otherwise,} \end{cases}$$

and we have

$$f^P \circ \Pi_I(x) = \begin{cases} -log\left(\frac{x}{3} + 1\right), & \text{if } x \in [-3, 0] \\ -log\left(-\frac{x}{3} + 1\right), & \text{if } x \in [0, 3] \\ \infty, & \text{otherwise,} \end{cases}$$

which is a convex function on $[-3,3]$ and therefore satisfies that $\text{co}(f^P \circ \Pi_I)(x) > 0$ for any nonzero $x \in [-3,3]$. By the same reasoning, we can work out that the composition function $f^P \circ \Pi_R$ is convex and takes on positive values for any nonzero $x \in [-20,40]$. Finally, by (3.25) and (3.26) again, we know that η_k for $k = 1,2,\ldots$ are uniformly and essentially bounded fuzzy random variables, and the random parameters $U_k, k = 1,2,\cdots$ of ξ_k ensure that $U_k \geq 3$ almost surely. As a consequence, from Theorems 3.4 and 3.7, we obtain the long-run expected number of elements replaced per month is

$$\lim_{t\to\infty} \frac{E[N(t)]}{t} = \frac{1}{E[U_1]} = 0.2,$$

and the long-run expected average cost of the replacement per month is

$$\lim_{t\to\infty} \frac{E[C(t)]}{t} = \frac{E[V_1]}{E[U_1]} = 12(\$).$$

Part II
Models

Chapter 4
System Reliability Optimization Models with Fuzzy Random Lifetimes

Reliability engineering plays a crucial role in a variety of systems. The primary goal of reliability engineering is to improve the reliability of a system. The redundancy allocation as a direct way of enhancing the system reliability involves the selection of the optimal combination of components and a system-level design configuration so as to maximize the reliability under the given cost and weight constraints, or alternatively, to meet reliability and weight constraints at a minimum cost. In conventional reliability models, an underlying assumption is that all the lifetimes of the components are characterized by random variables with probability distributions. Various kinds of stochastic reliability models have been proposed for different optimization purposes along this direction (see Barlow and Proschan [7], Elegbede et al. [33], Tavakkoli-Moghaddam et al. [137], and Yu et al. [170]).

Any probability distribution of lifetime that is reliable should be determined based on sufficient and precise lifetime data obtained via a large number of lifetime tests. Nevertheless, the real conditions for data capture (e.g., number of data, measurements, the ambient influence on the lifetime, data security,) vary largely depending on the types of devices. More often than not, such quantitative sufficiency or qualitative precision of the data, leave alone the both, sometimes is unable to be satisfied in real-life situations. For instance, to many long-life devices, never are there plenty of observations available, and for many devices (e.g., those with aluminum electrolytic capacitors inside), they are vulnerable to the ambient factors of working environment (e.g., temperature, humidity). So it is not easy to precisely predict the product lifetime in multifarious real working condition. Hence, given some working condition (a random scenario), the producer specialists are prone to recommend an estimation in terms of a safe lifetime range for the customers. Indubitably, this imprecise estimation together with the randomly changing working conditions leads to a sort of fuzzy probability distributions for product lifetime. Also, in some real-life circumstances, observations and measurements of lifetime are never precise (see [141]). Such imprecise observations or measurements could be engendered by the instability within the lifetime testing condition (e.g., the influence of surrounding environment on the tests), or caused by the intrinsic imprecision of tests (e.g., Accelerated Life Testing as a popular lifetime-value

S. Wang and J. Watada, *Fuzzy Stochastic Optimization: Theory, Models and Applications*, DOI 10.1007/978-1-4419-9560-5_4,

estimation method). In such a case, the lifetime data are therefore not precise but more or less fuzzy, and the real lifetime distribution should be also a kind of fuzzy probability distribution.

In many practical situations (as mentioned above), the stochastic variability (random working setting or data randomly tested in the experiments) and the fuzziness or imprecision (fuzzy estimation in each random scenario or imprecise measurement on each data) coexist in lifetime distributions. The reliability problems in those circumstances are unsuitable to be handled by probability models.

In this chapter, making use of fuzzy random variables as a tool to characterize the component lifetimes, we discuss system reliability optimization to a fuzzy random parallel-series system. For more studies related to this subject, see also references such as Kuo et al. [70], Kuo and Wan [71], and Mahapatra and Roy [107].

In Sect. 4.1, we introduce some necessary assumptions and the notation for the redundancy allocation problem. Then we define the reliability function with fuzzy random component lifetimes through the chance measure and build two fuzzy random redundancy allocation models (FR-RAMs I and II) which are tasks of fuzzy stochastic integer programming problems. In addition, some key difficulties for the models built are also indicated.

In Sect. 4.2, we discuss some mathematical properties of the models that include an analytical formula for reliability (with fuzzy random lifetimes) which is derived under a convexity condition for lifetime distributions, and some sensitivity conditions for the reliability w.r.t. the threshold lifetime.

In Sect. 4.3, we discuss the computation schemes for the reliability function that are critical to solving the models. As complexity increases, a formula is given for calculating the exact value of reliability with discrete lifetime distributions, a random-simulation algorithm is presented to compute the reliability with convex lifetime distributions, and a fuzzy-random-simulation algorithm is given to estimate the reliability value in a general case (with nonconvex lifetime distributions). Furthermore, the convergence properties for those computation schemes are discussed.

In Sect. 4.4, we introduce the solution approach for the models. Embedding the computation schemes into a genetic algorithm (GA), a hybrid solution approach is formed to solve the models, and some key procedures of this hybrid algorithm are initialization, solution evaluation or solution feasibility checking by computation schemes for reliability, solution selection, solution crossover and mutation, and solution improvement.

In Sect. 4.5, we conclude with several numerical experiments as well as comparisons for both models.

4.1 Problem Formulation

4.1.1 Mathematical Modeling

We consider a parallel-series system composed of s subsystems in parallel (see Fig. 4.1). Each subsystem i $(1 \leq i \leq s)$ is made up of actively redundant components

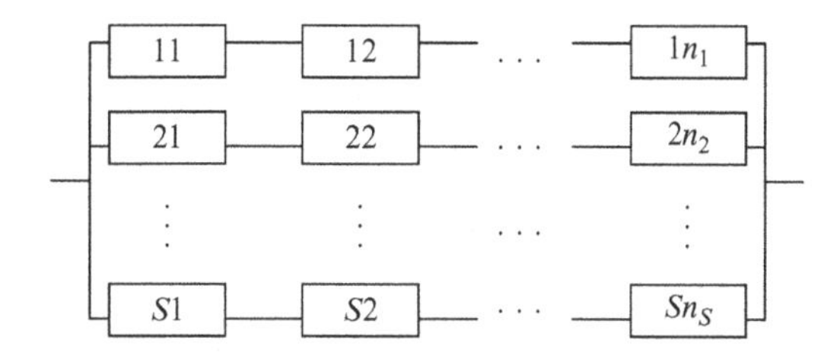

Fig. 4.1 A parallel-series system

in series. The lifetimes of the components are characterized by fuzzy random variables. The problem is how to find the optimal redundancy allocations to this fuzzy random parallel-series system so as to maximize system reliability, or to minimize the total cost of the system. Based on those two different objectives, two fuzzy random redundancy allocation models (FR-RAMs) will be built in this section, respectively.

Assumptions

1. All the lifetimes of components are treated as fuzzy random variables.
2. The redundancy level of subsystem i is bounded below by l_i and above by u_i.
3. The components of the same type have independent and identically distributed (i.i.d.) lifetimes.

Based on the above assumptions and notations, in this s-stage fuzzy random parallel-series system, if we use a fuzzy random vector

$$\xi = \begin{pmatrix} \xi_{1,1,1}, & \cdots, & \xi_{1,1,x_{1,1}}, & \cdots, & \xi_{1,n_1,1}, & \cdots, & \xi_{1,n_1,x_{1,n_1}}, \\ \cdots, & & & \cdots, & & & \cdots, \\ \xi_{s,1,1}, & \cdots, & \xi_{s,1,x_{s,1}}, & \cdots, & \xi_{s,n_s,1}, & \cdots, & \xi_{s,n_s,x_{s,n_s}} \end{pmatrix}$$

to characterize the fuzzy random lifetimes of the components, then the system lifetime at allocation x can be expressed as

$$T(x,\xi) = \bigvee_{i=1}^{s} \left[\bigwedge_{j=1}^{n_i} \left(\sum_{k=1}^{x_{i,j}} \xi_{i,j,k} \right) \right]. \tag{4.1}$$

Using the chance measure $\mathrm{Ch}\{\cdot\}$, the reliability of the fuzzy random parallel-series system can be characterized as follows:

$$\begin{aligned} R_{t^0}(x) &= \mathrm{Ch}\left\{T(x,\xi) \geq t^0\right\} \\ &= \int_{\Omega} \mathrm{Cr}\left\{ \bigvee_{i=1}^{s} \left[\bigwedge_{j=1}^{n_i} \left(\sum_{k=1}^{x_{i,j}} \xi_{i,j,k}(\omega) \right) \right] \geq t^0 \right\} \Pr(\mathrm{d}\,\omega), \end{aligned} \tag{4.2}$$

Notation

s	number of subsystems
i	index of subsystems, $1 \leq i \leq s$
n_i	number of different component types available for subsystem i
l_i	lower bounds on the number of redundant components in subsystem i
u_i	upper bounds on the number of redundant components in subsystem i
$x_{i,j}$	number of components of type j in subsystem i
x	decision vector $(x_{1,1}, \cdots, x_{1,n_1}, \cdots, x_{s,1}, \cdots, x_{s,n_s})$
$\xi_{i,j,k}$	the fuzzy random lifetime of component k of type j in the subsystem i $1 \leq k \leq x_{ij}, 1 \leq j \leq n_i$
t^0	preselected threshold system lifetime
$R_{t^0}(x)$	system reliability for a decision x at the threshold lifetime t^0
c_{ij}	the cost of each component of type j in subsystem i
c^0	the maximum capital available
r^0	the target overall reliability of system

Remark 4.1. If the fuzzy random vector ξ reduces to a random vector, therefore $T(x, \xi(\omega))$ is a crisp number for any $\omega \in \Omega$. By the definition of the chance measure, we have

$$\begin{aligned}
R_{t^0}(x) &= \mathrm{Ch}\{T(x,\xi) \geq t^0\}, \\
&= \int_\Omega \mathrm{Cr}\{T(x,\xi(\omega)) \geq t^0\} \Pr(\mathrm{d}\,\omega), \\
&= \int_\Omega I_{\left\{\omega | T(x,\xi(\omega)) \geq t^0\right\}}(\omega) \Pr(\mathrm{d}\,\omega), \\
&= \Pr\{T(x,\xi) \geq t^0\},
\end{aligned}$$

where $I_{\{\cdot\}}$ is the indicator function of set $\{\cdot\}$. Hence, the fuzzy random reliability (4.2) degenerates to the reliability in the probabilistic reliability theory.

Remark 4.2. If the fuzzy random vector ξ reduces to a fuzzy vector, then clearly the reliability (4.2) degenerates to

$$R_{t^0}(x) = \mathrm{Cr}\left\{T(x,\xi) \geq t^0\right\}$$

which is the reliability in the fuzzy system reliability model (see [179]).

Maximizing the overall system reliability under the given cost c^0 and redundancy level constraints, we obtain the first model([149]):
[FR-RAM I]

$$
\left.\begin{array}{l}
\max R_{t^0}(x) = \displaystyle\int_{\Omega} \mathrm{Cr}\left\{ \bigvee_{i=1}^{s} \left[\bigwedge_{j=1}^{n_i} \left(\sum_{k=1}^{x_{i,j}} \xi_{i,j,k}(\omega) \right) \right] \geq t^0 \right\} \Pr(\mathrm{d}\,\omega) \\
\text{subject to} \\
\qquad \sum_{i=1}^{s} \sum_{j=1}^{n_i} c_{ij} x_{i,j} \leq c^0, \\
\qquad l_i \leq \sum_{j=1}^{n_i} x_{i,j} \leq u_i, \text{for } i = 1, \cdots, s, \\
\qquad x_{i,j} \in \mathbb{N}, \text{for } j = 1, \cdots, n_i, i = 1, \cdots, s,
\end{array}\right\} \quad (4.3)
$$

where $\mathbb{N}$ is the set of all positive integers.

Alternatively, if we minimize the total cost meeting the overall system target reliability r^0 and the redundancy level constraints, the second model ([149]) can be built as follows:

[FR-RAM II]

$$
\left.\begin{array}{l}
\min \displaystyle\sum_{i=1}^{s} \sum_{j=1}^{n_i} c_{ij} x_{i,j} \\
\text{subject to} \\
\qquad R_{t^0}(x) \geq r^0, \text{for } i = 1, \cdots, s, \\
\qquad l_i \leq \displaystyle\sum_{j=1}^{n_i} x_{i,j} \leq u_i, \text{for } i = 1, \cdots, s, \\
\qquad x_{i,j} \in \mathbb{N}, \text{for } j = 1, \cdots, n_i, i = 1, \cdots, s,
\end{array}\right\} \quad (4.4)
$$

where

$$
R_{t^0}(x) = \int_{\Omega} \mathrm{Cr}\left\{ \bigvee_{i=1}^{s} \left[\bigwedge_{j=1}^{n_i} \left(\sum_{k=1}^{x_{i,j}} \xi_{i,j,k}(\omega) \right) \right] \geq t^0 \right\} \Pr(\mathrm{d}\,\omega).
$$

4.1.2 Difficulties

A difficult problem to tackle in the above FR-RAMs I and II is the computation of reliability $R_{t^0}(x)$ with fuzzy random lifetimes. In contrast to the probability reliability function which can be usually written equivalently as an analytical formula with the probability density functions of the lifetimes, to FR-RAMs however, it is not easy to find a generic formula to analytically calculate the reliability. This is on account of a complexity nature of the fuzzy random lifetime distribution being the combination of the probability distribution and the membership function. Such a difficulty on reliability (objective value) computation yields directly the difficulty to model solution.

4.2 Model Analysis

This section focuses on some theoretical properties of the FR-RAMs. In the following Theorem 4.1, we derive an analytical formula for the overall reliability $R_{t^0}(x)$ of the system, provided all the lifetimes of components have convex distributions. This formula is helpful to the computation of reliability with convex lifetimes, which will be discussed in Sect. 4.3. What's more, Theorems 4.2–4.4 discuss the sensitivity of reliability $R_{t^0}(x)$ in FR-RAMs w.r.t. the threshold lifetime t^0.

4.2.1 An Analytical Formula for Reliability

Theorem 4.1 ([149]). *Assume that in the FR-RAMs, the lifetimes $\xi_{i,j,k}$ of components for $i = 1,2,\cdots,s; j = 1,2,\cdots,n_i; andk = 1,2,\cdots,x_{ij}$ are fuzzy random variables on probability space $(\Omega,\Sigma,\Pr)$. Suppose for almost every $\omega \in \Omega$, that $\xi_{i,j,k}(\omega)$ is a convex fuzzy variable with membership function $\mu^{\omega}_{i,j}$, and $\mu^{\omega}_{i,1}(v^{\omega}_{i,1}) = \mu^{\omega}_{i,2}(v^{\omega}_{i,2}) = \cdots = \mu^{\omega}_{i,n_i}(v^{\omega}_{i,n_i}) = 1$ with $x_{i,1}v^{\omega}_{i,1} \le x_{i,2}v^{\omega}_{i,2} \le \cdots \le x_{i,n_i}v^{\omega}_{i,n_i}$. If we denote μ^{ω}_T the membership function of fuzzy variable $T(x,\xi(\omega))$ for any $\omega \in \Omega$, then given any allocation x, the system reliability is*

$$R_{t^0}(x) = \int_{\left\{\omega | t^0 \le x_{s,1}v^{\omega}_{s,1}\right\}} 1 - \frac{\mu^{\omega}_T(t^0)}{2}\Pr(\mathrm{d}\,\omega) + \int_{\left\{\omega | t^0 > x_{s,1}v^{\omega}_{s,1}\right\}} \frac{\mu^{\omega}_T(t^0)}{2}\Pr(\mathrm{d}\,\omega), \quad (4.5)$$

where for almost every $\omega \in \Omega$, $\mu^{\omega}_T(t)$ is given by

$$\mu^{\omega}_T(t) = \begin{cases} \bigwedge_{i=1}^{s} \mu^{\omega}_i(t), & t < x_{1,1}v^{\omega}_{1,1} \\ \bigwedge_{i=l+1}^{s} \mu^{\omega}_i(t), & x_{l,1}v^{\omega}_{l,1} \le t < x_{l+1,1}v^{\omega}_{l+1,1}, 1 \le l \le s-1 \\ \bigvee_{i=1}^{s} \mu^{\omega}_i(t), & t \ge x_{s,1}v^{\omega}_{s,1}; \end{cases} \quad (4.6)$$

here we assume that $x_{1,1}v^{\omega}_{1,1} \le x_{2,1}v^{\omega}_{2,1} \le \cdots \le x_{s,1}v^{\omega}_{s,1}$ without losing any generality, and $\mu^{\omega}_i(t)$ is calculated by

$$\mu^{\omega}_i(t) = \begin{cases} \bigvee_{j=1}^{n_i} \mu^{\omega}_{i,j}\left(\frac{t}{x_{i,j}}\right), & t < x_{i,1}v^{\omega}_{i,1} \\ \bigwedge_{j=1}^{l} \mu^{\omega}_{i,j}\left(\frac{t}{x_{i,j}}\right), & x_{i,l}v^{\omega}_{i,l} \le t < x_{i,l+1}v^{\omega}_{i,l+1}, 1 \le l \le n_i - 1 \\ \bigwedge_{j=1}^{n_i} \mu^{\omega}_{i,j}\left(\frac{t}{x_{i,j}}\right), & t \ge x_{i,n_i}v^{\omega}_{i,n_i}. \end{cases} \quad (4.7)$$

Proof. First of all, recalling that all lifetimes are convex and the lifetimes of the same type of components are identically distributed, given any i, j, and $t > 0$, we have

$$\mathrm{Pos}\left\{\sum_{k=1}^{x_{i,j}} \xi_{i,j,k}(\omega) = t\right\} = \mathrm{Pos}\left\{\xi_{i,j,1}(\omega) = \frac{t}{x_{i,j}}\right\} = \mu^{\omega}_{i,j}\left(\frac{t}{x_{i,j}}\right)$$

for almost every $\omega \in \Omega$, which also is a convex fuzzy number. Furthermore,

$$\mathrm{Pos}\left\{\sum_{k=1}^{x_{i,j}} \xi_{i,j,k}(\omega) = x_{i,j}v_{i,j}^{\omega}\right\} = \mu_{i,j}^{\omega}\left(v_{i,j}^{\omega}\right) = 1$$

for $i = 1,2,\cdots,s; j = 1,2,\cdots,n_i$ and

$$x_{i,1}v_{i,1}^{\omega} \le x_{i,2}v_{i,2}^{\omega} \le \cdots \le x_{i,n_i}v_{i,n_i}^{\omega}.$$

Therefore, by the minimum t-norm operation of convex fuzzy numbers (refer to Theorem B.5 in Appendix B), we obtain the membership function $\mu_i^{\omega}(t)$ of

$$\bigwedge_{j=1}^{n_i}\sum_{k=1}^{x_{i,j}} \xi_{i,j,k}(\omega)$$

is

$$\mu_i^{\omega}(t) = \begin{cases} \bigvee_{j=1}^{n_i} \mu_{i,j}^{\omega}\left(\frac{t}{x_{i,j}}\right), & t < x_{i,1}v_{i,1}^{\omega} \\ \bigwedge_{j=1}^{l} \mu_{i,j}^{\omega}\left(\frac{t}{x_{i,j}}\right), & x_{i,l}v_{i,l}^{\omega} \le t < x_{i,l+1}v_{i,l+1}^{\omega}, 1 \le l \le n_i - 1 \\ \bigwedge_{j=1}^{n_i} \mu_{i,j}^{\omega}\left(\frac{t}{x_{i,j}}\right), & t \ge x_{i,n_i}v_{i,n_i}^{\omega}, \end{cases}$$

for $i = 1,2,\cdots,s$. Note that $\mu_i^{\omega}(t)$ is also a convex fuzzy number, which is nondecreasing in $[-\infty, x_{i,1}v_{i,1}]$ and nonincreasing in $[x_{i,1}v_{i,1}^{\omega}, \infty]$ for all i, and

$$x_{1,1}v_{1,1}^{\omega} \le x_{2,1}v_{2,1}^{\omega} \le \cdots \le x_{s,1}v_{s,1}^{\omega},$$

making use of the maximum t-norm operation of convex fuzzy numbers (refer to Theorem B.4 in Appendix B), the membership function $\mu_T^{\omega}(t)$ of the system lifetime

$$T\left(x,\xi(\omega)\right) = \bigvee_{i=1}^{s}\left[\bigwedge_{j=1}^{n_i}\left(\sum_{k=1}^{x_{i,j}} \xi_{i,j,k}(\omega)\right)\right]$$

can be given in the following form

$$\mu_T^{\omega}(t) = \begin{cases} \bigwedge_{i=1}^{s} \mu_i^{\omega}(t), & t < x_{1,1}v_{1,1}^{\omega} \\ \bigwedge_{i=l+1}^{s} \mu_i^{\omega}(t), & x_{l,1}v_{l,1}^{\omega} \le t < x_{l,1}v_{l+1,1}^{\omega}, 1 \le l \le s-1 \\ \bigvee_{i=1}^{s} \mu_i^{\omega}(t), & t \ge x_{s,1}v_{s,1}^{\omega}. \end{cases}$$

Since $\mu_T^{\omega}(t)$ is nondecreasing in $[-\infty, x_{s,1}v_{s,1}^{\omega}]$ and nonincreasing in $[x_{s,1}v_{s,1}^{\omega}, \infty]$, by the definition, we can calculate

$$\mathrm{Cr}\left\{T\left(x,\xi(\omega)\right) \ge t^0\right\} = \frac{1}{2}\left[1 + \sup_{t \ge t^0} \mu_T^{\omega}(t) - \sup_{t < t^0} \mu_T^{\omega}(t)\right],$$

Table 4.1 Lifetime of each component in Example 4.1

$\omega=\omega_1$ (Probability $=0.4$)	$\omega=\omega_2$ (Probability $=0.6$)
$\xi_{1,1}(\omega_1)=(2,3,4)$	$\xi_{1,1}(\omega_2)=(3,4,6)$
$\xi_{1,2}(\omega_1)=(3,5,8)$	$\xi_{1,2}(\omega_1)=(5,8,10)$
$\xi_{2,1}(\omega_1)=(6,7,8)$	$\xi_{2,1}(\omega_2)=(6,8,10)$

which implies

$$\begin{aligned}&\mathrm{Cr}\left\{T\left(x,\xi(\omega)\right)\geq t^0\right\}\\ &=I_{\{\omega|t^0\leq x_{s,1}v_{s,1}^{\omega}\}}(\omega)\left(1-\frac{\mu_T^{\omega}(t^0)}{2}\right)+I_{\{\omega|t^0>x_{s,1}v_{s,1}^{\omega}\}}(\omega)\left(\frac{\mu_T^{\omega}(t^0)}{2}\right),\end{aligned}$$

for almost every $\omega\in\Omega$, where $I_{\{\cdot\}}$ is the indicator function of set $\{\cdot\}$. Integrating w.r.t. ω on the both sides of the above equation deduces the required result (4.5). The proof of the theorem is completed. □

Example 4.1. Consider a parallel-series system with two subsystems, and there are two types of component in the first subsystem and one type in the second, and the redundancy allocation is $x=(x_{1,1},x_{1,2},x_{2,1})=(2,1,1)$. Suppose that the fuzzy random lifetimes $\xi_{1,1,1}$, $\xi_{1,2,1}$ and $\xi_{2,1,1}$ are characterized by a discrete random variable which takes on values $\omega=\omega_1$ with probability 0.4 and $\omega=\omega_2$ with probability 0.6, and they have the distributions in Table 4.1. Set $t^0=7$, we calculate the reliability $R_7(x)=\mathrm{Ch}\{T(x,\xi)\geq 7\}$ of the system.

Since the lifetime of the system

$$T(x,\xi)=[(\xi_{1,1,1}+\xi_{1,1,2})\wedge\xi_{1,2,1}]\vee\xi_{2,1,1},$$

we have

$$T\left(x,\xi(\omega_k)\right)=\left[\left(\xi_{1,1,1}(\omega_k)+\xi_{1,1,2}(\omega_k)\right)\wedge\xi_{1,2,1}(\omega_k)\right]\vee\xi_{2,1,1}(\omega_k)$$

for $k=1,2$. That is, $T(x^0,\xi)$ takes on the fuzzy values $[(4,6,8)\wedge(3,5,8)]\vee(6,7,8)$ with probability 0.4, and $[(6,8,12)\wedge(5,8,10)]\vee(6,8,10)$ with probability 0.6. For $\omega=\omega_1$, from (4.7) in Theorem 4.1, we can calculate that

$$(4,6,8)\wedge(3,5,8)=(3,5,8).$$

Furthermore, making use of (4.6), we obtain

$$T(x,\xi(\omega_1))=(3,5,8)\vee(6,7,8)=(6,7,8)$$

and

$$\mu_T^{\omega_1}(t)=\begin{cases}t-6, & 6\leq t<7\\ 8-t, & 7\leq t<8\\ 0, & \text{otherwise.}\end{cases}$$

We note that $x_{2,1} = 1$ and $t^0 = 7 = x_{2,1}\nu_{2,1}^{\omega_1}$; therefore,

$$\mathrm{Cr}\left\{T\left(x,\xi(\omega_1)\right) \geq t^0\right\} = 1 - \frac{\mu_T^{\omega_1}(7)}{2} = 0.5.$$

Similarly, we obtain

$$\mu_T^{\omega_2}(t) = \begin{cases} \dfrac{t-6}{2}, & 6 \leq t < 8 \\ \dfrac{10-t}{2}, & 8 \leq t < 10 \\ 0, & \text{otherwise.} \end{cases}$$

Since $t^0 = 7 < x_{2,1}\nu_{2,1}^{\omega_2} = 8$, we have

$$\mathrm{Cr}\left\{T\left(x,\xi(\omega_2)\right) \geq t^0\right\} = 1 - \frac{\mu_T^{\omega_2}(7)}{2} = 0.75.$$

Consequently from (4.5), we have

$$\begin{aligned} R_7(x) &= \int_\Omega \mathrm{Cr}\left\{T\left(x,\xi(\omega)\right) \geq 7\right\} \Pr(\mathrm{d}\omega) \\ &= \int_\Omega 1 - \frac{\mu_T^{\omega}(7)}{2} \Pr(\mathrm{d}\omega) \\ &= 0.5 \times 0.4 + 0.75 \times 0.6 = 0.65. \end{aligned}$$

4.2.2 Sensitivity Analysis

In the FR-RAMs I and II, we note that the reliability $R_{t^0}(x)$ is predetermined by the decision maker, and any changes of the threshold lifetime t^0 may influence the objective value of FR-RAM I and the reliability constraint of FR-RAM II, and finally influence the allocation decision. Therefore, it is important to analyze the sensitivity of $R_{t^0}(x)$ w.r.t. the threshold lifetime t^0.

Theorem 4.2 ([149]). *Given fuzzy random lifetimes $\xi_{i,j,k}$ of components in the FR-RAMs for $i = 1,2,\cdots,s$; $j = 1,2,\cdots,n_i$; and $k = 1,2,\cdots,x_{i,j}$ on a probability space $(\Omega,\Sigma,\Pr)$ such that for almost every $\omega \in \Omega$, $\xi_{i,j,k}(\omega)$ is a convex fuzzy variable. If $\xi_{i,j,k}(\omega)$ for each $i,j,$andk is left continuous or upper semicontinuous, then the reliability $R_{t^0}(x)$ is left continuous w.r.t. the threshold lifetime $t^0 \in \Re$.*

Proof. First of all, we deal with the case that $\xi_{i,j,k}(\omega)$ is left continuous. Denote $\mu_{i,j}^{\omega}$ the membership function of $\xi_{i,j,k}(\omega)$ for any $\omega \in \Omega$, and $i = 1,2,\cdots,s$; $j =$

$1,2,\cdots,n_i; and k=1,2,\cdots,x_{i,j}$. In the following, we prove that the membership function of fuzzy variable $\bigwedge_{j=1}^{n_i}\sum_{k=1}^{x_{i,j}}\xi_{i,j,k}(\omega)$ denoted by μ_i^ω is left continuous for almost every $\omega\in\Omega$.

Without losing any generality, we assume that $x_{1,1}v_{1,1}^\omega\le x_{2,1}v_{2,1}^\omega\le\cdots\le x_{s,1}v_{s,1}^\omega$, recall that fuzzy variable $\xi_{i,j,k}(\omega)$ is convex for almost every $\omega\in\Omega$, and then from Theorem 4.1, we have

$$\mu_i^\omega(t)=\begin{cases}\bigvee_{j=1}^{n_i}\mu_{i,j}^\omega\left(\frac{t}{x_{i,j}}\right), & t<x_{i,1}v_{i,1}^\omega\\ \bigwedge_{j=1}^{l}\mu_{i,j}^\omega\left(\frac{t}{x_{i,j}}\right), & x_{i,l}v_{i,l}^\omega\le t<x_{i,l+1}v_{i,l+1}^\omega, 1\le l\le n_i-1\\ \bigwedge_{j=1}^{n_i}\mu_{i,j}^\omega\left(\frac{t}{x_{i,j}}\right), & t\ge x_{i,n_i}v_{i,n_i}^\omega.\end{cases} \tag{4.8}$$

For almost every $\omega\in\Omega$, from the assumption that fuzzy variable $\xi_{i,j,k}(\omega)$ is left continuous, that is, the membership function $\mu_{i,j}^\omega$ is a left continuous real-valued function for $i=1,2,\cdots,s$, $j=1,2,\cdots,n_i$, therefore, for any $t_0\in\Re$,

$$\begin{aligned}\lim_{t\to t_0-}\mu_{i,1}^\omega(t)\bigvee\mu_{i,2}^\omega(t) &= \lim_{t\to t_0-}\mu_{i,1}^\omega(t)\bigvee\lim_{t\to t_0-}\mu_{i,2}^\omega(t)\\ &= \mu_{i,1}^\omega(t_0)\bigvee\mu_{i,2}^\omega(t_0),\end{aligned}$$

and

$$\begin{aligned}\lim_{t\to t_0-}\mu_{i,1}^\omega(t)\bigwedge\mu_{i,2}^\omega(t) &= \lim_{t\to t_0-}\mu_{i,1}^\omega(t)\bigwedge\lim_{t\to t_0-}\mu_{i,2}^\omega(t)\\ &= \mu_{i,1}^\omega(t_0)\bigwedge\mu_{i,2}^\omega(t_0).\end{aligned}$$

Hence, $\mu_{i,1}^\omega\vee\mu_{i,2}^\omega$ and $\mu_{i,1}^\omega\wedge\mu_{i,2}^\omega$ are left continuous. By the method of induction, we can obtain that $\bigvee_{j=1}^{n}\mu_{i,j}^\omega$ and $\bigwedge_{j=1}^{n}\mu_{i,j}^\omega$ are left continuous real-valued functions for any finite positive integer n. Therefore, from (4.8), we have $\mu_i^\omega(t)$ is left continuous in $(-\infty,x_{i,1}v_{i,1}^\omega)$, $[x_{i,l}v_{i,l}^\omega,x_{i,l+1}v_{i,l+1}^\omega)$ for $1\le l\le n_i-1$, and $[x_{i,n_i}v_{i,n_i}^\omega,\infty)$, respectively. Thus, to prove the leftcontinuity of μ_i^ω, it suffices to prove μ_i^ω is left continuous at $x_{i,1}v_{i,1}^\omega,x_{i,2}v_{i,2}^\omega,\cdots,x_{i,n_i}v_{i,n_i}^\omega$ for $i=1,2,\cdots,s$. Given each i, for $x_{i,1}v_{i,1}^\omega$, we have

$$\begin{aligned}\lim_{t\to x_{i,1}v_{i,1}^\omega-}\mu_i^\omega(t) &= \lim_{t\to x_{i,1}v_{i,1}^\omega-}\bigvee_{j=1}^{n_i}\mu_{i,j}^\omega\left(\frac{t}{x_{i,j}}\right)\\ &= 1=\mu_i^\omega(x_{i,1}v_{i,1}^\omega),\end{aligned} \tag{4.9}$$

which implies that μ_i^ω is left continuous at $x_{i,1}v_{i,1}^\omega$. Next, for $x_{i,l}v_{i,l}^\omega$, $2\le l\le n_i$, we have

$$\begin{aligned}\lim_{t\to x_{i,l}v_{i,l}^\omega-}\mu_i^\omega(t) &= \lim_{t\to x_{i,l}v_{i,l}^\omega-}\bigwedge_{j=1}^{l-1}\mu_{i,j}^\omega\left(\frac{t}{x_{i,j}}\right)\\ &= \bigwedge_{j=1}^{l-1}\mu_{i,j}^\omega\left(v_{i,l}^\omega\right)=\mu_i^\omega(x_{i,l}v_{i,l}^\omega).\end{aligned} \tag{4.10}$$

That is, μ_i^ω is left continuous at $x_{i,l}v_{i,l}^\omega$ for $2 \le l \le n_i$. As a consequence, we have proved that μ_i^ω is left continuous for any $i = 1,2,\cdots,s$ and almost every $\omega \in \Omega$. Furthermore, by the same reasoning, from the expression (4.6) of the membership function μ_T^ω of $T(x,\xi(\omega))$, we can prove that μ_T^ω is also left continuous for almost every $\omega \in \Omega$.

In the following, we prove μ_T^ω is upper semicontinuous provided $\xi_{i,j,k}(\omega)$ is upper semicontinuous. Before this, we shall prove that given i, $\bigvee_{j=1}^n \mu_{i,j}^\omega$ and $\bigwedge_{j=1}^n \mu_{i,j}^\omega$ are upper semicontinuous real-valued functions for any finite positive integer n provided $\mu_{i,j}^\omega$ for $j = 1,2,\cdots,n$ are upper semicontinuous. In fact, when $n = 2$, by the upper semicontinuity of $\mu_{i,1}^\omega$ and $\mu_{i,2}^\omega$, we have for any $\varepsilon > 0$, there exists a $\delta > 0$ such that

$$\begin{aligned}
&\mu_{i,1}^\omega(t) \bigvee \mu_{i,2}^\omega(t) \\
&\quad < \left(\mu_{i,1}^\omega(t^0) + \varepsilon\right) \bigvee \left(\mu_{i,2}^\omega(t^0) + \varepsilon\right) \\
&\quad = \left(\mu_{i,1}^\omega(t^0) \bigvee \mu_{i,2}^\omega(t^0)\right) + \varepsilon,
\end{aligned}$$

and

$$\begin{aligned}
&\mu_{i,1}^\omega(t) \bigwedge \mu_{i,2}^\omega(t) \\
&\quad < \left(\mu_{i,1}^\omega(t^0) + \varepsilon\right) \bigwedge \left(\mu_{i,2}^\omega(t^0) + \varepsilon\right) \\
&\quad = \left(\mu_{i,1}^\omega(t^0) \bigwedge \mu_{i,2}^\omega(t^0)\right) + \varepsilon,
\end{aligned}$$

for any $t^0 \in \Re$. That is, $\mu_{i,1}^\omega \vee \mu_{i,2}^\omega$ and $\mu_{i,1}^\omega \wedge \mu_{i,2}^\omega$ are upper semicontinuous. By the method of induction, it is not difficult to show $\bigvee_{j=1}^n \mu_{i,j}^\omega$ and $\bigwedge_{j=1}^n \mu_{i,j}^\omega$ are upper semicontinuous real-valued functions for any finite positive integer n. Furthermore, similarly as in the case that $\xi_{i,j,k}(\omega)$ is left continuous, to prove the upper semicontinuity of μ_i^ω, from (4.8), it suffices to prove that μ_i^ω is upper semicontinuous at $x_{i,1}v_{i,1}^\omega, x_{i,2}v_{i,2}^\omega, \cdots, x_{i,n_i}v_{i,n_i}^\omega$ for $i = 1,2,\cdots,s$. Given $i = 1,2,\cdots,s$, for $x_{i,1}v_{i,1}^\omega$, we have

$$\limsup_{t \to x_{i,1}v_{i,1}^\omega} \mu_i^\omega(t) \le 1 = \mu_i^\omega(x_{i,1}v_{i,1}^\omega),$$

which implies μ_i^ω is upper semicontinuous at $x_{i,1}v_{i,1}^\omega$ for each i. As to $x_{i,l}v_{i,l}^\omega$, $2 \le l \le n_i$, we note from the proof of Theorem 4.1 that μ_i^ω is nonincreasing in $[x_{i,1}v_{i,1}^\omega, \infty]$ for all i; hence, we have

$$\limsup_{t \to x_{i,l}v_{i,l}^\omega} \mu_i^\omega(t) \le \limsup_{t \to x_{i,l}v_{i,l}^\omega -} \mu_i^\omega(t) = \limsup_{t \to x_{i,l}v_{i,l}^\omega -} \bigwedge_{j=1}^{l-1} \mu_{i,j}^\omega\left(\frac{t}{x_{i,j}}\right). \tag{4.11}$$

It follows from the upper semicontinuity of $\bigwedge_{j=1}^{l-1} \mu_{i,j}^{\omega}$ that

$$\limsup_{t \to x_{i,l} v_{i,l}^{\omega}-} \bigwedge_{j=1}^{l-1} \mu_{i,j}^{\omega}\left(\frac{t}{x_{i,j}}\right) \leq \bigwedge_{j=1}^{l-1} \mu_{i,j}^{\omega}\left(v_{i,l}^{\omega}\right) = \mu_i^{\omega}(x_{i,l} v_{i,l}^{\omega}). \tag{4.12}$$

Combining (4.11) and (4.12), we have μ_i^{ω} is upper semicontinuous at $x_{i,l}v_{i,l}^{\omega}$, for $2 \leq l \leq n_i$. So far, we have proved that μ_i^{ω} is upper semicontinuous for any $i = 1, 2, \cdots, s$ and almost every $\omega \in \Omega$. Based on this fact, by the same reasoning, we can prove that μ_T^{ω} is also upper semicontinuous for almost every $\omega \in \Omega$.

Therefore, we have μ_T^{ω} is left continuous or upper semicontinuous according that $\xi_{i,j,k}(\omega)$ for each $i, j, and k$ is left continuous or upper semicontinuous. Thus, by the left-continuity condition for the distribution functions of fuzzy random variable (refer to [147]), we have

$$\lim_{t \to t^0 -} \mathrm{Ch}\{T(x,\xi) \geq t\} = \mathrm{Ch}\{T(x,\xi) \geq t^0\}$$

for any $t^0 \in \Re$. That is, the reliability $R_{t^0}(x)$ is a left continuous function of the threshold lifetime $t^0 \in \Re$. □

Example 4.2. Consider a parallel-series system with two subsystems, and the redundancy allocation is $x = (x_{1,1}, x_{1,2}, x_{2,1}) = (1, 1, 2)$. The components have the distributions as follows: fuzzy random lifetimes $\xi_{1,1}$, $\xi_{1,2}$ and $\xi_{2,1}$ are characterized by a discrete random variable which takes on values $\omega = \omega_1$ with probability 0.8 and $\omega = \omega_2$ with probability 0.2, and the membership functions are left continuous and upper semicontinuous which are given as below:

$$\mu_{\xi_{1,1}(\omega_1)}(r) = \begin{cases} r-2, & 2 \leq r \leq 3 \\ \dfrac{4-r}{2}, & 3 < r \leq 4 \\ 0, & \text{otherwise}, \end{cases}$$

$$\mu_{\xi_{1,1}(\omega_2)}(r) = \begin{cases} r-3, & 3 \leq r \leq 4 \\ \dfrac{5-r}{2}, & 4 < r \leq 5 \\ 0, & \text{otherwise}, \end{cases}$$

$$\mu_{\xi_{1,2}(\omega_1)}(r) = \begin{cases} \dfrac{r-3}{2}, & 3 \leq r \leq 5 \\ \dfrac{7-r}{4}, & 5 < r \leq 7 \\ 0, & \text{otherwise}, \end{cases}$$

$$\mu_{\xi_{1,2}(\omega_2)}(r) = \begin{cases} r-5, & 5 \le r \le 6 \\ \dfrac{7-r}{2}, & 6 < r \le 7 \\ 0, & \text{otherwise,} \end{cases}$$

$$\mu_{\xi_{2,1}(\omega_1)}(r) = \begin{cases} r-6, & 6 \le r \le 7 \\ \dfrac{8-r}{2}, & 7 < r \le 8 \\ 0, & \text{otherwise,} \end{cases}$$

and

$$\mu_{\xi_{2,1}(\omega_2)}(r) = \begin{cases} r-7, & 7 \le r \le 8 \\ \dfrac{9-r}{2}, & 8 < r \le 9 \\ 0, & \text{otherwise.} \end{cases}$$

Now, we verify the left continuity of $R_t(x)$ w.r.t. t.

First of all, from Theorem 4.1, we have the membership function $\mu_T^{\omega_1}(t)$ of

$$T\Big(x,\xi(\omega_1)\Big) = \Big(\xi_{1,1,1}(\omega_1) \wedge \xi_{1,2,1}(\omega_1)\Big) \bigvee \Big(\xi_{2,1,1}(\omega_1) + \xi_{2,1,2}(\omega_1)\Big)$$

is

$$\mu_T^{\omega_1}(r) = \begin{cases} \dfrac{r-12}{2}, & 12 < r \le 14 \\ \dfrac{16-r}{4}, & 14 < r \le 16 \\ 0, & \text{otherwise.} \end{cases}$$

Furthermore, we have

$$\mathrm{Cr}\Big\{T\Big(x,\xi(\omega_1)\Big) \ge t\Big\} = \begin{cases} 1, & t \le 12 \\ \dfrac{16-t}{4}, & 12 < t \le 14 \\ \dfrac{16-t}{8}, & 14 < t \le 16 \\ 0, & \text{otherwise.} \end{cases}$$

Similarly, we can obtain

$$\mathrm{Cr}\Big\{T\Big(x,\xi(\omega_2)\Big) \ge t\Big\} = \begin{cases} 1, & t \le 14 \\ \dfrac{18-t}{4}, & 14 < t \le 16 \\ \dfrac{18-t}{8}, & 16 < t \le 18 \\ 0, & \text{otherwise.} \end{cases}$$

As a consequence, by the definition, we have

$$\mathrm{Ch}\{T(x,\xi) \geq t\} = \begin{cases} 1, & t \leq 12 \\ \dfrac{17-t}{5}, & 12 < t \leq 14 \\ \dfrac{50-3t}{20}, & 14 < t \leq 16 \\ \dfrac{18-t}{20}, & 16 < t \leq 18 \\ 0, & \text{otherwise,} \end{cases}$$

which is a left continuous function of $t \in \Re$. This result coincides with that of Theorem 4.2.

By the similar reasoning used in the Theorem 4.2, it is not difficult to obtain the following results.

Theorem 4.3 ([149]). *Suppose that the lifetimes $\xi_{i,j,k}$ of components in the FR-RAMs for $i = 1,2,\cdots,s; j = 1,2,\cdots,n_i; andk = 1,2,\cdots,x_{i,j}$ are fuzzy random variables on probability space $(\Omega,\Sigma,\Pr)$ such that for almost every $\omega \in \Omega$, $\xi_{i,j,k}(\omega)$ is a convex fuzzy variable. If $\xi_{i,j,k}(\omega)$ for each $i,j,andk$ is right continuous and lower semicontinuous, then the reliability $R_{t^0}(x)$ is right continuous w.r.t. the threshold lifetime $t^0 \in \Re$.*

Theorem 4.4 ([149]). *Assume that the lifetimes $\xi_{i,j,k}$ of components in the FR-RAMs for $i = 1,2,\cdots,s; j = 1,2,\cdots,n_i; andk = 1,2,\cdots,x_{i,j}$ are fuzzy random variables on probability space $(\Omega,\Sigma,\Pr)$ such that for almost every $\omega \in \Omega$, $\xi_{i,j,k}(\omega)$ is a convex fuzzy variable. If $\xi_{i,j,k}(\omega)$ for each $i,j,andk$ is continuous, then we have the reliability $R_{t^0}(x)$ is continuous w.r.t. the threshold lifetime $t^0 \in \Re$.*

4.3 Computing Reliability

We note from the FR-RAMs I and II that the reliability function $R_{t^0}(x)$ is the objective of FR-RAM I and a constraint of FR-RAM II, therefore, in order to solve the FR-RAMs, we have to cope with the computation of the system reliability $R_{t^0}(x)$. However, due to the fuzzy random parameters, the reliability function $R_{t^0}(x)$ in general cannot be calculated directly.

Let

$$\xi = \begin{pmatrix} \xi_{1,1,1}, & \cdots, & \xi_{1,1,x_{1,1}}, & \cdots, & \xi_{1,n_1,1}, & \cdots, & \xi_{1,n_1,x_{1,n_1}}, \\ \cdots, & & & \cdots, & & & \cdots, \\ \xi_{s,1,1}, & \cdots, & \xi_{s,1,x_{s,1}}, & \cdots, & \xi_{s,n_s,1}, & \cdots, & \xi_{s,n_s,x_{s,n_s}} \end{pmatrix}$$

be the lifetime vector involved in the fuzzy random parallel-series system, where each lifetime $\xi_{i,j,k}$ can be any general positive fuzzy random variable for $i =$

$1,2,\cdots,s; j = 1,2,\cdots,n_i$; and $k = 1,2,\cdots,x_{i,j}$. The discussion is divided into three cases: reliability with discrete lifetimes, reliability with convex lifetimes, and reliability with nonconvex (continuous) lifetimes.

4.3.1 Reliability with Discrete Lifetimes

Suppose that ξ is a fuzzy random vector whose randomness is characterized by a discrete random vector ω assuming finite number of values ω_k, $k=1,2,\cdots,N$, with probability p_k, $k=1,2,\cdots,N$, respectively, and for each k,

$$\xi(\omega_k) = \begin{pmatrix} \xi_{1,1,1}(\omega_k), & \cdots, & \xi_{1,1,x_{1,1}}(\omega_k), & \cdots, & \xi_{1,n_1,1}(\omega_k), & \cdots, & \xi_{1,n_1,x_{1,n_1}}(\omega_k), \\ \cdots, & & & \cdots, & & & \cdots, \\ \xi_{s,1,1}(\omega_k), & \cdots, & \xi_{s,1,x_{s,1}}(\omega_k), & \cdots, & \xi_{s,n_s,1}(\omega_k), & \cdots, & \xi_{s,n_s,x_{s,n_s}}(\omega_k) \end{pmatrix}$$

is a discrete fuzzy vector taking on M_k values

$$\widehat{\xi}^{k,j} = \begin{pmatrix} \widehat{\xi}^{k,j}_{1,1,1}, & \cdots, & \widehat{\xi}^{k,j}_{1,1,x_{1,1}}, & \cdots, & \widehat{\xi}^{k,j}_{1,n_1,1}, & \cdots, & \widehat{\xi}^{k,j}_{1,n_1,x_{1,n_1}}, \\ \cdots, & & & \cdots, & & & \cdots, \\ \widehat{\xi}^{k,j}_{s,1,1}, & \cdots, & \widehat{\xi}^{k,j}_{s,1,x_{s,1}}, & \cdots, & \widehat{\xi}^{k,j}_{s,n_s,1}, & \cdots, & \widehat{\xi}^{k,j}_{s,n_s,x_{s,n_s}} \end{pmatrix}$$

with membership degree $\mu_k^j > 0$, $k=1,2,\cdots,N, j=1,2,\cdots,M_k$, and

$$\max_{j=1}^{M_k} \mu_k^j = 1.$$

In this case, the support Ξ of ξ can be expressed as

$$\Xi = \left\{ \widehat{\xi}^{k,j} \mid k=1,2,\cdots,N, j=1,2,\cdots,M_k \right\},$$

which is a finite set. Hence, from the definition, we have

$$R_{t^0}(x) = \sum_{k=1}^{N} p_k Q_k(x), \tag{4.13}$$

where

$$Q_k(x) = \mathrm{Cr}\{T(x,\xi(\omega_k)) \geq t^0\}$$

is calculated by

$$Q_k(x) = \frac{1}{2}\left[\max\left\{\mu_k^j \mid T\left(x,\widehat{\xi}^{k,j}\right) \geq t^0\right\} + 1 - \max\left\{\mu_k^j \mid T\left(x,\widehat{\xi}^{k,j}\right) < t^0\right\}\right]. \tag{4.14}$$

For the simplicity, we abbreviate the formula (4.13)–(4.14) of the reliability with discrete lifetimes to RDL.

4.3.2 Reliability with Convex Lifetimes

In The preceding Theorem 4.1 already supplies us with an analytical expression (4.5)–(4.7) of the reliability when all the lifetimes of components have convex distributions. It enables us to compute the reliability with convex lifetimes by the following random simulation.

First of all, given a decision x, for any $\omega \in \Omega$, we calculate $\mu_T^{\omega}(t^0)$ by formula (4.6)–(4.7). Furthermore, we note that formula (4.5) for reliability $R_{t^0}(x)$ with convex lifetimes can be rewritten as

$$R_{t^0}(x) = E\left[I_{\{\omega | t^0 \le x_{s,1} \nu_{s,1}^{\omega}\}}(\omega) \left(1 - \frac{\mu_T^{\omega}(t^0)}{2}\right) + I_{\{\omega | t^0 > x_{s,1} \nu_{s,1}^{\omega}\}}(\omega) \left(\frac{\mu_T^{\omega}(t^0)}{2}\right)\right],$$

where $E[\cdot]$ is the expected value operator of a random variable and $I_{\{\cdot\}}$ is the indicator function of set $\{\cdot\}$. Therefore, making using of random simulation (see [36]), we can compute the reliability by

$$R_{t^0}(x) \leftarrow \frac{1}{M} \sum_{i=1}^{M} R_{t^0}(x, \omega_i), \quad (M \to \infty), \tag{4.15}$$

where

$$R_{t^0}(x, \omega_i) = I_{\{\omega | t^0 \le x_{s,1} \nu_{s,1}^{\omega}\}}(\omega_i) \left(1 - \frac{\mu_T^{\omega_i}(t^0)}{2}\right) + I_{\{\omega | t^0 > x_{s,1} \nu_{s,1}^{\omega}\}}(\omega_i) \left(\frac{\mu_T^{\omega_i}(t^0)}{2}\right). \tag{4.16}$$

Here, ω_i for $i = 1, 2, \cdots, n$ are the random samples generated from the distribution of the random parameter involved in the fuzzy random vector ξ. It is well known that the random simulation (4.15) converges with probability 1 as $M \to \infty$, which is ensured by the strong law of large numbers. The above computation procedure is summarized as the following algorithm.

Algorithm 4.1.

Step 1. *Set* $R = 0$.
Step 2. *Randomly generate a sample point* $\widehat{\omega}$ *from the distribution of the random vector involved in* ξ.
Step 3. *Compute the* $R_{t^0}(x, \widehat{\omega})$ *through (4.16).*
Step 4. $R \leftarrow R + R_{t^0}(x, \widehat{\omega})$.
Step 5. *Repeat the Steps 2-4 M times.*
Step 6. *Return the value of* $R_{t^0}(x) = R/M$.

4.3.3 Reliability with Nonconvex Lifetimes

In order to compute the system reliability with nonconvex lifetimes, here we present a fuzzy random simulation approach. Suppose the randomness of ξ is characterized by a continuous random vector, and for any random realization $\omega \in \Omega$, $\xi(\omega)$ is a nonconvex continuous fuzzy vector with an infinite support denoted by

$$\Xi = \prod_{k=1}^{K} [a_i, b_i], \tag{4.17}$$

where $K = \sum_{i=1}^{s} \sum_{j=1}^{n_i} x_{i,j}$, $[a_k, b_k]$ is the support of ξ_k for $k = 1, 2, \cdots, K$.

First of all, we employ a discretization method (see [92]) to generate a sequence $\{\zeta_l\}$ of discrete fuzzy random vectors which converges to the original continuous ξ. To simplify it a little, we denote the original fuzzy random lifetime vector $\xi = (\xi_1, \xi_2 \cdots, \xi_K)$. For each positive integer l, $\zeta_l = (\zeta_{l,1}, \zeta_{l,2}, \cdots, \zeta_{l,K})$ is constructed by the following method: define $\zeta_{l,i} = g_{l,i}(\xi_i)$ for $i = 1, 2, \cdots, K$, where the functions $g_{l,i}$'s are given by

$$g_{l,i}(v_i) = \begin{cases} a_i, & v_i \in [a_i, a_i + \frac{1}{l}) \\ \sup\left\{\frac{k_i}{l} \mid k_i \in Z, \text{ s.t. } \frac{k_i}{l} \le v_i\right\}, & v_i \in [a_i + \frac{1}{l}, b_i] \end{cases} \tag{4.18}$$

and Z is the set of integers. In what follows, the sequence $\{\zeta_l\}$ of discrete fuzzy random vectors generated by (4.18) is referred to as the *discretization* of original ξ. It can been proved that

$$\begin{aligned} &\|\zeta_l(\omega)(\gamma) - \xi(\omega)(\gamma)\| \\ &= \sqrt{\sum_{j=1}^{K} \left(\zeta_{l,i}(\omega)(\gamma) - \xi_i(\omega)(\gamma)\right)^2} \le \frac{\sqrt{K}}{l}, \end{aligned}$$

for all $(\omega, \gamma) \in \Omega \times \Gamma$, which implies that the discretization $\{\zeta_l\}$ converges to ξ uniformly (see [96]). The above process is usually called fuzzy simulation.

Next, we estimate $R_{t^0}(x)$. recalling from the definition that the reliability function

$$R_{t^0}(x) = \int_{\Omega} \mathrm{Cr}\{T(x, \xi(\omega)) \ge t^0\} \Pr(\mathrm{d}\omega).$$

For every $\omega_i \in \Omega$, we first replace the ξ with its discretization $\{\zeta_l\}$ generated by (4.18), and estimate

$$\mathrm{Cr}\{T(x, \xi(\omega_i)) \ge t^0\}$$

by

$$Q(x, \omega_i) = \mathrm{Cr}\{T(x, \zeta_l(\omega_i)) \ge t^0\}$$

which can be determined by (4.14). Then, the reliability $R_{t^0}(x)$ can be estimated by a random simulation

$$R_{t^0}(x) \leftarrow \frac{1}{M}\sum_{i=1}^{M} Q(x,\omega_i), \quad (M \to \infty) \tag{4.19}$$

with probability 1.

The fuzzy random simulation procedure for the $R_{t^0}(x)$ with nonconvex lifetimes is summarized as following algorithm.

Algorithm 4.2.

Step 1. *Set $R = 0$.*
Step 2. *Randomly, generate a sample point $\omega \in \Omega$ through the corresponding probability distribution.*
Step 3. *Generate $\zeta_l(\omega)$ from the support Ξ of $\xi(\omega)$ through formula (4.18).*
Step 4. *Calculate*

$$Q(x,\omega) = \mathrm{Cr}\{T(x,\zeta_l(\omega)) \geq t^0\}$$

through formula (4.14).
Step 5. *$R \leftarrow R + Q(x,\omega)$.*
Step 6. *Repeat Steps 2-5 M times.*
Step 7. *Return the value of $R_{t^0}(x) = R/M$.*

4.3.4 Convergence

The reliability with discrete lifetimes can be determined exactly by formula (4.13)–(4.14) (RDL), and Algorithm 4.1 (random simulation) owns the convergence with probability 1 when calculating the reliability with convex lifetimes. Regarding Algorithm 4.2 for the reliability with nonconvex lifetimes, the convergence can be proved.

In fact, the following Theorem 4.5 shows that $R_{t^0,\zeta_l}(x)$ converges to $R_{t^0}(x)$ for almost every $t^0 > 0$, as $l \to \infty$. As a consequence, the original reliability function $R_{t^0}(x)$ can be well approximated by $R_{t^0,\zeta_l}(x)$ through Algorithm 4.2, provided l is sufficiently large.

Theorem 4.5 ([149]). *Consider FR-RAMs I and II for a parallel-series system. Let ξ be the continuous fuzzy random lifetime vector of the components in the system, which has the compact interval support (4.17), $\{\zeta_l\}$ be the discretization of ξ, and t^0 the preselected threshold system lifetime. Then, for any feasible decision x, the approximating system reliability function $R_{t^0,\zeta_l}(x)$ converges to the original system reliability function, i.e.,*

$$\lim_{l\to\infty} R_{t^0,\zeta_l}(x) = R_{t^0}(x),$$

provided $R_t(x)$ is continuous at $t = t^0$.

Proof. Recall that the lifetime of parallel-series system

$$T(x,\xi)=\bigvee_{i=1}^{s}\left[\bigwedge_{j=1}^{n_i}\left(\sum_{k=1}^{x_{i,j}}\xi_{i,j,k}\right)\right],$$

is a continuous function w.r.t. ξ, for any given

$$x=(x_{1,1},\cdots,x_{1,n_1},\cdots,x_{s,1},\cdots,x_{s,n_s}).$$

Since the support of ξ, i.e.,

$$\Xi=\prod_{i=1}^{K}[a_i,b_i]$$

is a compact set in $\Re^K$, where

$$K=\sum_{i=1}^{s}\sum_{j=1}^{n_i}x_{i,j},$$

$T(x,\xi)$ is uniformly continuous on Ξ. Hence, given a feasible x, for any $\varepsilon>0$, there is a $\delta>0$ such that

$$\left|T(x,\widehat{\xi}')-T(x,\widehat{\xi}'')\right|<\varepsilon \tag{4.20}$$

whenever $\widehat{\xi}',\widehat{\xi}''\in\Xi$, and

$$\left\|\widehat{\xi}'-\widehat{\xi}''\right\|=\sqrt{\sum_{i=1}^{s}\sum_{j=1}^{n_i}\sum_{k=1}^{xi,j}\left(\widehat{\xi}'_{i,j,k}-\widehat{\xi}''_{i,j,k}\right)^2}<\delta.$$

Noting that the discretization $\{\zeta_l\}$ is a sequence of fuzzy random vectors which converges uniformly to ξ on $\Omega\times\Gamma$, for the above δ, there exists a positive integer L such that for all $(\omega,\gamma)\in\Omega\times\Gamma$,

$$\begin{aligned}&\|\zeta_l(\omega)(\gamma)-\xi(\omega)(\gamma)\|\\&=\sqrt{\sum_{i=1}^{s}\sum_{j=1}^{n_i}\sum_{k=1}^{x_{i,j}}\left(\zeta^l_{i,j,k}(\omega)(\gamma)-\xi_{i,j,k}(\omega)(\gamma)\right)^2}<\delta\end{aligned}$$

provided $l\geq L$. Combining (4.20), for all $(\omega,\gamma)\in\Omega\times\Gamma$,

$$\left|T\left(x,\zeta_l(\omega)(\gamma)\right)-T\left(x,\xi(\omega)(\gamma)\right)\right|<\varepsilon$$

whenever $l\geq L$. That is, the sequence $\{T(x,\zeta_l)\}$ of fuzzy random variables converges uniformly to $T(x,\xi)$ on $\Omega\times\Gamma$. It follows that, for any $\varepsilon>0$, we have

$$\lim_{l\to\infty}\mathrm{Ch}\left\{\left|T(x,\zeta_l)-T(x,\xi)\right|\geq\varepsilon\right\}=0.$$

Since convergence in chance implies convergence in distribution (see [98]), we obtain

$$\lim_{l\to\infty} \mathrm{Ch}\{T(x,\zeta_l) \geq t^0\} = \mathrm{Ch}\{T(x,\xi) \geq t^0\} \tag{4.21}$$

provided $\mathrm{Ch}\{T(x,\xi) \geq t\}$ is continuous at $t = t^0$. The proof of the theorem is completed. □

4.4 A GA-Based Hybrid Approach

It is easy to see that the FR-RAMs I and II are tasks of fuzzy random integer programming problems. Since genetic algorithm (GA), as an evolutionary search method, has been applied successfully to many practical optimization problems with integer decisions (see [39,48,86,109]), in this section, we incorporate the preceding three computation schemes (formula RDL, Algorithm 4.1, and Algorithm 4.2) of the reliability $R_{t^0}(x)$ in different cases into the mechanism of GA to search for the approximately optimal solution of the FR-RAMs I and II. In the hybrid algorithm, the GA is used to search for the best redundancy allocation, and the computation methods are used to calculate the objective value of the FR-RAM I and to check the feasibility of each chromosome in the FR-RAM II (Fig. 4.2). Such a hybrid solution mechanism, i.e., "computation schemes for uncertain functions plus metaheuristics" will be also utilized in the models of later chapters.

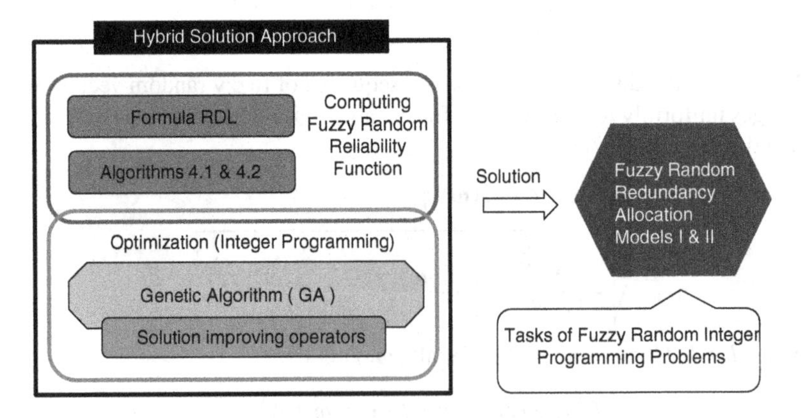

Fig. 4.2 Hybrid algorithm structure for FR-RAMs I and II

4.4.1 Solution Representation and Initialization

A positive integer vector $C = (C_1, C_2, \cdots, C_N)$ is used as a chromosome to represent a solution $x = (x_{1,1}, \cdots, x_{1,n_1}, \cdots x_{s,1}, \cdots, x_{s,n_s})$ of the FR-RAMs I and II, where $N = n_1 + n_2 + \cdots + n_s$.

To initialize the population, we first generate randomly an integer vector $C = (C_1, C_2, \cdots, C_N)$ from a positive integer set $\{1, 2, \cdots, K\}^N$, where K is a sufficiently large integer. If C is feasible, it is taken as an initial chromosome, otherwise, regenerate the vector C from $\{1, 2, \cdots, K\}^N$ until the C is proved to be feasible.

Here, for the FR-RAM I, the feasibility of the chromosome

$$C = (C_1, C_2, \cdots, C_N)$$

is checked by

$$\sum_{i=1}^{s} \sum_{j=1}^{n_i} c_{ij} C_{ij} \leq c^0 \tag{4.22}$$

$$l_i \leq \sum_{j=1}^{n_i} C_{ij} \leq u_i, \text{for } i = 1, \cdots, s. \tag{4.23}$$

Here, $C_{ij} \in \{C_1, C_2, \cdots, C_N\}$. While in the FR-RAM II, to check the feasibility of the chromosome C, we should compute the reliability $R_{t^0}(C)$ by the formula RDL, Algorithm 4.1, or Algorithm 4.2, according to the different distributions of lifetimes. Then, the following constraints are checked:

$$R_{t^0}(C) = \mathrm{Ch}\{T(C, \xi) \geq t^0\} \geq r^0 \tag{4.24}$$

$$l_i \leq \sum_{j=1}^{n_i} C_{ij} \leq u_i, \text{for } i = 1, \cdots, s. \tag{4.25}$$

Repeating the above process *pop_size* times, we get *pop_size* initial chromosomes $C_1, C_2, \cdots, C_{pop\_size}$.

4.4.2 Selection Process

The selection process is done based on *elitist strategy* and *spinning roulette wheel*. Before spinning the roulette wheel, we first calculate the objective function for each chromosome, i.e.,

$$R_{t^0}(C) \text{ for FR-RAM I,}$$

and

$$\sum_{i=1}^{s} \sum_{j=1}^{n_i} c_{ij} C_{ij} \text{ for FR-RAM II,}$$

respectively, and the *pop_size* chromosomes are rearranged in a descending order of fitness based on the values of their objective functions. Here, the objective of the

FR-RAM I is computed by using the formula RDL, Algorithm 4.1, or Algorithm 4.2 for the different cases of lifetimes, respectively.

In order to ensure that the best chromosome C_1 of current population can always be selected successfully as an offspring, that is, to move C_1 directly into the next generation (elitist strategy), then, we operate the selection process to the rest $pop\_size - 1$ chromosomes as follows: Employing the *evaluation function*, we assign a probability of reproduction to each chromosome $C_k, k = 2,3,\cdots,pop\_size$, so that the chromosome with the higher fitness will have more chance to be reproduced. There are several kinds of evaluation functions; here we adopt a popular one, *rank-based evaluation function*, which is defined as below:

$$eval(C_k) = a(1-a)^{k-2}, k = 2,3,\cdots,pop\_size,$$

where $a \in (0,1)$ is a system parameter, and $k = 1$ means the best individual, while $k = pop\_size$ the worst one. Next, we calculate the cumulative probability p_k for each chromosome $C_k, k = 2,3,\cdots,pop\_size$ as follows:

$$p_1 = 0, p_k = eval(C_2) + eval(C_3) + \cdots + eval(C_k),$$

and then normalize all p_k's dividing each $p_k, k = 2,3,\cdots,pop\_size$ by p_{pop_size} such that $p_{pop\_size} = 1$. After that, generate a random number $r \in (0,1]$, the probability of $p_{k-1} < r \le p_k$ is the probability that the kth chromosome will be selected for the new population for $k = 2,3,\cdots,pop\_size$. Repeating the following process $pop\_size - 1$ times, we can select $pop\_size - 1$ copies of chromosomes: generate a random number $r \in (0,1]$, and select the kth chromosome C_k for $2 \le k \le pop\_size$ if $p_{k-1} < r \le p_k$. Combining the previous C_1, we obtain pop_size offspring.

4.4.3 Crossover Operation

In this process, a system parameter $p_c \in (0,1)$ is predetermined as the probability of crossover. We repeat the following process pop_size times to determine the parents for the crossover operation: generate a random number r from interval $(0,1]$; the chromosome C_k is selected as a parent for crossover provided $r < p_c$, where $k = 1,2,\cdots,pop\_size$. Denote $C_1', C_2', C_3', \cdots$ the selected parents. They are divided into pairs:

$$(C_1', C_2'), (C_3', C_4'), (C_5', C_6'), \cdots.$$

The crossover operation on each pair (C_1', C_2') is done in the following way: Let

$$C_1' = \left(c_1^{(1)}, c_2^{(1)}, \cdots, c_N^{(1)}\right),$$

and

$$C_2' = \left(c_1^{(2)}, c_2^{(2)}, \cdots, c_N^{(2)}\right).$$

We randomly choose an integer N_c between 1 and N as the crossover point. Then, exchange the genes of the chromosomes C_1' and C_2' and produce two children as follows:

$$C_1'' = \left(c_1^{(2)}, c_2^{(2)}, \cdots, c_{N_c-1}^{(2)}, c_{N_c}^{(2)}, c_{N_c+1}^{(1)}, \cdots, c_N^{(1)}\right)$$

and

$$C_2'' = \left(c_1^{(1)}, c_2^{(1)}, \cdots, c_{N_c-1}^{(1)}, c_{N_c}^{(1)}, c_{N_c+1}^{(2)}, \cdots, c_N^{(2)}\right).$$

If both children are feasible, then the parents are replaced by them. Otherwise, keep the feasible one if it exists, and then repeat the crossover process by generating a new crossover points until two feasible children are obtained.

4.4.4 *Mutation Operation*

Similar to the crossover operation, a parameter $p_m \in (0,1)$ is predetermined as the probability of mutation. We repeat the following process *pop_size* times: randomly generate a real number r from $(0,1]$; the chromosome C_k is selected as parents for mutation provided $r < p_m$, where $k = 1, 2, \cdots, pop\_size$. On each selected parent, denoted $C = (C_1, C_2, \cdots, C_N)$, the mutation is done in the following way. We first randomly choose a mutation position N_m between 1 and N, i.e., $N_m = \text{(int)rand}(1, N)$, where $\text{(int)rand}(a, b)$ is a random integer generated from interval (a, b). Then, initialize $C_1', C_2', \cdots, C_{N_m-1}', C_{N_m}'$ from integer set $\{1, 2, \cdots, K\}$, and produce a new chromosome

$$C' = \left(C_1', C_2', \cdots, C_{N_m-1}', C_{N_m}', C_{N_m+1}, \cdots, C_N\right).$$

If C' is feasible for the constraints, then replace C with it. Otherwise, repeat this process until a feasible child is obtained.

4.4.5 *Chromosome Improvement*

We note that in FR-RAM I, the more components allocated, the larger the objective value (reliability level) is, while in FR-RAM II, the less components allocated, the smaller the objective value (total cost) is. So in both processes of crossover and mutation, we can improve the quality of the new offspring C' by increasing the component number to FR-RAM I, while reducing the component number to FR-RAM II, under the constraint conditions (4.22)–(4.23) and (4.24)–(4.25), respectively. That is, in the crossover and mutation processes to FR-RAM I, we do

```
if(Constrint (4.22)–(4.23) is satisfied)
{   while(Constrint (4.22)–(4.23) is satisfied)
    {
      M = (int)rand(1,N);                                      (4.26)
      C'_M ++;
    } C'_M --;
}
```

for each generated chromosome $C' = (C'_1, C'_2, \cdots, C'_N)$, and to FR-RAM II, we do

```
if(Constrint (4.24)–(4.25) is satisfied)
{   while(Constrint (4.24)–(4.25) is satisfied)
    {
      M = (int)rand(1,N);                                      (4.27)
      C'_M --;
    } C'_M ++;
}
```

for each generated chromosome.

4.4.6 Algorithm Procedure

The algorithm for solving the FR-RAMs I and II is summarized as follows (see also the flowchart Fig. 4.3).

Algorithm 4.3.

Step 1. *Input the parameters: pop_size, p_c, p_m, and a.*

Step 2. *Initialize pop_size chromosomes from the positive integer set $\{1,2,\cdots,K\}^N$. Here, the feasibility of the chromosomes in FR-RAM II is checked by the formula RDL, Algorithm 4.1, or Algorithm 4.2.*

Step 3. *Compute the objective values of all chromosomes. Here, for FR-RAM I, the objective values are computed by formula RDL, Algorithm 4.1, or Algorithm 4.2.*

Step 4. *Calculate the rank-based evaluation function for all the chromosomes according to their objective values.*

Step 5. *Select the chromosomes by spinning the roulette wheel with elitist strategy.*

Step 6. *Update the chromosomes by crossover and mutation operations with chromosome improvement. Again, for FR-RAM II, the feasibility of the chromosomes is checked by the formula RDL, Algorithm 4.1, or Algorithm 4.2.*

Step 7. *Repeat Step 3 to Step 6 for a given number of cycles.*

Step 8. *Return the best chromosome as the optimal solution.*

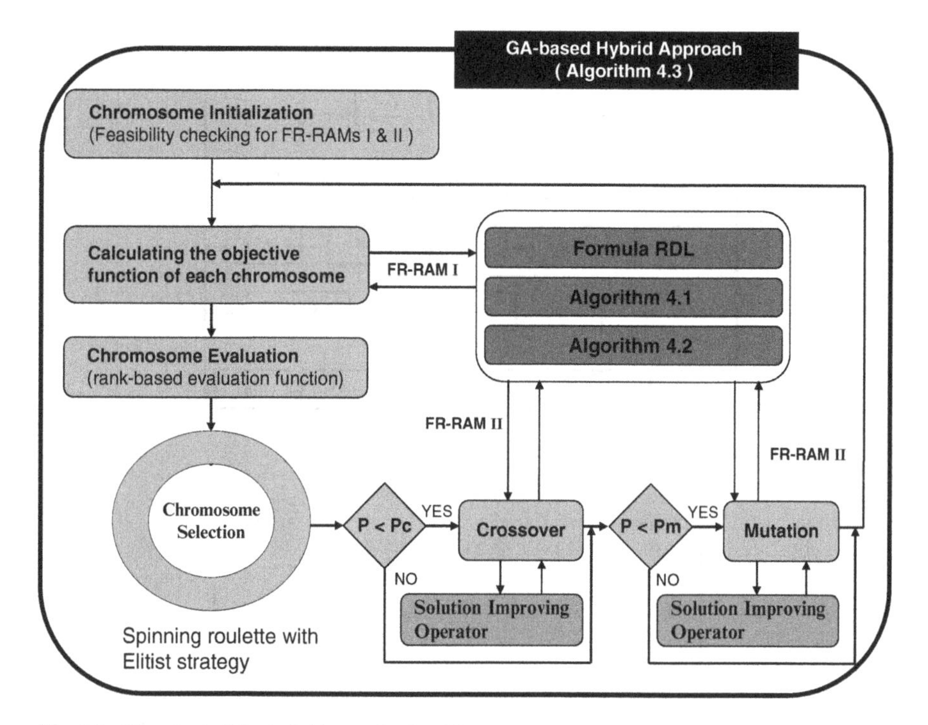

Fig. 4.3 Flowchart of the hybrid genetic algorithm

4.5 Numerical Experiments

In this section, two numerical examples of the FR-RAMs I and II are provided (all the numerical experiments of the book are performed on a personal computer, Intel(R) Core(TM) 2 Duo CPU 2.00 GHz, 1.0 GB memory, and Visual C++).

Example 4.3. Consider a parallel-series system with six subsystems, where there are 3 types of components in the first subsystem, 2 types in the second subsystem, 4 types in the third subsystem, 5 types in the third subsystem, 3 types in the third subsystem, and 6 types in the third subsystem (see Fig. 4.4).

The redundancy allocation decision vector is

$$x = \begin{pmatrix} x_{1,1},\ x_{1,2},\ x_{1,3}, \\ x_{2,1},\ x_{2,2}, \\ x_{3,1},\ x_{3,2},\ x_{3,3}, \\ x_{4,1},\ x_{4,2},\ x_{4,3},\ x_{4,4},\ x_{4,5}, \\ x_{5,1},\ x_{5,2},\ x_{5,3}, \\ x_{6,1},\ x_{6,2},\ x_{6,3},\ x_{6,4},\ x_{6,5},\ x_{6,6}; \end{pmatrix},$$

the vector of fuzzy random lifetimes is

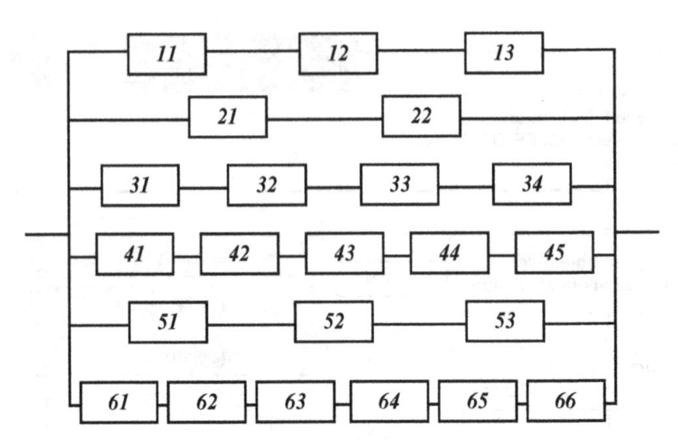

Fig. 4.4 A parallel-series system with six subsystems

$$\xi = \begin{pmatrix} \xi_{1,1,1}, \cdots, \xi_{1,1,x_{1,1}}, & \cdots, & \xi_{1,3,1}, & \cdots, & \xi_{1,3,x_{1,3}}, \\ \xi_{2,1,1}, \cdots, \xi_{2,1,x_{2,1}}, \xi_{2,2,1}, & \cdots, & \xi_{2,2,x_{2,2}}, & & \\ \xi_{3,1,1}, \cdots, \xi_{3,1,x_{3,1}}, & \cdots, & \xi_{3,3,1}, & \cdots, & \xi_{3,3,x_{3,3}}, \\ \xi_{4,1,1}, \cdots, \xi_{4,1,x_{4,1}}, & \cdots, & \xi_{4,5,1}, & \cdots, & \xi_{4,5,x_{4,5}}, \\ \xi_{5,1,1}, \cdots, \xi_{5,1,x_{5,1}}, & \cdots, & \xi_{5,3,1}, & \cdots, & \xi_{5,3,x_{5,3}}, \\ \xi_{6,1,1}, \cdots, \xi_{6,1,x_{6,1}}, & \cdots, & \xi_{6,1,x_{6,1}}, & \cdots, & \xi_{6,6,x_{6,6}} \end{pmatrix}$$

in which fuzzy random lifetimes and the cost of each component are given in Table 4.2 (we use $\xi_{i,j}$ and c_{ij} to represent the distribution and the cost of all lifetimes $\xi_{i,j,k}$ for $k = 1,2,\cdots,x_{i,j}$, since they are the same type of components). The threshold system lifetime $t^0 = 6$. The total available capital $c^0 = 700$. For each subsystem $i, i = 1,2,\cdots,6$, the lower and upper bounds of the number of the redundant components are given in Table 4.3, respectively.

Maximizing the reliability and making use of FR-RAM I, we can build a redundancy allocation model for this system as follows:

$$\left.\begin{array}{l} \max R_6(x) = \displaystyle\int_\Omega \mathrm{Cr}\left\{ \bigvee_{i=1}^{6} \left[\bigwedge_{j=1}^{n_i} \left(\sum_{k=1}^{x_{i,j}} \xi_{i,j,k}(\omega) \right) \right] \geq 6 \right\} \Pr(\mathrm{d}\omega) \\ \text{subject to} \\ \qquad 10x_{1,1} + 12x_{1,2} + 14x_{1,3} + 10x_{2,1} + 12x_{2,2} + 16x_{3,1} \\ \qquad +12x_{3,2} + 14x_{3,3} + 15x_{3,4} + 12x_{4,1} + 18x_{4,2} + 16x_{4,3} \\ \qquad +12x_{4,4} + 10x_{4,5} + 18x_{5,1} + 11x_{5,2} + 14x_{5,3} + 12x_{6,1} \\ \qquad +10x_{6,2} + 11x_{6,3} + 13x_{6,4} + 18x_{6,5} + 16x_{6,6} \leq 700, \\ \qquad 3 \leq \sum_{j=1}^{3} x_{1,j} \leq 12, \\ \qquad 2 \leq \sum_{j=1}^{2} x_{2,j} \leq 6, \\ \qquad 4 \leq \sum_{j=1}^{4} x_{3,j} \leq 24, \\ \qquad 5 \leq \sum_{j=1}^{5} x_{4,j} \leq 32, \\ \qquad 3 \leq \sum_{j=1}^{3} x_{5,j} \leq 18, \\ \qquad 6 \leq \sum_{j=1}^{6} x_{6,j} \leq 38, \\ \qquad x_{i,j} \in \mathbb{N}, \text{for } j = 1,\cdots,n_i, i = 1,\cdots,6. \end{array}\right\} \quad (4.28)$$

Table 4.2 Lifetime and cost of each component in Example 5

Component ij	Cost c_{ij}	Lifetime ξ_{ij}	Random parameter Y_{ij}
11	10	$(2+Y_{11},3+Y_{11},5+Y_{11})$	$Y_{11}\sim\mathscr{U}(2,3)$
12	12	$(3+Y_{12},4+Y_{12},6+Y_{12})$	$Y_{12}\sim\mathscr{U}(2,3)$
13	14	$(4+Y_{13},5+Y_{13},6+Y_{13})$	$Y_{13}\sim\mathscr{U}(1,2)$
21	10	$(1+Y_{21},3+Y_{21},4+Y_{21})$	$Y_{21}\sim\mathscr{U}(2,4)$
22	12	$(2+Y_{22},4+Y_{22},5+Y_{22})$	$Y_{22}\sim\mathscr{U}(1,3)$
31	16	$(4+Y_{31},6+Y_{31},8+Y_{31})$	$Y_{31}\sim\mathscr{U}(0,2)$
32	12	$(3+Y_{32},4+Y_{32},5+Y_{32})$	$Y_{32}\sim\mathscr{U}(1,3)$
33	14	$(4+Y_{33},5+Y_{33},6+Y_{33})$	$Y_{33}\sim\mathscr{U}(2,3)$
34	15	$(4+Y_{34},6+Y_{34},8+Y_{34})$	$Y_{34}\sim\mathscr{U}(1,2)$
41	12	$(2+Y_{41},4+Y_{41},6+Y_{41})$	$Y_{41}\sim\mathscr{U}(0,2)$
42	18	$(5+Y_{42},7+Y_{42},9+Y_{42})$	$Y_{42}\sim\mathscr{U}(1,3)$
43	16	$(4+Y_{43},8+Y_{43},10+Y_{43})$	$Y_{43}\sim\mathscr{U}(2,3)$
44	12	$(3+Y_{44},6+Y_{44},8+Y_{44})$	$Y_{44}\sim\mathscr{U}(1,3)$
45	10	$(2+Y_{45},4+Y_{45},7+Y_{45})$	$Y_{45}\sim\mathscr{U}(0,1)$
51	18	$(6+Y_{51},8+Y_{51},10+Y_{51})$	$Y_{51}\sim\mathscr{U}(0,2)$
52	11	$(3+Y_{52},4+Y_{52},5+Y_{52})$	$Y_{52}\sim\mathscr{U}(1,3)$
53	14	$(4+Y_{53},7+Y_{53},9+Y_{53})$	$Y_{53}\sim\mathscr{U}(1,2)$
61	12	$(4+Y_{61},6+Y_{61},8+Y_{61})$	$Y_{61}\sim\mathscr{U}(0,2)$
62	10	$(2+Y_{62},4+Y_{62},5+Y_{62})$	$Y_{62}\sim\mathscr{U}(0,1)$
63	11	$(3+Y_{63},5+Y_{63},6+Y_{63})$	$Y_{63}\sim\mathscr{U}(2,3)$
64	13	$(4+Y_{64},6+Y_{64},8+Y_{64})$	$Y_{64}\sim\mathscr{U}(0,2)$
65	18	$(6+Y_{65},8+Y_{65},10+Y_{65})$	$Y_{65}\sim\mathscr{U}(1,3)$
66	16	$(5+Y_{66},7+Y_{66},9+Y_{66})$	$Y_{66}\sim\mathscr{U}(1,2)$

Table 4.3 Lower and upper bounds for subsystems in Example 4.4

Lower bounds l_i	Upper bounds u_i
$l_1=3$	$u_1=12$
$l_2=2$	$u_2=6$
$l_3=4$	$u_3=24$
$l_4=5$	$u_4=32$
$l_5=3$	$u_5=18$
$l_6=6$	$u_6=38$

Noting that all the lifetimes of components have the convex distributions, therefore, at each given allocation decision x, the system reliability

$$R_6(x)=\int_{\Omega}\mathrm{Cr}\left\{\bigvee_{i=1}^{6}\left[\bigwedge_{j=1}^{n_i}\left(\sum_{k=1}^{x_{i,j}}\xi_{i,j,k}(\omega)\right)\right]\geq 6\right\}\Pr(\mathrm{d}\omega) \tag{4.29}$$

can be calculated by Algorithm 4.1. Incorporating Algorithm 4.1 into the GA, we use Algorithm 4.3 to search for the optimal solution of problem (4.28).

Algorithm 4.3 has been run with 3,000 times of random simulation (4.15) in Algorithm 4.1, and 200 generations in GA, and in Table 4.4, we compare solutions by careful variations of parameters of GA (the probability of crossover p_c, the probability of mutation p_m, and the parameter a in the rank-based evaluation function.) with the same stopping rule. The parameters are given in Table 4.4 from

Table 4.4 The parameters and comparison solutions of Example 4.3

pop_size	p_c	p_m	a	Optimal solution	Objective	Error(%)
20	0.2	0.4	0.05	(1,1,5,4,1,1,1,3,2,4,1,3,4,1,2,2,3,2,5,1,4,1,1)	0.868	2.58
20	0.3	0.2	0.05	(1,1,5,4,1,1,1,3,2,4,1,3,4,1,2,2,3,2,5,1,4,1,1)	0.868	2.58
20	0.2	0.2	0.05	(2,1,1,1,1,1,4,2,2,3,1,2,4,1,2,2,3,2,5,1,4,1,1)	0.862	3.25
20	0.4	0.2	0.10	(1,1,4,1,1,1,1,1,1,8,1,2,4,1,4,2,3,3,5,1,4,2,1)	0.876	1.68
20	0.4	0.3	0.10	(1,1,2,1,1,1,1,1,1,2,1,5,1,1,4,5,3,3,5,1,4,3,1)	0.883	0.90
30	0.2	0.4	0.05	(1,1,1,2,1,1,1,1,2,1,1,1,3,1,1,3,5,3,3,1,4,1,1)	0.871	2.24
30	0.3	0.2	0.05	(1,1,6,1,1,2,1,3,2,5,1,2,4,1,2,2,3,2,5,1,5,1,1)	0.868	2.58
30	0.2	0.2	0.05	(1,1,5,1,1,1,1,1,1,4,1,3,1,1,4,4,2,5,1,2,4,4,1)	0.891	0.00
30	0.4	0.2	0.10	(1,1,3,2,1,1,2,1,1,3,1,2,1,1,4,4,2,3,4,3,2,1,1)	0.886	0.56
30	0.4	0.3	0.10	(1,1,1,1,1,1,1,1,1,3,1,1,1,1,5,1,1,4,4,2,7,1,3)	0.889	0.22

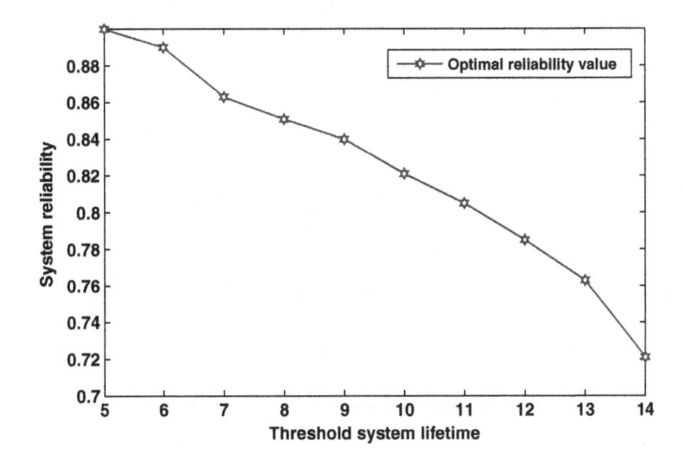

Fig. 4.5 The trend line of the reliability with respect to t^0

the first to the fifth column, and the computational results are provided in the sixth and seventh columns. Moreover, the *relative error* is given in the last column, which is defined by

$$Error = \frac{optimal\ value - objective\ value}{optimal\ value} \times 100\%.$$

It follows from Table 4.4 that the relative error does not exceed 3.25% when different parameters of GA are selected. In addition, the trend line of the reliability with respect to the threshold system lifetime t^0 is provided in Fig. 4.5. The performance implies the solution algorithm is robust to the parameter settings and effective to solve the FR-RAM I.

Example 4.4. In Example 4.3, if the decision maker intends to minimize the total cost to meet some reliability constraint, i.e., $R_{t^0}(x) \geq r^0$, then the problem can be modeled by FR-RAM II. Here, we suppose the lifetimes of components have the same costs as those in Example 4.3 but have partly different distributions as listed in Table 4.5, where positive triangular fuzzy random variables, positive normal fuzzy

Table 4.5 Lifetime and cost of each component in Example 4.4

Component ij	Cost c_{ij}	Lifetime ξ_{ij}	Random parameter Y_{ij}
11	10	$(2+Y_{11},3+Y_{11},5+Y_{11})$	$Y_{11} \sim \mathscr{U}(2,3)$
12	12	$\mathscr{N}^{+}_{FR}(Y_{12},4)$	$Y_{12} \sim \mathscr{U}(2,4)$
13	14	$\nabla(4+Y_{13},5+Y_{13},6+Y_{13})$	$Y_{13} \sim \mathscr{U}(1,3)$
21	10	$\nabla(1+Y_{21},3+Y_{21},4+Y_{21})$	$Y_{21} \sim \mathscr{U}(1,2)$
22	12	$(2+Y_{22},4+Y_{22},5+Y_{22})$	$Y_{22} \sim \mathscr{U}(1,3)$
31	16	$\mathscr{N}^{+}_{FR}(Y_{31},6)$	$Y_{31} \sim \mathscr{U}(1,3)$
32	12	$\nabla(3+Y_{32},4+Y_{32},5+Y_{32})$	$Y_{32} \sim \mathscr{U}(0,2)$
33	14	$(4+Y_{33},5+Y_{33},6+Y_{33})$	$Y_{33} \sim \mathscr{U}(2,3)$
34	15	$(4+Y_{34},5+Y_{34},6+Y_{34})$	$Y_{34} \sim \mathscr{U}(1,2)$
41	12	$(2+Y_{41},4+Y_{41},6+Y_{41})$	$Y_{31} \sim \mathscr{U}(0,2)$
42	18	$(5+Y_{42},7+Y_{42},9+Y_{42})$	$Y_{32} \sim \mathscr{U}(1,3)$
43	16	$(4+Y_{43},8+Y_{43},10+Y_{43})$	$Y_{33} \sim \mathscr{U}(2,3)$
44	12	$\mathscr{N}^{+}_{FR}(Y_{44},6)$	$Y_{44} \sim \mathscr{U}(1,3)$
45	10	$(2+Y_{45},4+Y_{45},7+Y_{45})$	$Y_{45} \sim \mathscr{U}(0,1)$
51	18	$(6+Y_{51},8+Y_{51},10+Y_{51})$	$Y_{51} \sim \mathscr{U}(0,2)$
52	11	$\nabla(4+Y_{52},6+Y_{52},8+Y_{52})$	$Y_{52} \sim \mathscr{U}(1,3)$
53	14	$(4+Y_{53},7+Y_{53},9+Y_{53})$	$Y_{53} \sim \mathscr{U}(1,2)$
61	12	$(4+Y_{61},6+Y_{61},8+Y_{61})$	$Y_{61} \sim \mathscr{U}(0,2)$
62	10	$\nabla(3+Y_{62},6+Y_{62},8+Y_{62})$	$Y_{62} \sim \mathscr{U}(0,1)$
63	11	$(3+Y_{63},5+Y_{63},6+Y_{63})$	$Y_{63} \sim \mathscr{U}(2,3)$
64	13	$\mathscr{N}^{+}_{FR}(Y_{64},8)$	$Y_{64} \sim \mathscr{U}(0,2)$
65	18	$(6+Y_{65},8+Y_{65},10+Y_{65})$	$Y_{65} \sim \mathscr{U}(1,3)$
66	16	$(5+Y_{66},7+Y_{66},9+Y_{66})$	$Y_{66} \sim \mathscr{U}(1,2)$

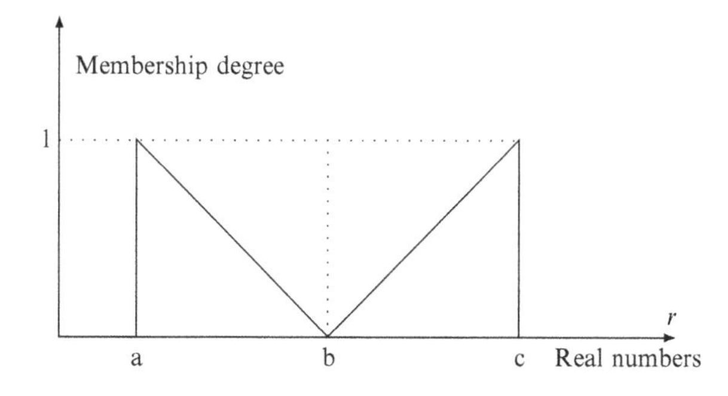

Fig. 4.6 The membership function of an inverse triangular fuzzy variable $X = \nabla(a,b,c)$

random variables, and inverse triangular fuzzy random variables are mixed together in the distributions of the component lifetimes. Here, the membership function (see Fig. 4.6) of an inverse triangular fuzzy variable which is denoted by $\nabla(a,b,c)$ is given by

$$\mu_{\nabla(a,b,c)}(r) = \begin{cases} \dfrac{b-r}{b-a}, & a \le r < b \\ \dfrac{x-b}{c-b}, & b \le r \le c \\ 0, & \text{otherwise.} \end{cases}$$

Table 4.6 Lower and upper bounds for subsystems in Example 4.4

Lower bounds l_i	Upper bounds u_i
$l_1 = 3$	$u_1 = 12$
$l_2 = 2$	$u_2 = 6$
$l_3 = 8$	$u_3 = 24$
$l_4 = 10$	$u_4 = 32$
$l_5 = 3$	$u_5 = 18$
$l_6 = 12$	$u_6 = 38$

Taking the target reliability $r^0 = 0.8$ and the threshold lifetime $t^0 = 6$, and the lower and upper bounds of numbers of the redundant components that are provided in Table 4.6, a cost minimization-based FR-RAM can be formed by

$$\left.\begin{array}{l}
\min\ 10x_{1,1}+12x_{1,2}+14x_{1,3}+10x_{2,1}+12x_{2,2}+16x_{3,1} \\
\quad +12x_{3,2}+14x_{3,3}+15x_{3,4}+12x_{4,1}+18x_{4,2}+16x_{4,3} \\
\quad +12x_{4,4}+10x_{4,5}+18x_{5,1}+11x_{5,2}+14x_{5,3}+12x_{6,1} \\
\quad +10x_{6,2}+11x_{6,3}+13x_{6,4}+18x_{6,5}+16x_{6,6} \\
\text{subject to} \\
\quad R_6(x) \geq 0.8, \\
\quad 3 \leq \sum_{j=1}^{3} x_{1,j} \leq 12, \\
\quad 2 \leq \sum_{j=1}^{2} x_{2,j} \leq 6, \\
\quad 8 \leq \sum_{j=1}^{4} x_{3,j} \leq 24, \\
\quad 10 \leq \sum_{j=1}^{5} x_{4,j} \leq 32, \\
\quad 3 \leq \sum_{j=1}^{3} x_{5,j} \leq 18, \\
\quad 12 \leq \sum_{j=1}^{6} x_{6,j} \leq 38, \\
\quad x_{i,j} \in \mathbb{N}, \text{for } j = 1, \cdots, n_i, i = 1, \cdots, 6,
\end{array}\right\} \quad (4.30)$$

where

$$R_6(x) = \int_{\Omega} \mathrm{Cr}\left\{ \bigvee_{i=1}^{6} \left[\bigwedge_{j=1}^{n_i} \left(\sum_{k=1}^{x_{i,j}} \xi_{i,j,k}(\omega) \right) \right] \geq 6 \right\} \Pr(d\omega),$$

When using the Algorithm 4.3 (for FR-RAM II) to solve the problem (4.30). In the processes of initialization, crossover, and mutation, we need to check the feasibility of each chromosome, which means to compute $R_6(x)$ in each checking

Table 4.7 The parameters and comparison solutions of Example 4.4

pop_size	p_c	p_m	a	Optimal solution	Objective	Error(%)
20	0.2	0.4	0.05	(1,1,1,1,1,1,2,4,1,2,1,1,3,3,1,1,1,1,3,2,3,1,2)	489.00	2.73
20	0.3	0.2	0.05	(1,1,1,1,1,1,2,4,1,2,1,1,2,4,1,1,1,1,3,2,3,1,2)	487.00	2.31
20	0.2	0.2	0.05	(1,1,1,1,1,1,3,3,1,2,2,1,1,4,1,1,1,2,2,4,1,1,2)	489.00	2.73
20	0.4	0.2	0.10	(1,1,1,1,1,1,4,2,1,2,1,1,2,4,1,1,1,2,5,1,1,2,1)	480.00	0.84
20	0.4	0.3	0.10	(1,1,1,1,1,1,2,4,1,2,1,1,2,4,1,1,1,1,3,2,3,1,2)	487.00	2.31
30	0.2	0.4	0.05	(1,1,1,1,1,1,3,1,3,4,1,1,2,2,1,1,1,2,3,2,3,1,1)	487.00	2.31
30	0.3	0.2	0.05	(1,1,1,1,1,1,4,2,1,2,1,1,2,4,1,1,1,2,2,2,3,1,2)	485.00	1.89
30	0.2	0.2	0.05	(1,1,1,1,1,1,4,2,1,2,1,1,2,4,1,1,1,2,5,1,1,2,1)	480.00	0.84
30	0.4	0.2	0.10	(1,1,1,1,1,1,4,2,1,2,1,1,1,5,1,1,1,2,5,1,1,1,2)	476.00	0.00
30	0.4	0.3	0.10	(1,1,1,1,1,1,5,1,1,5,1,1,1,2,1,1,1,2,3,4,1,1,1)	477.00	0.21

process. We note from Table 4.5 that the lifetimes

$$\begin{aligned} &\nabla(4+Y_{13},5+Y_{13},6+Y_{13}) \quad \text{in subsystem 1,} \\ &\nabla(1+Y_{21},3+Y_{21},4+Y_{21}) \quad \text{in subsystem 2,} \\ &\nabla(3+Y_{32},4+Y_{32},5+Y_{32}) \quad \text{in subsystem 3,} \\ &\nabla(4+Y_{52},6+Y_{52},8+Y_{52}) \quad \text{in subsystem 5,} \end{aligned}$$

and

$$\nabla(3+Y_{62},6+Y_{62},8+Y_{62}) \quad \text{in subsystem 6,}$$

are nonconvex fuzzy random variables; Algorithm 4.1, therefore, is inapplicable to computing the system reliability $R_6(x)$. As a consequence, we use the fuzzy random simulation (Algorithm 4.2) to compute $R_6(x)$, and embed it into GA to search for the approximately best solution of the problem (4.30).

We run Algorithm 4.3 with 200 generations, where we execute 3,000 times of random simulation (4.19) and generate 1,000 discrete fuzzy sample points in Algorithm 4.2; the comparison solutions with different parameters are collected in Table 4.7, where the *relative error* is defined by

$$Error = \frac{objective\ value - optimal\ value}{optimal\ value} \times 100\%.$$

We see from Table 4.7 that the relative error dose not exceed 2.73% which shows that Algorithm 4.3 is robust to the parameter settings and effective to solve the FR-RAM II.

In order to further testify the effectiveness of the designed hybrid GA approach, we compare it with a hybrid tabu search (TS) algorithm which was proposed in [167] for a fuzzy integer programming problem. As for the details of TS, one may refer to Appendix D. In the Table 4.8, we provide the comparison results (here in TS, the neighbor number is set as 29, and the tabu tenure TT = (int)rand(8, 10) is selected from different experiments) when dealing with Examples 4.3 and 4.4, respectively. What's more, the convergence comparisons in Examples 4.3 and 4.4 are given in

Table 4.8 The solution result comparison

Example	Approach	Optimal solution	Objective value
4.3	Hybrid GA	(1,1,5,1,1,1,1,1,1,4,1,3,1,1,4,4,2,5,1,2,4,4,1)	0.891 (reliability)
	Hybrid TS	(1,1,1,1,1,1,5,1,1,2,1,3,1,1,2,1,2,4,2,1,1,1,4)	0.873 (reliability)
4.4	Hybrid GA	(1,1,1,1,1,1,4,2,1,2,1,1,1,5,1,1,1,2,5,1,1,1,2)	476.00 (cost)
	Hybrid TS	(1,1,1,1,1,1,5,1,1,2,1,1,2,4,1,1,1,1,5,1,1,1,3)	480.00 (cost)

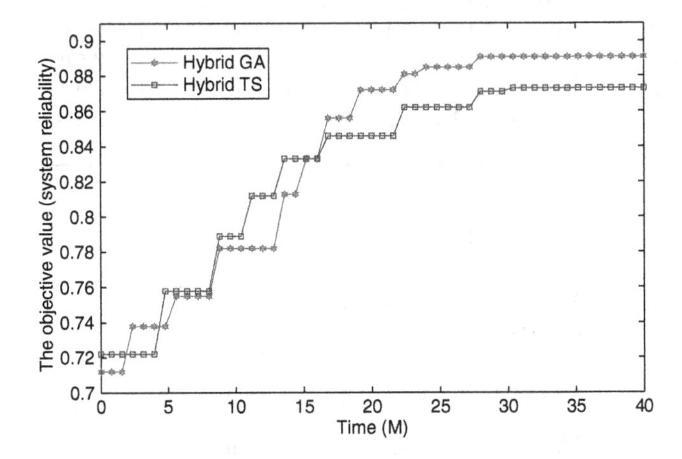

Fig. 4.7 The convergence comparison of different approaches for Example 4.3

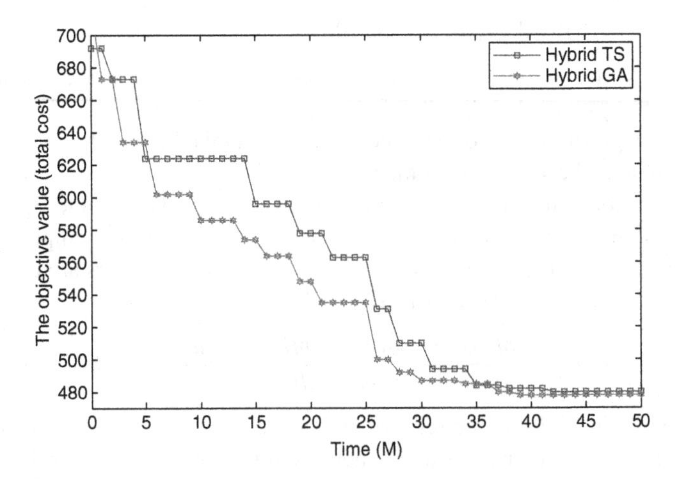

Fig. 4.8 The convergence comparison of different approaches for Example 4.4

Figs. 4.7 and 4.8, respectively, within the same time duration (40 minutes (Ms) for Example 4.3 and 50 Ms for Example 4.4, the iterations of algorithms have become stable within both time durations). From those results, we can see that the hybrid GA exhibits better performance than hybrid TS when dealing with the FR-RAMs I and II.

Chapter 5
Recourse-Based Fuzzy Random Facility Location Model with Fixed Capacity

Facility location selection is one of the most critical and strategic issues in supply chain design and management; it exhibits a significant impact on market share and profitability. Roughly speaking, the objective of a facility location strategy is to maximize the profit or minimize the costs, by determining which plants to open given a set of potential plant locations. Depending on whether or not taking the uncertainty into consideration, the location problems can be classified into two groups: deterministic location problems, and location problems with uncertain parameters. To the former, several qualitative techniques of nonlinear programming methods as well as heuristics have been proposed, such as those presented by Akinc and Khumawala [3], Badri [5], Dupont [31], Ernst and Krishnamoorthy [34], Schutz et al. [130], and Lozano et al. [102].

As for the location problems in an uncertain environment, a large number of stochastic models for location selection have been discussed intensively in the literature(see Berman and Drezner [8], Laporte et al. [76], Logendran and Terrell [99], Louveaux and Peeters [100], and Zeng and Ward [178]). Exerting the probability theory and the stochastic programming as the mainstay, the stochastic location model treats the uncertain parameters (e.g., customers' demands and operating costs of plants) as random variables, and computation depends fully on the probability distributions of the random parameters.

Nevertheless, the complex nature of practical uncertain circumstances makes it unrealistic to use only random variable to characterize the customers' demands or operating costs of plants in many real-world location problems:

Case I. When the products of plants are new to the customers, no sale record can be utilized to serve as the historic data for the client demands. So the demand data can only be captured by conducting a market survey to customers directly. In this circumstance, chances are pretty good that customers have too scarce information (purchase records) on the new products to surely determine the demand, but rather, they prefer to provide an estimation, say, demand range within which a lot of possible values could be the real purchase quantity. Such kind of demand data

S. Wang and J. Watada, *Fuzzy Stochastic Optimization: Theory, Models and Applications*, DOI 10.1007/978-1-4419-9560-5_5, © Springer Science+Business Media New York 2012

is essentially vague or fuzzy values instead of crisp values, and apparently, the distribution identified statistically from them should no question be imprecise or fuzzy probability distribution.

Case II. As for the variable costs on the products, since in real situations , more often than not, the potential sites for new plants to build in are some new regions for the firm; hence, it is not possible to obtain the historical variable cost data on in each potential sites. As a consequence, the variable costs are provided by the experts in terms of an imprecise estimation based on the cost record of old plants of this firm in other areas. Furthermore, the volatile variable costs differ a lot in different real sales situations which are affected largely by the whole marketing condition. No sooner is a prediction for the sale scenarios (random scenarios) found via the statistical analysis with the historical marketing data, than the experts are able to give more realistic estimations of variable costs in different marketing scenarios. No doubt, the variable costs in this case also make a fuzzy probability distribution.

Case III. Another sort of uncertainty mixture is that the stochastic variability and the imprecision exist amid different types of parameters, separately, for instance, when the demands are random variables but the variable costs are some fuzzy values estimated by the experts, or vice versa.

Undoubtedly, due to the existence of imprecision or fuzziness in terms of different forms, the random variable in probability theory is not capable to handle any of the situations explicated in the above Cases. Actually, the imprecise probability distributions of the demands and variable costs in Cases I and II, respectively, are intrinsically the distributions of fuzzy random variables. Likewise in the Case III, since the fuzzy random variable owes a generalized distribution structure which can characterize not only random variable but also fuzzy number, there is no theoretical obstacle to utilize fuzzy random variable as a fusion vehicle to holistically deal with the random parameters and fuzzy parameters altogether in Case III.

This chapter presents a fuzzy random facility location model with fixed capacity in the context of the fuzzy random variable. In addition, for more on the location selection models considering fuzziness, see Ishii, Lee, and Yeh [55],Wang, Watada, and Pedrycz [152], and Wen and Iwamura [163].

In Sect. 5.1, we give the mathematical formulation of the fuzzy random facility location model with recourse and fixed capacity (FR-FLM-RFC) which is mathematically a two-stage fuzzy stochastic binary (0-1) integer programming. The two-stage structure of the model is elaborated, and the solution difficulties are discussed.

In Sect. 5.2, we discuss the mathematical properties of the two-stage FR-FLM-RFC. The value of fuzzy information in the decision-making process of the location model is analyzed in terms of the difference of fuzzy random solution and random solution. Along with that, two lower and upper bounds for the solution for the two-stage FR-FLM-RFC are derived.

In Sect. 5.3, we introduce the computation schemes for the value of recourse function $\mathcal{Q}(x)$ at a given location decision x of the two-stage FR-FLM-RFC.

Two different computation schemes are designed for cases of discrete and continuous fuzzy random parameters (general case), where for the latter, the convergence of the fuzzy random simulation is theoretically discussed.

In Sect. 5.4, we discuss the solution approach. A hybrid modified binary ant colony optimization (MBACO) algorithm is designed to search for the approximately optimal solution of the two-stage FR-FLM-RFC. In the hybrid mechanism, the simplex algorithm is utilized to determine the second-stage optimal value for a given location decision and each fuzzy random realization, the computation schemes is used to calculate the values of recourse function, and the MBACO's job is to deal with the 0-1 integer optimization, i.e., to search for the optimal (approximately) location solution.

In Sect. 5.5, we test the hybrid MBACO algorithm for the two-stage FR-FLM-RFC with two numerical examples as well as some comparisons with other binary metaheuristics.

5.1 Problem Formulation

5.1.1 Mathematical Modeling

Problem settings: a firm intends to open a facility in n potential sites, the cost of each facility consists of fix opening and operating cost and variable operating cost V, the latter being a fuzzy random variable. There are m customers having fuzzy random demands D for some commodity. Each customer can be supplied from an open facility where the commodity is made available. The distribution pattern from facilities to customers is not predetermined but adapted to the realization of the fuzzy random event with respect to the demands and variable operating costs. The objective of the firm is to maximize the expected profit by choosing the optimal number of facilities to open and their locations in the market areas. To proceed with the detailed discussion, it would be advantageous to introduce some useful notation as well as assumptions.

Assumptions

1. The capacity for each facility is fixed.
2. Each customer's demand cannot be overserved, but it is possible that the demand is not fully satisfied.
3. The total supply from one facility to all clients cannot exceed the capacity of the facility.
4. Fuzzy random demand-cost vector $\xi = (D_1, \cdots, D_m, V_1, \cdots, V_n)$ is defined from a probability space $(\Omega, \Sigma, \Pr)$ to a collection of fuzzy vectors on possibility space $(\Gamma, \mathscr{A}, \mathrm{Pos})$.

Notation	
i	the index of facilities, $1 \leq i \leq n$
j	the index of clients, $1 \leq j \leq m$
D_j	fuzzy demand of client j
r_j	the unit price charged to client j
s_i	the capacity of facility i
c_i	the fixed cost for opening and operating facility i
V_i	the unit variable operating cost of facility i, which is a fuzzy random variable
ξ	fuzzy random demand-cost vector $\xi = (D_1, \cdots, D_m, V_1, \cdots, V_n)$
x_i	binary decision variable equal to one if facility i, is open and zero otherwise
x	decision vector which is $x = (x_1, x_2, \cdots, x_n)$
y_{ij}	the quantity supplied to client j from i
t_{ij}	unit transportation cost from i to j

Given the above assumptions and making use of the notation, a two-stage fuzzy random facility location model with recourse and fixed capacity (FR-FLM-RFC [154]) can be formulated concisely as follows:

$$\left.\begin{array}{ll} \max & \mathscr{Q}(x) - \sum\limits_{i=1}^{n} c_i x_i \\ \text{subject to } x_i \in \{0,1\}, i = 1,2,\cdots,n, & \end{array}\right\} \tag{5.1}$$

where $\mathscr{Q}(x) = E\left[Q(x,\xi)\right]$, and

$$\left.\begin{array}{l} Q\Big(x, \xi(\omega,\gamma)\Big) = \max \sum\limits_{i=1}^{n} \sum\limits_{j=1}^{m} \left(r_j - V_i(\omega,\gamma) - t_{ij}\right) y_{ij}^{(\omega,\gamma)} \\ \qquad \text{subject to} \\ \qquad\qquad \sum\limits_{i=1}^{n} y_{ij}^{(\omega,\gamma)} \leq D_j(\omega,\gamma), j = 1,2,\cdots,m, \\ \qquad\qquad \sum\limits_{j=1}^{m} y_{ij}^{(\omega,\gamma)} \leq s_i x_i, i = 1,2,\cdots,n, \\ \qquad\qquad y_{ij}^{(\omega,\gamma)} \geq 0, i = 1,2,\cdots,n, j = 1,2,\cdots,m. \end{array}\right\} \tag{5.2}$$

Here, $D_j(\omega,\gamma)$ and $V_i(\omega,\gamma)$ are the realizations of the fuzzy random demand D_j and fuzzy random cost V_i, respectively, for any $(\omega,\gamma) \in \Omega \times \Gamma$.

In the two-stage fuzzy random facility location problem (7.1)–(7.3) with recourse and fixed capacity, a distinction is made between the first stage and the second stage of processing (see Fig. 5.1). The location decision vector x is the first-stage decision which must be considered before the realizations of fuzzy random demand-cost vector

$$\xi(\omega,\gamma) = \Big(D_1(\omega,\gamma), \cdots, D_m(\omega,\gamma), V_1(\omega,\gamma), \cdots, V_n(\omega,\gamma)\Big) \tag{5.3}$$

coming out. At the second stage, the fuzzy random demands and costs are known (the realizations $\xi(\omega,\gamma)$ are observed); therefore, the second-stage decision

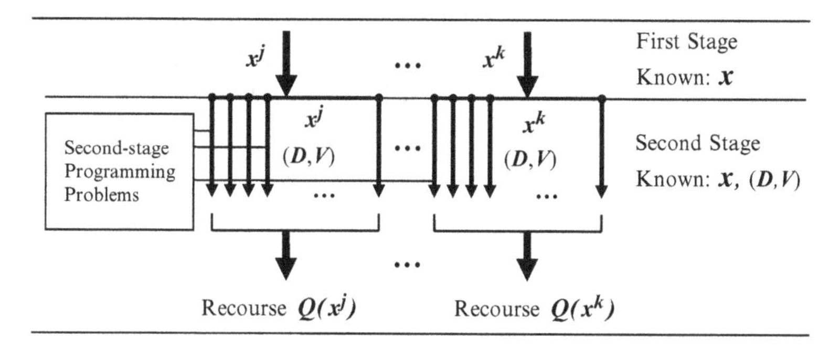

Fig. 5.1 Two-stage process of fuzzy random facility location problem

variables (also called recourse decisions) $y_{ij}, 1 \leq i \leq n, 1 \leq j \leq m$, which represent the distribution pattern from facilities to customers, are determined to maximize the return

$$\sum_{i=1}^{n} \sum_{j=1}^{m} \left(r_j - V_i(\omega,\gamma) - t_{ij}\right) y_{ij}^{(\omega,\gamma)}$$

corresponding to the current outcome of the fuzzy random event (ω,γ). The decisions $y_{ij}^{(\omega,\gamma)}$'s are determined as soon as x and ξ are known by solving a linear programming (7.3), which is called a *second-stage problem*. Since $y_{ij}^{(\omega,\gamma)}, 1 \leq i \leq n, 1 \leq j \leq m$ are completely determined by the selection of x and the realized value $\xi(\omega,\gamma)$ of ξ, the real decision of problem (7.1)–(7.3) is only the location decision x.

Furthermore, given x and $\xi(\omega,\gamma)$, the optimal value of the second-stage problem (7.3), i.e., $Q(x,\xi(\omega,\gamma))$, is usually referred to as a second-stage value function, and the expected value of $Q(x,\xi)$, $\mathcal{Q}(x) = E\left[Q(x,\xi)\right]$ is called the *recourse function* of the two-stage fuzzy random facility location problem.

From the above descriptions, we see that in order to obtain the objective value $\mathcal{Q}(x) - \sum_{i=1}^{n} c_j x_j$ at fixed decision x of problem (7.1)–(7.3), we first need to calculate the recourse function $\mathcal{Q}(x)$ which contains a complex process of solving the second-stage problems (7.3) to obtain $Q(x,\xi(\omega,\gamma))$ for all realizations $\xi(\omega,\gamma), (\omega,\gamma) \in \Omega \times \Gamma$. The solution process to the problem (7.1)–(7.3) can be illustrated by a simple example.

Example 5.1. Consider a two-stage fuzzy random facility location problem as follows: $n = 2$, $m = 1, r = 6, (s_1,s_2) = (3,5), (t_1,t_2) = (4,2), (c_1,c_2) = (1,2)$. Suppose that the fuzzy random costs $V_1 \equiv 1, V_2 \equiv 2$, and fuzzy random demand D have the following distribution:

$$D(\omega) = \begin{cases} X_1, & \text{with probability } 1/3 \\ X_2, & \text{with probability } 2/3, \end{cases}$$

where fuzzy variable X_1 takes on values $6, 8$, and 20 with membership degree $2/5, 1$, and $3/5$, respectively, and the fuzzy variable X_2 assumes the values 3 and 5 with membership degree $4/5$ and 1, respectively. The model can be built as

$$\left.\begin{array}{ll} \max & \mathscr{Q}(x) - x_1 - 2x_2 \\ \text{subject to } x_i \in \{0,1\}, i = 1,2. & \end{array}\right\} \tag{5.4}$$

where $\mathscr{Q}(x) = E\left[Q(x,\xi)\right]$ and

$$\left.\begin{array}{rl} Q\Big(x, \xi(\omega,\gamma)\Big) = \max & y_1 + 6y_2 \\ \text{subject to} & y_1 + y_2 \le D(\omega,\gamma), \\ & 0 \le y_1 \le 3x_1, \\ & 0 \le y_2 \le 5x_2. \end{array}\right\} \tag{5.5}$$

To solve this problem, we need to compare the objective values of all the (first-stage) decisions, i.e., $x_1 = (1,1), x_2 = (0,1), x_3 = (1,0)$, and $x_4 = (0,0)$, respectively, which requires calculating recourse function $\mathscr{Q}(x)$. We note that the fuzzy random demand-cost vector $\xi = (D, V_1, V_2) \equiv (D, 1, 2)$ has discrete realizations, denoted $\widehat{\xi}^{ij}, i = 1,2; j = 1,2, N_i (N_1 = 2, N_2 = 3)$, which are determined by a discrete random variable ω and two discrete fuzzy variables X_1 and X_2. That is, for $\omega = \omega_1$ with probability $p_1 = 1/3$, fuzzy vector $\xi(\omega_1) = (X_1, 1, 2)$ takes on the realizations $\widehat{\xi}^{11} = (6,1,2)$ with membership degree $\mu_{11} = 2/5$, $\widehat{\xi}^{12} = (8,1,2)$ with $\mu_{12} = 1$, and $\widehat{\xi}^{13} = (20,1,2)$ with $\mu_{13} = 3/5$, while for $\omega = \omega_2$ with probability $p_2 = 2/3$, $\xi(\omega_2) = (X_2, 1, 2)$ takes on the realizations $\widehat{\xi}^{21} = (3,1,2)$ with $\mu_{21} = 4/5$ and $\widehat{\xi}^{22} = (5,1,2)$ with $\mu_{22} = 1$. For this discrete fuzzy random vector ξ, the recourse function $\mathscr{Q}(x)$ can be calculated from the definition as follows:

$$\mathscr{Q}(x) = \sum_{i=1}^{2} p_i \mathscr{Q}(x, \omega_i), \tag{5.6}$$

where

$$\mathscr{Q}(x, \omega_i) = \sum_{j=1}^{N_i} q_{ij} Q(x, \widehat{\xi}^{ij});$$

each $Q\left(x, \widehat{\xi}^{ij}\right)$ is obtained by solving the second-stage problem (5.5) for each realization $\widehat{\xi}^{ij}$ and, all of them are ranked as

$$Q\left(x, \widehat{\xi}^{i1}\right) \le Q\left(x, \widehat{\xi}^{i2}\right) \le \cdots \le Q\left(x, \widehat{\xi}^{iN_i}\right);$$

the corresponding weights q_{ij}'s are given by the following formulae:

$$q_{ij} = \frac{1}{2}\left(\max_{k=1}^{j} \mu_{ik} - \max_{k=0}^{j-1} \mu_{ik}\right) + \frac{1}{2}\left(\max_{k=j}^{N_i} \mu_{ik} - \max_{k=j+1}^{N_i+1} \mu_{ik}\right) \tag{5.7}$$

$(\mu_{i0} = 0, \mu_{i,N_i+1} = 0)$ for $i = 1,2; j = 1,2,\cdots,N_i$. The detailed procedure is depicted below.

When $x = x_1 = (1,1)$, the second-stage problem (5.5) can be rewritten as

$$\left.\begin{array}{rl} Q\Big(x, \xi(\omega,\gamma)\Big) = & \max\ y_1 + 6y_2 \\ & \text{subject to } y_1 + y_2 \leq D(\omega,\gamma), \\ & 0 \leq y_1 \leq 3, \\ & 0 \leq y_2 \leq 5. \end{array}\right\} \tag{5.8}$$

Let $\omega = \omega_1$ (with probability $1/3$). For $\widehat{\xi}^{11} = (6,1,2)$ with membership degree $2/5$, by solving the second-stage problem, we obtain the optimal second-stage decision is $(y_1^*, y_2^*) = (1,5)$. Hence, we have $Q(x_1, \widehat{\xi}^{11}) = 31$ with the membership degree of $2/5$. In the same way, we obtain $Q(x_1, \widehat{\xi}^{12}) = 33$ with membership degree 1, and $Q(x_1, \widehat{\xi}^{13}) = 33$ with membership degree $3/5$. We note that

$$Q(x_1, \widehat{\xi}^{11}) < Q(x_1, \widehat{\xi}^{12}) = Q(x_1, \widehat{\xi}^{13}).$$

Then making use of (5.7), we determine

$$\begin{aligned} q_{11} &= \frac{1}{2} \times \frac{2}{5} + \frac{1}{2} \times \left(\max\left\{\frac{2}{5}, 1, \frac{3}{5}\right\} - \max\left\{1, \frac{3}{5}\right\}\right) = \frac{1}{5}, \\ q_{12} &= \frac{1}{2} \times \left(1 - \frac{2}{5}\right) + \frac{1}{2} \times \left(1 - \frac{3}{5}\right) = \frac{1}{2}, \\ q_{13} &= \frac{1}{2} \times (1-1) + \frac{1}{2} \times \frac{3}{5} = \frac{3}{10}. \end{aligned}$$

Therefore,

$$\mathcal{Q}(x_1, \omega_1) = \frac{1}{5} \times 31 + \frac{1}{2} \times 33 + \frac{3}{10} \times 33 = 32.6.$$

Next, we let $\omega = \omega_2$ (with probability $2/3$). Similarly as in the above case, we can obtain

$$18 = Q\left(x_1, \widehat{\xi}^{21}\right) < Q\left(x_1, \widehat{\xi}^{22}\right) = 30,$$

$q_{21} = 2/5, q_{22} = 3/5$, and $\mathcal{Q}(x_1, \omega_2) = 25.2$. As a consequence, by (5.6) we get

$$\mathcal{Q}(x_1) = \frac{1}{3} \times 32.6 + \frac{2}{3} \times 25.2 = 27.67,$$

and the objective value at x_1 is

$$Q(x_1) - c_1 - c_2 = 27.67 - 3 = 24.67.$$

When $x = x_2 = (0,1)$, for $\omega = \omega_1$, we calculate

$$30 = Q\left(x_2, \widehat{\xi}^{11}\right) = Q\left(x_2, \widehat{\xi}^{12}\right) = Q\left(x_2, \widehat{\xi}^{13}\right) = 30.$$

Therefore, $\mathscr{Q}(x_2, \omega_1) = 30$. On the other hand, for $\omega = \omega_2$, we obtain

$$18 = Q\left(x_2, \widehat{\xi}^{21}\right) < Q\left(x_2, \widehat{\xi}^{22}\right) = 30,$$

$q_{21} = 2/5, q_{22} = 3/5$, and $\mathscr{Q}(x_2, \omega_2) = \frac{2}{5} \times 18 + \frac{3}{5} \times 30 = 25.2$. Thus,

$$\mathscr{Q}(x_2) = \frac{1}{3} \times 30 + \frac{2}{3} \times 25.2 = 26.8,$$

and the corresponding objective value is

$$\mathscr{Q}(x^2) - c_2 = 26.8 - 2 = 24.8.$$

Similarly, in the case of $x = x_3 = (1,0)$, we obtain $\mathscr{Q}(x_3) = 3$, and the objective value is $\mathscr{Q}(x_3) - c_1 = 2$, while for $x = x_4 = (0,0)$, we obtain $\mathscr{Q}(x_4) = 0$, and the objective value also is 0. Comparing all the objective values, we can find the optimal solution is $x^* = (0,1)$ with 24.4 as the optimal value of the objective function.

5.1.2 *Difficulties*

We see from Example 5.1 that the basic procedure to solve the two-stage FR-FLM-RFC (7.1)–(7.3) comes as follows: firstly, calculate the recourse function $\mathscr{Q}(x)$ to obtain the objective value for each first-stage decision x. Secondly, compare all the objective values to determine the optimal solution x^*. Nevertheless, neither of the above two steps is easy.

On the one hand, since the recourse function $\mathscr{Q}(x)$ is the expected value of $Q(x,\xi)$, from Example 5.1, to obtain $\mathscr{Q}(x)$, we need to compute $Q(x,\xi(\omega,\gamma))$ by solving a second-stage (linear) programming (5.2) for each (ω,γ). In Example 5.1, since the fuzzy random demand-cost vector ξ comes with finite realizations, $Q(x,\xi)$ also takes on finite values. So we can obtain the recourse function $\mathscr{Q}(x)$ through finite calculations. However, generally speaking, for each ω, $\xi(\omega)$ could be a continuous fuzzy vector which has infinite numbers of realizations. In such a case, calculating $\mathscr{Q}(x)$ becomes an infinite-dimensional optimization problem which cannot be solved analytically, since it requires solving infinite linear programming problems. Furthermore, the complexity of the problem rapidly increases if the random parameter involved in ξ is also a continuous random vector, which means we have an infinite number of $\omega's$ to deal with in addition to the process to an infinite number of $\gamma's$.

On the other hand, although we can compare the objective values for all the decisions in Example 5.1, such an exhaustive comparison is not suitable when the number of decision variables becomes moderately large. Moreover, from the above discussion, we see that in general the recourse function $\mathcal{Q}(x)$ cannot be expressed analytically; therefore, the two-stage FR-FLM-RFC (5.1)–(5.2) cannot be solved by the classical mathematical programming methods.

The above two difficulties will be dealt with in Sects. 5.3 and 5.4.

5.2 Model Analysis

This section is intended to present some theoretical properties for the two-stage FR-FLM-RFC. To clarify the discussion a little, we write the following original problem:

$$\left.\begin{array}{ll}\max\limits_{x} & z(x,\xi) = \max\limits_{y} \sum\limits_{i=1}^{n}\sum\limits_{j=1}^{m}(r_j - V_i - t_{ij})y_{ij} - \sum\limits_{i=1}^{n} c_i x_i \\ \text{subject to} & \\ & x_i \in \{0,1\}, i = 1,2,\cdots,n, \\ & \sum\limits_{i=1}^{n} y_{ij} \leq D_j, j = 1,2,\cdots,m, \\ & \sum\limits_{j=1}^{m} y_{ij} \leq s_i x_i, i = 1,2,\cdots,n, \\ & y_{ij} \geq 0, i = 1,2,\cdots,n, j = 1,2,\cdots,m,\end{array}\right\} \tag{5.9}$$

where $\xi = (D_1,\cdots,D_m,V_1,\cdots,V_n)$ is the fuzzy random vector of parameters in (5.3).

5.2.1 *Value of Fuzzy Information*

Recall that in our two-stage FR-FLM-RFC, the location decision x is made before the realization of the ξ, and the optimal solution is determined by considering all the fuzzy random uncertainties; this *fuzzy random solution*(FRS), based on (5.9), can be rewritten as

$$FRS = \max_{x} E\left[z\left(x,\xi\right)\right]. \tag{5.10}$$

Now, let us consider a simplified approach to the two-stage FR-FLM-RFC. First of all, suppose the decision maker does not take fuzzy or imprecise information into account when dealing with the location problems in a hybrid uncertain environment. He/she replaces the fuzzy value $\xi(\omega)$ with some crisp one, without losing any generality, denoted $\bar{\xi}(\omega)$ for any $\omega \in \Omega$. Hence, the original problem is converted into a stochastic programming problem, and an optimal solution $x^*(\bar{\xi})$ can be obtained in such simplified situation:

$$x^*(\bar{\xi}) = \arg\max_{x} E_{\omega}\left[z\left(x, \bar{\xi}(\omega)\right)\right], \tag{5.11}$$

where $E_{\omega}[\cdot]$ represents the expected value of a random variable. Furthermore, the solution $x^*(\bar{\xi})$ will be taken to deal with the different scenarios in real situations with fuzzy random uncertainty. This solution is called *random solution* (RS) to fuzzy random problems, and it can be expressed as follows:

$$RS = E\left[z\left(x^*(\bar{\xi}), \xi\right)\right], \tag{5.12}$$

where $x^*(\bar{\xi})$ is given by (6.16). Furthermore, the value of fuzzy information (VFI) of the solution in the two-stage FR-FLM-RFC is defined as

$$VFI = FRS - RS.$$

Proposition 5.1. *Assume that FRS and RS are the fuzzy random solution and random solution defined in (6.15) and (6.17), respectively, then we have*

$$VFI \geq 0.$$

Proof. Recalling that

$$RS = E\left[z\left(x^*(\bar{\xi}), \xi\right)\right],$$

we note that $x^*(\bar{\xi}) \in \{0,1\}^n$ is only one solution to the problem

$$\max_{x} E\left[z\left(x, \xi\right)\right].$$

Hence, we have

$$RS \leq FRS,$$

which is

$$VFI \geq 0.$$

The proof is completed. □

The above result, though obvious, is of much significance. It indicates a fact that when dealing with the location problems under hybrid uncertainty containing randomness and fuzziness simultaneously, neglecting the fuzziness may lead to risky results of value loss. This can be concretely shown by the following example.

Example 5.2. We assume that a firm plans to open new plants in two potential sites ($n = 2$) where nearby there is only one client ($m = 1$); in order to determine the optimal locations (which site or both) for the new plants to open, it needs to acquire the information for the parameters involved in the location problem. After the investigation, the fixed parameter values have been easily obtained: $n = 2, m = 1$; the unit price charged to client from plants 1 and 2: $(b_1, b_2) = (15, 14)$; the capacity of plants 1 and 2: $(s_1, s_2) = (10, 8)$; the unit transportation cost from plants to the client: $(t_1, t_2) = (3, 6)$; and the fixed cost for opening and operating the plants: $(c_1, c_2) = (2, 4.5)$. As for the variable parameters, we have got the exact information for the demands of the client ($D(\omega) \equiv 3$), and the variable cost for operating the plant 2 ($V_2(\omega) \equiv 4$). Unfortunately, since the historical data are inadequate for the variable cost (V_2) of operating the plant 2, the firm can only obtain the following imprecise distribution provided by a group of experts:

$$V_1 = \begin{cases} \widetilde{V_1(\omega_1)}, \ \omega = \omega_1 \text{ with probability } 0.8; \\ \widetilde{V_1(\omega_2)}, \ \omega = \omega_2 \text{ with probability } 0.2, \end{cases} \tag{5.13}$$

where

$$\widetilde{V_1(\omega_1)} = \Big\{ \langle 1, 0.6 \rangle; \langle 3, 1 \rangle; \langle 5, 0.8 \rangle \Big\},$$

$$\widetilde{V_1(\omega_2)} = \Big\{ \langle 8, 0.2 \rangle; \langle 10, 0.6 \rangle; \langle 12, 1 \rangle \Big\},$$

which means that the V_1 takes on discrete imprecise or fuzzy values $\widetilde{V_1(\omega_1)}$ and $\widetilde{V_1(\omega_2)}$ with probability 0.8 and 0.2, respectively, and fuzzy variable $\widetilde{V_1(\omega_1)}$ ($\widetilde{V_1(\omega_2)}$ (can be similarly illustrated) assumes values $V_1(\omega_1, \gamma_1) = 1$, $V_1(\omega_1, \gamma_2) = 3$, and $V_1(\omega_1, \gamma_3) = 5$ with membership degrees 0.6, 1, and 0.8, respectively; here the $\omega_i, i = 1, 2$, and $\gamma_j, j = 1, 2, 3$ represent the random events and fuzzy events, respectively.

Solution (FRS) with Fuzzy Information. Based on the above circumstances, we first determine the solution to this problem without losing any uncertainty information. This problem can be formulated as a location model as follows:

$$\begin{aligned} \max \quad & E\left[Q(x, V_1)\right] - c_1 x_1 - c_2 x_2 \\ \text{subject to } & x_i \in \{0, 1\}, i = 1, 2, \end{aligned} \tag{5.14}$$

where $E[\cdot]$ is denoted as the expected value operator, and

$$\begin{aligned} Q\Big(x, V_1(\omega, \gamma)\Big) = \max \sum_{i=1}^{2} \Big(b_i - t_i - V_i(\omega, \gamma)\Big) y_i \\ \text{subject to } y_1 + y_2 \le D, \\ 0 \le y_1 \le s_1 x_1, \\ 0 \le y_2 \le s_2 x_2. \end{aligned} \tag{5.15}$$

For location decision $x^I = (x_1^I, x_2^I) = (1,1)$. In the scenario $\omega = \omega_1$ with probability 0.8, from the distribution (5.13) of variable cost V_1, for each realization $(\omega_1, \gamma_j), j = 1,2,3$, we can calculate the second-stage optimal value

$$Q\left(x^I, V_1(\omega_1, \gamma_j)\right), j = 1,2,3$$

respectively as follows. For $V_1(\omega_1, \gamma_1) = \langle 1, 0.6 \rangle$, we have

$$\begin{aligned} Q\left(x^I, V_1(\omega_1, \gamma_1)\right) &= \max 11y_1 + 4y_2 \\ &\text{subject to } y_1 + y_2 \leq 3, \\ &\qquad 0 \leq y_1 \leq 2, \\ &\qquad 0 \leq y_2 \leq 2, \end{aligned}$$

with possibility 0.6; hence, we obtain $Q\left(x^I, V_1(\omega_1, \gamma_1)\right) = 26$ with possibility 0.6, or equivalently,

$$Q\left(x^I, V_1(\omega_1, \gamma_1)\right) = \langle 26, 0.6 \rangle.$$

Similarly, we can calculate that

$$Q\left(x^I, V_1(\omega_1, \gamma_2)\right) = \langle 22, 1 \rangle,$$

and

$$Q\left(x^I, V_1(\omega_1, \gamma_3)\right) = \langle 18, 0.8 \rangle.$$

Therefore, using the expected value of a discrete fuzzy variable, we have

$$E\left[Q\left(x^I, \widetilde{V_1(\omega_1)}\right)\right] = 29.4. \tag{5.16}$$

As for the scenario $\omega = \omega_2$ with probability 0.2, we can similarly work out that

$$E\left[Q\left(x^I, \widetilde{V_1(\omega_2)}\right)\right] = 8.8. \tag{5.17}$$

In sum of the above calculations, we can obtain overall expected value at solution x^I is

$$\begin{aligned} E\left[Q\left(x^I, V_1\right)\right] &= \int_{\Omega} E\left[Q\left(x^I, \widetilde{V_1(\omega)}\right)\right] \Pr(\mathrm{d}\omega) \\ &= 0.8 \times 29.4 + 0.2 \times 8.8 = 25.28, \end{aligned} \tag{5.18}$$

and the objective value $z_{FR}\left(x^I\right) = 25.28 - 2 - 4.5 = 18.78$.

As to other cases for the location decision $x^{II} = (1,0)$, $x^{III} = (0,1)$, and $x^{IV} = (0,0)$, we can obtain the solutions by the similar calculations. They are $z_{FR}\left(x^{II}\right) = 15.28, z_{FR}\left(x^{III}\right) = 3.5$, and $z_{FR}\left(x^{IV}\right) = 0$.

With the comparison of all the solutions, we get that the optimal solution with fuzzy random information that is $x^* = x^I = (1,1)$ with objective value that is $FRS = 18.78$ (fuzzy random solution).

Solution (RS) Without Considering Fuzzy Information. Next, we show that what solution will be obtained if the decision maker only takes account of the random information in this location problem. A natural attempt which is also supposed to be reasonable is to replace the imprecise realizations

$$\widetilde{V_1(\omega_1)} = \left\{\langle 1,0.6\rangle;\langle 3,1\rangle;\langle 5,0.8\rangle\right\},$$

and

$$\widetilde{V_1(\omega_2)} = \left\{\langle 8,0.2\rangle;\langle 10,0.6\rangle;\langle 12,1\rangle\right\}$$

with their expected values

$$E\left[\widetilde{V_1(\omega_1)}\right] = 3.2,$$

and

$$E\left[\widetilde{V_1(\omega_2)}\right] = 11.2,$$

respectively. Hence, the fuzzy random cost V_1 degenerates to a random variable whose distribution becomes accordingly as follows:

$$V_1 = \begin{cases} V_1(\omega_1) = 3.2, & \omega = \omega_1 \text{ with probability } 0.8; \\ V_1(\omega_2) = 11.2, & \omega = \omega_2 \text{ with probability } 0.2. \end{cases}$$

As a consequence, the location problem becomes

$$\begin{array}{ll} \max & E\left[Q(x,V_1)\right] - 2x_1 - 4.5x_2 \\ \text{subject to } x_i \in \{0,1\}, i = 1,2, & \end{array} \tag{5.19}$$

where

$$\begin{aligned} Q\Big(x,V_1(\omega)\Big) = \max\ & 12y_1 + 4y_2 - V_1(\omega)y_1 \\ \text{subject to } & y_1 + y_2 \le 3, \\ & 0 \le y_1 \le 2x_1, \\ & 0 \le y_2 \le 2x_2. \end{aligned} \tag{5.20}$$

Now we determine the best solution x^* with the reduction of the uncertainty.

As to location decision $x^I = (x_1^I, x_2^I) = (1,1)$. For the scenario $\omega = \omega_1$ with probability 0.8, we can get $Q(x^I, V_1(\omega_1)) = 21.6$. Similarly, as for the scenario $\omega = \omega_2$ with probability 0.2, we can calculate that $Q(x^I, V_1(\omega_2)) = 8.8$.

As a consequence, we can obtain

$$E\left[Q(x^I, V_1)\right] = 0.8 \times 21.6 + 0.2 \times 8.8 = 19.4, \tag{5.21}$$

and the objective value is $19.4 - 6.5 = 12.54$. As for the other location decisions $x^{II} = (1,0), x^{III} = (0,1)$, and $x^{IV} = (0,0)$, we can work out that the corresponding objective values are 15.28, 3.5, and 0, respectively. In sum, the optimal solution without considering the imprecise or fuzzy information is $x^* = x^{II} = (1,0)$. At last, we utilize this solution $x^* = (1,0)$ to deal with the real situations (with completely hybrid uncertain information) and get the objective value that is $RS = 15.28$ (random solution).

Solution Comparison. Comparing the two different solutions obtained by considering and not considering the imprecise information arise in this problem, clearly,

$$z{*}_{FR} = 18.78 > RS = 15.28,$$

which implies the firm makes the decision without considering the fuzziness in the problem with hybrid uncertainty, it loses a mount of value of $18.78 - 15.28 = 3.50$.

5.2.2 Solution Bounds

Proposition 5.2 ([154]). *Consider the two-stage FR-FLM-RFC (7.1)–(7.3), suppose the fuzzy random cost-vector $V = (V_1, \cdots, V_n)$ is bounded by two fixed vectors $V_{\min}$ and $V_{\max}$, i.e., $V_{\min} \leq V \leq V_{\max}$, and the demand-vector $D = (D_1, \cdots, D_m)$ is bounded by $D_{\min} \leq D \leq D_{\max}$. If we denote $x_{\min}$ and $x_{\max}$ the optimal solutions to the problems $z(x, D_{\min}, V_{\max})$ and $z(x, D_{\max}, V_{\min})$ in (5.9), respectively, then we have*

$$z\left(x_{\min}, D_{\min}, V_{\max}\right) \leq FRS \leq z\left(x_{\max}, D_{\max}, V_{\min}\right).$$

Proof. We first prove the inequality shown on the left hand side. Let us write the feasible sets as

$$\mathscr{H}\left(x, D\right) = \left\{ y \geq 0 \mid \begin{array}{l} \sum\limits_{i=1}^{n} y_{ij} \leq D_j, j = 1, 2, \cdots, m, \\ \sum\limits_{j=1}^{m} y_{ij} \leq s_i x_i, i = 1, 2, \cdots, n \end{array} \right\}$$

for any location $x \in \{0,1\}^n$ and demand $D(\omega,\gamma),(\omega,\gamma) \in \Omega \times \Gamma$. Hence, we have

$$\mathscr{H}\left(x, D_{\min}\right) \subset \mathscr{H}\left(x, D(\omega,\gamma)\right)$$

for any location decision x and $(\omega,\gamma) \in \Omega \times \Gamma$. Therefore,

$$z\left(x, D_{\min}, V(\omega,\gamma)\right) \le z\left(x, D(\omega,\gamma), V(\omega,\gamma)\right)$$

for any location decision x and $(\omega,\gamma) \in \Omega \times \Gamma$. Moreover, observing the objective function of (5.9), the inequality $V_{\max} \ge V$ implies

$$z\left(x, D_{\min}, V_{\max}\right) \le z\left(x, D(\omega,\gamma), V(\omega,\gamma)\right),$$

which follows that

$$z\left(x, D_{\min}, V_{\max}\right) \le \mathscr{E}(x,\omega) \tag{5.22}$$

for any location decision x and $\omega \in \Omega$, where

$$\begin{aligned}\mathscr{E}(x,\omega) = & \int_0^\infty \mathrm{Cr}\left\{z\left(x, D(\omega), V(\omega)\right) \ge r\right\} \mathrm{d}r \\ & - \int_{-\infty}^0 \mathrm{Cr}\left\{z\left(x, D(\omega), V(\omega)\right) \le r\right\} \mathrm{d}r.\end{aligned}$$

Integrating (5.22) with respect to ω, we obtain

$$z\left(x, D_{\min}, V_{\max}\right) \le E\left[z\left(x, D.V\right)\right].$$

Furthermore,

$$z\left(x, D_{\min}, V_{\max}\right) \le \max_x E\left[z\left(x, D, V\right)\right]$$

for any decision x. Thus,

$$z\left(x_{\min}, D_{\min}, V_{\max}\right) \le \max_x E\left[z\left(x, D, V\right)\right].$$

In a similar approach, we can obtain

$$\max_x E\left[z\left(x, D, V\right)\right] \le z\left(x_{\max}, D_{\max}, V_{\min}\right).$$

The proof of the theorem is completed. □

5.3 Computing Recourse Function

This section focuses on the computation of the recourse function $\mathscr{Q}(x)$. The discussion breaks into two cases of discrete and continuous fuzzy random parameters.

5.3.1 Discrete Case

For discrete case, the fuzzy random demand-cost vector ξ involved in the problem (7.1)–(7.3) is a discrete one such that ω is a discrete random vector taking on a finite number of values ω_i with probability p_i, $i = 1, 2, \cdots, N$, respectively, and for each i, $\xi(\omega_i)$ is a discrete fuzzy vector which takes on the following values:

$$
\begin{array}{ll}
\widehat{\xi}^{i1} = \left(\widehat{D}_1^{i1}, \cdots, \widehat{D}_m^{i1}, \widehat{V}_1^{i1}, \cdots, \widehat{V}_n^{i1}\right) & \text{with membership degree } \mu_{i1} > 0 \\
\widehat{\xi}^{i2} = \left(\widehat{D}_1^{i2}, \cdots, \widehat{D}_m^{i2}, \widehat{V}_1^{i2}, \cdots, \widehat{V}_n^{i2}\right) & \text{with membership degree } \mu_{i2} > 0 \\
\cdots & \cdots \\
\widehat{\xi}^{iN_i} = \left(\widehat{D}_1^{iN_i}, \cdots, \widehat{D}_m^{iN_i}, \widehat{V}_1^{iN_i} \cdots, \widehat{V}_n^{iN_i}\right) & \text{with membership degree } \mu_{iN_i} > 0,
\end{array}
$$

where for each i, $\max_{j=1}^{N_i} \mu_{ij} = 1$. Without any loss of generality, we assume that for each i and fixed x, the second-stage value function satisfies the condition

$$
Q\left(x, \widehat{\xi}^{i1}\right) \le Q\left(x, \widehat{\xi}^{i2}\right) \le \cdots \le Q\left(x, \widehat{\xi}^{iN_i}\right);
$$

then as illustrated in Example 5.1, the value of the recourse function $\mathscr{Q}(x)$ at x is computed as follows:

$$
\mathscr{Q}(x) = \sum_{i=1}^{N} p_i \mathscr{Q}(x, \omega_i), \tag{5.23}
$$

where

$$
\mathscr{Q}(x, \omega_i) = \sum_{j=1}^{N_i} q_{ij} Q\left(x, \widehat{\xi}^{ij}\right), \tag{5.24}
$$

for each pair (i, j). Here, the second-stage value $Q(x, \widehat{\xi}^{ij})$ is obtained by solving the second-stage programming (5.2), and the corresponding weights q_{ij}'s are given in the form

$$
q_{ij} = \frac{1}{2}\left(\max_{k=1}^{j} \mu_{ik} - \max_{k=0}^{j-1} \mu_{ik}\right) + \frac{1}{2}\left(\max_{k=j}^{N_i} \mu_{ik} - \max_{k=j+1}^{N_i+1} \mu_{ik}\right) \tag{5.25}
$$

$(\mu_{i0} = 0, \mu_{i,N_i+1} = 0)$ for $i = 1,2,\cdots,N; j = 1,2,\cdots,N_i$, and satisfy the following constraints

$$q_{ij} \geq 0, \text{ and } \sum_{j=1}^{N_i} q_{ij} = \max_{j=1}^{N_i} \mu_{ij} = 1, i = 1,2,\cdots,N.$$

The computation procedure for the recourse function $\mathscr{Q}(x)$ with discrete fuzzy random demand-cost vector can be summarized in the following manner.

Algorithm 5.1.

Step 1. *Set i from* 1 *to N, and repeat the following Steps* 2–5.
Step 2. *Compute the* $Q(x,\widehat{\xi}^{ij})$ *by solving the second-stage programming (5.2) for* $j = 1,2,\cdots,N_i$.
Step 3. *Rearrange the subscript j of* $Q(x,\widehat{\xi}^{ij})$ *such that*

$$Q\left(x,\widehat{\xi}^{i1}\right) \leq Q\left(x,\widehat{\xi}^{i2}\right) \leq \cdots \leq Q\left(x,\widehat{\xi}^{iN_i}\right).$$

Step 4. *Determine the weight* q_{ij} *of* $Q\left(x,\widehat{\xi}^{ij}\right)$ *by (5.25) for* $j = 1,\cdots,N_i$.
Step 5. *Calculate* $\mathscr{Q}(x,\omega_i)$ *via (5.24).*
Step 6. *Return the value of* $\mathscr{Q}(x)$ *through (5.23).*

5.3.2 Continuous Case

When demand-cost vector ξ is a continuous fuzzy random vector, i.e., the randomness of ξ is determined by a continuous random vector, and for any ω, fuzzy vector $\xi(\omega)$ is also a continuous one which takes on values in the following compact support

$$\Xi = \prod_{j=1}^{m+n} [a_j, b_j]. \tag{5.26}$$

To handle the recourse function $\mathscr{Q}(x)$ with the continuous fuzzy random demand-cost vector ξ, the following two steps (fuzzy random simulation) are realized:

(a) For any random realization ω, the continuous fuzzy vector $\xi(\omega)$ is discretized by the discretization method as we have introduced in Sect. 4.3.3. With the discretization method, we construct a discrete fuzzy vector for approximation:

$$\zeta_l(\omega) = \left(\mathscr{D}_{l,1}(\omega),\ldots,\mathscr{D}_{l,m}(\omega),\mathscr{V}_{l,1}(\omega),\ldots,\mathscr{V}_{l,n}(\omega)\right) \to \xi(\omega), \tag{5.27}$$

where l is some large positive integer.

(b) For the continuous random parameter ω, we first randomly generate $\widehat{\omega}_i$ for $1 \le i \le M$ from the distribution of the ω. In virtue of the discretization in (a), we replace $\xi(\widehat{\omega}_i)$ with the discrete fuzzy vector $\zeta_l(\widehat{\omega}_i)$ for a sufficiently large integer l, and compute the recourse function with ζ_l by using random simulation. That is, for each $\widehat{\omega}_i$, we calculate the

$$Q\left(x, \zeta_l(\widehat{\omega}_i)^j\right), q_{\widehat{\omega}_i, j}, 1 \le i \le M, 1 \le j \le N_{\widehat{\omega}_i},$$

and then determine $\mathscr{Q}(x, \widehat{\omega}_i), 1 \le i \le M$ by the method as discussed in Sect. 5.3.1. Making use of the random simulation technique, the recourse function $\mathscr{Q}(x)$ can be approximated as

$$\frac{\mathscr{Q}(x, \widehat{\omega}_1) + \mathscr{Q}(x, \widehat{\omega}_2) + \cdots + \mathscr{Q}(x, \widehat{\omega}_M)}{M} \to \mathscr{Q}(x). \tag{5.28}$$

The above random simulation (5.28) is characterized by convergence with probability 1 as $M \to \infty$. Summarizing the overall discussion, the computing procedure of $\mathscr{Q}(x)$ is given as below, where we use $\widehat{\zeta}_l^j$ to represent $\zeta_l(\widehat{\omega})^j$.

Algorithm 5.2.

Step 1. *Set $Q = 0$ and $l = L$, where L is a sufficiently large integer.*

Step 2. *Randomly generate a simple point $\widehat{\omega}$ from the distribution of the continuous random vector.*

Step 3. *Generate sample points*

$$\widehat{\zeta}_l^j = \left(\widehat{\mathscr{D}}_{l,1}^j, \cdots, \widehat{\mathscr{D}}_{l,m}^j, \cdots, \widehat{\mathscr{V}}_{l,1}^j, \widehat{\mathscr{D}}_{l,n}^j\right)$$

uniformly via discretization method (4.18) from the support Ξ of $\xi(\widehat{\omega})$ for $j = 1, 2, \cdots, N_{\widehat{\omega}}$.

Step 4. *Compute the $Q\left(x, \widehat{\zeta}_l^j\right)$ through solving the second-stage programming (5.2for $j = 1, 2, \cdots, N_{\widehat{\omega}}$.*

Step 5. *Rearrange the subscript j of $Q\left(x, \widehat{\zeta}_l^j\right)$ such that*

$$Q\left(x, \widehat{\zeta}_l^1\right) \le Q\left(x, \widehat{\zeta}_l^2\right) \le \cdots \le Q\left(x, \widehat{\zeta}_l^{N_{\widehat{\omega}}}\right).$$

Step 6. *Determine the weight q_j of $Q\left(x, \widehat{\zeta}_l^j\right)$ by (5.25) for $j = 1, \cdots, N_{\widehat{\omega}}$.*

Step 7. *Calculate $\mathscr{Q}(x, \widehat{\omega})$ by (5.24).*

Step 8. *$Q \leftarrow Q + Q(x, \widehat{\omega})$.*

Step 9. *Repeat the Steps 2–8 M times.*

Step10. *Return the value of $\mathscr{Q}(x) = Q/M$.*

5.3.3 Convergence

For the discrete case, the convergence of Algorithm 5.1 is ensured automatically since the computation formulas (5.23)–(5.25) provides the exact value of the recourse function $\mathcal{Q}(x)$. As to the continuous cases, the convergence of Algorithm 5.2 can be proved.

Theorem 5.1 ([154]). *Consider the two-stage fuzzy random facility location problem (5.1)–(5.2). For the continuous case, suppose the fuzzy random demand-cost vector ξ has a compact support (5.26), and $\mathscr{D}_{l_1}$ and $\mathscr{V}_{l_2}$ are the discretizations of $D=(D_1,\cdots,D_m)$ and $V=(V_1,\cdots,V_n)$, respectively. If we denote $\mathcal{Q}(x,\mathscr{D}_{l_1},\mathscr{V}_{l_2})$ the recourse function with discretizations $\mathscr{D}_{l_1}$ and $\mathscr{V}_{l_2}$, and $\mathcal{Q}(x)$ the recourse function for original $\xi=(D,V)$, then for any given feasible decision x, we have*

$$\lim_{l_1\to\infty}\lim_{l_2\to\infty}\mathcal{Q}(x,\mathscr{D}_{l_1},\mathscr{V}_{l_2})=\mathcal{Q}(x).$$

Proof. We note that the second-stage programming (7.3) of problem (7.1)–(7.3) can be expressed as

$$\left.\begin{array}{r}\max U(\omega,\gamma)^T y\\ \text{subject to } Wy=H(\omega,\gamma)+Bx\\ y\geq 0,\end{array}\right\}\tag{5.29}$$

where $U(\omega,\gamma)$ is a vector which contains r_j,t_{ij} and fuzzy random costs $V_i(\omega,\gamma)$, B is a fixed vector of s_i, $H(\omega,\gamma)$ is a vector which is made up of fuzzy random demands $D_j(\omega,\gamma)$ and 0, and a fixed matrix W.

Since the recourse matrix M is fixed, taking advantage of the knowledge of parametric programming (see [13]), we can obtain that the second-stage value function $Q(x,D,V)$ is concave with respect to(w.r.t.) D and convex w.r.t. V. Hence, $Q(x,D,V)$ is continuous w.r.t. D and V respectively. Furthermore, we note that the supports Ξ_D of D and Ξ_V of V are compact; therefore, $Q(x,D,V)$ is uniformly continuous w.r.t. D on Ξ_D and V on Ξ_V, respectively.

Recall that the discrete fuzzy random vector $\mathscr{V}_{l_2}$ converges uniformly to V, combining the uniform continuity of $Q(x,D,V)$ w.r.t. V on Ξ_V, we have that for fixed $\mathscr{D}_{l_1}$,

$$\lim_{l_2\to\infty}Q\Big(x,\mathscr{D}_{l_1},\mathscr{V}_{l_2}(\omega,\gamma)\Big)=Q\Big(x,\mathscr{D}_{l_1},V(\omega,\gamma)\Big)$$

for all $(\omega,\gamma)\in\Omega\times\Gamma$. That is, $Q(x,\mathscr{D}_{l_1},\mathscr{V}_{l_2})$ converges uniformly to $Q(x,\mathscr{D}_{l_1},V)$ for fixed $\mathscr{D}_{l_1}$. Furthermore, noting that the support Ξ_V is bounded, and making use of the convergence properties of fuzzy random variables (see [98]), we have

$$\lim_{l_2\to\infty}E\Big[Q(x,\mathscr{D}_{l_1},\mathscr{V}_{l_2})\Big]=E\Big[Q(x,\mathscr{D}_{l_1},V)\Big],\tag{5.30}$$

or, equivalently,

$$\lim_{l_2\to\infty} \mathscr{Q}(x,\mathscr{D}_{l_1},\mathscr{V}_{l_2}) = \mathscr{Q}(x,\mathscr{D}_{l_1}), \tag{5.31}$$

for fixed $\mathscr{D}_{l_1}$, where $\mathscr{Q}(x,\mathscr{D}_{l_1})$ is the recourse function with discretization $\mathscr{D}_{l_1}$ and original cost vector V.

Next, similarly as in the above case, since the discrete fuzzy random vector $\mathscr{D}_{l_1}$ converges uniformly to D, the uniform continuity of $Q(x,D,V)$ w.r.t. D on Ξ_D implies

$$\lim_{l_1\to\infty} Q\Big(x,\mathscr{D}_{l_1}(\omega,\gamma),V\Big) = Q\Big(x,D,V\Big)$$

for all $(\omega,\gamma)\in\Omega\times\Gamma$. Furthermore, we have

$$\lim_{l_1\to\infty} E\Big[Q(x,\mathscr{D}_{l_1},V)\Big] = E\Big[Q(x,D,V)\Big], \tag{5.32}$$

which combining with (5.31) implies

$$\lim_{l_1\to\infty}\lim_{l_2\to\infty} \mathscr{Q}(x,\mathscr{D}_{l_1},\mathscr{V}_{l_2}) = \lim_{l_1\to\infty} \mathscr{Q}(x,\mathscr{D}_{l_1}) = \mathscr{Q}(x).$$

The proof of the theorem has been completed. □

5.4 A Hybrid Binary ACO Algorithm

So far, we can compute the value of recourse function $\mathscr{Q}(x)$ with demand-cost vector ξ for each location decision x through Algorithm 5.1 (discrete case) and Algorithm 5.1 (continuous case). Since the simplex algorithm is one of the most efficient algorithms for solving the linear programming problems (see [21]), here we utilize the simplex algorithm to solve each second-stage problem (5.2) in the process of fuzzy random simulation. In order to solve the two-stage FR-FLM-RFC (5.1)–(5.2), in this section, we design a hybrid mutation-based binary ant colony optimization (MBACO) algorithm, which integrates the fuzzy random simulation and the simplex algorithm (Fig. 5.2).

Ant colony optimization (ACO), proposed in the early 1990s (see [27, 28]), is a metaheuristic method for difficult discrete optimization problems. ACO was inspired by the behavior of real ants. The individual ants communicate information with others via a medium, i.e., pheromone. When searching for food, a moving ant deposits a chemical pheromone trail on the ground. The pheromone is reinforced as other ants use the same trail. The quantity of pheromone depending on the quantity and quality of the food will guide other ants to the food source. In general, the ACO includes the following characteristics: 1. The construction of solutions which uses the pheromone trails. 2. A method to reinforce the pheromone. 3. A local (neighborhood) search to improve solutions. For more successful applications of ACO, one may refer to [14, 20, 46, 51, 108, 135].

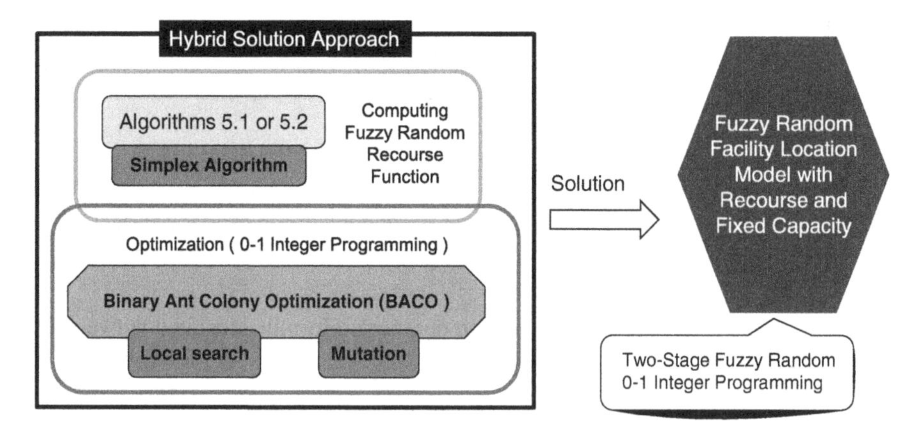

Fig. 5.2 The structure of hybrid algorithm for the two-stage FR-FLM-RFC

To solve the two-stage FR-FLM-RFC which is essentially a two-stage 0-1 fuzzy random programming problem more effectively with the aid of ACO, some modifications are considered:

(a) A 0-1 pheromone updating operator is incorporated to form the binary ACO (BACO).
(b) In the hybrid mechanism, the simplex algorithm is used to solve the second-stage linear programming problems, and Algorithm 5.1 or Algorithm 5.2 (fuzzy random simulation) is embedded into the BACO for the corresponding cases to compute the recourse with fuzzy random parameters.
(c) A mutation operator is employed to extend the search space of BACO so as to decrease the probability of getting trapped in a local optimum.

5.4.1 Solution Encoding and Construction

In the MBACO algorithm, N_a ants $x_1, x_2, \cdots, x_{N_a}$ are used to search for the optimal location. Each ant $x = (x_1, x_2, \cdots, x_n)$ is a binary integer vector which represents a location solution to the two-stage FR-FLM-RFC (7.1)–(7.3), where n is the number of the potential sites of the new facilities.

In the construction phase, all the ants construct solutions through two categories of pheromone trail intensity, called 1-pheromone and 0-pheromone which are denoted by τ_{1i} and τ_{0i}, respectively, where $i = 1, 2, \cdots, n$. A solution is constructed according to the following rule:

$$\texttt{if}(\text{rand}() < \mathbf{P}_i)\ \texttt{then}\ x_i = 1;\ \texttt{otherwise}\ x_i = 0; \tag{5.33}$$

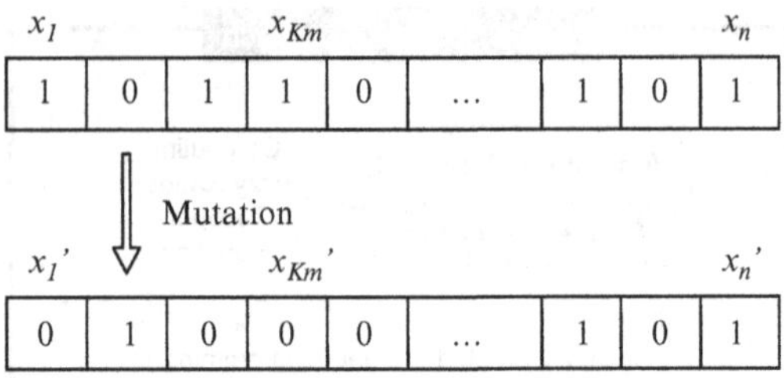

Fig. 5.3 Mutation operation

x_1			x_{Km}					x_n
1	0	1	1	0	...	1	0	1

Mutation

x_1'			x_{Km}'					x_n'
0	1	0	0	0	...	1	0	1

for $i = 1,2,\cdots,n$, where rand() is random number uniformly generated between 0 and 1, and $\mathbf{P}_i$ is the transition probability for each component of the solution, which is defined by

$$\mathbf{P}_i = \frac{\tau_{1i}}{\tau_{1i} + \tau_{0i}}, \quad i = 1,2,\cdots,n. \tag{5.34}$$

5.4.2 Fitness Function

Let the fitness of each ant x be its objective value, i.e.,

$$\mathbf{F}(x) = \mathscr{Q}(x) - c^T x, \tag{5.35}$$

where $\mathbf{F}(x)$ is the fitness function and $c = (c_1, c_2, \cdots, c_n)$. Therefore, the particles of larger objective values come with higher fitness. Here, for each ant x, the fitness value $\mathbf{F}(x)$ is calculated by Algorithm 5.1 or Algorithm 5.2 together with the simplex algorithm.

5.4.3 Mutation

A mutation operation is applied to enlarge the search space of the colony. We predetermine a parameter $p_m \in (0,1)$ as the probability of mutation. For each ant x, if running the mutation, we first randomly choose a mutation position K_m between 1 and n, and then exchange the values of $x_1, x_2, \cdots, x_{K_m}$ between 0 and 1. This is shown in Fig. 5.3.

5.4.4 Local Search for Improving Solutions

After a colony has been generated and mutated (if mutation is selected), each ant uses a local search to improve the solution. We generate a neighborhood $\mathbf{N}(x)$ for each ant x by the following method: for each ant $x = (x_1, x_2, \cdots, x_n)$, we first generate

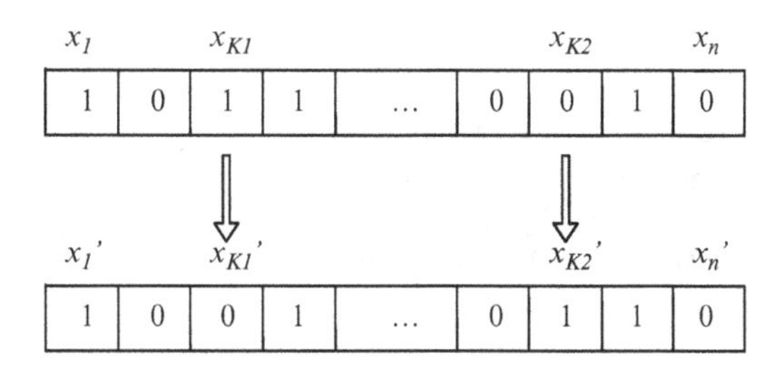

Fig. 5.4 Generating a neighbor for an ant

two positions K_1 and K_2 between 1 and n, and then for each K_1 and K_2, exchange its value between 0 and 1 (see Fig. 5.4). Repeating the above two steps M_N times, we obtain M_N neighbors of ant x. After the neighborhood of each ant is generated, we evaluate each ant and its neighbors via computing their fitness values $\mathbf{F}(x)$ by using the fuzzy random simulation together with the simplex algorithm, and update the position of each ant by

$$x^j = \arg \max_{x \in \mathbf{N}(x^j)} \mathbf{F}(x) \tag{5.36}$$

for $j = 1, 2, \cdots, N_a$. Then, we find the ant x_{best} being the best solution so far.

5.4.5 *Online–Offline Update of Pheromone Intensity*

The pheromone trail intensity update in ACO consists of online (ant-to-ant) update and offline (colony) update. The online update is performed after each solution has been constructed, whose goal is mainly to decay the pheromone intensity of the components of the solution just constructed to encourage exploration the choices of other components, so as to diversify the search performed by the subsequent ants during an iteration. Online updating is expressed in the form

$$\tau_{ki} = \varphi \cdot \tau_{ki} + (1 - \varphi) \cdot \tau^0, \tag{5.37}$$

where $k = 0, 1$, $i = 1, 2, \cdots, n$, $\varphi \in (0, 1)$ is the pheromone decay coefficient, and τ^0 is the initial value of the 1- and 0-pheromone which is set over the component number, i.e., $1/n$.

After all the solutions of the colony have been constructed, mutated, and improved by the local search, the offline pheromone update is applied to each component x_i by the following rule:

$$\tau_{ki} = \begin{cases} \rho \cdot \tau_{ki} + (1 - \rho) \cdot \sum_{j=1}^{N_a} \Delta \tau_{ki}^j, & \text{if } x_i = k \text{ belongs to the best ant } x_{best} \\ \tau_{ki}, & \text{otherwise,} \end{cases} \tag{5.38}$$

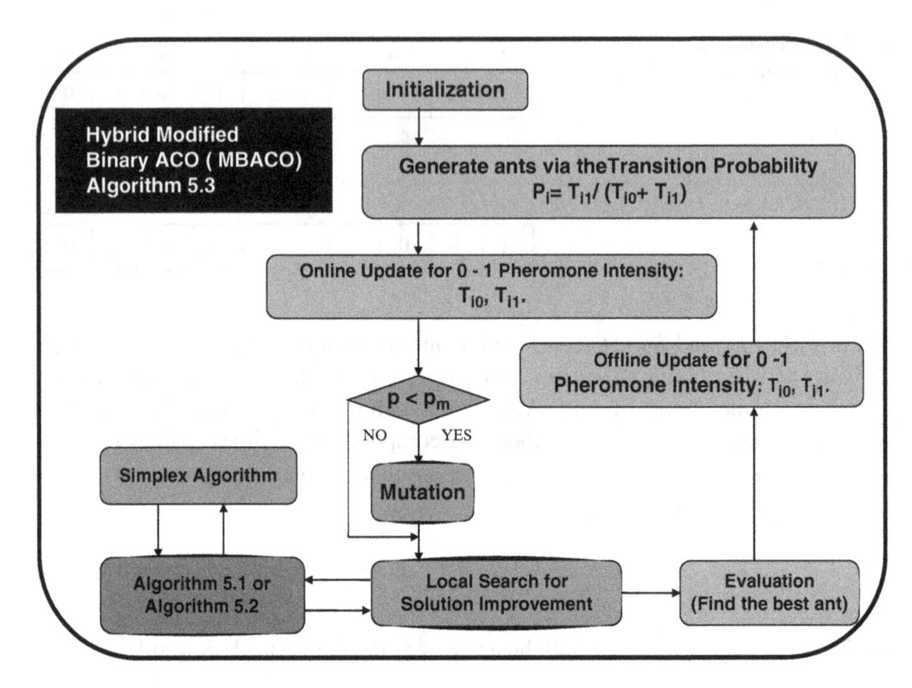

Fig. 5.5 Flowchart of the hybrid MBACO algorithm

where $\rho \in (0,1)$ is the evaporation rate and $\Delta\tau_{ki}^{j}$ is the quantity of the pheromone contributed by the ant x^j to the trial intensity of component i, which is defined as

$$\Delta\tau_{ki}^{j} = \begin{cases} \frac{w \cdot \mathbf{F}(x^j)}{n}, & \text{if } x_i^j = k \\ 0, & \text{otherwise}, \end{cases} \tag{5.39}$$

for $k = 0, 1$ and $i = 1, 2, \cdots, n$; here the weight w is a constant.

5.4.6 *Hybrid Algorithm Procedure*

The hybrid MBACO algorithm for solving the two-stage FR-FLM-RFC is summarized in Algorithm 5.3, and its flowchart is given in Fig. 5.5.

Algorithm 5.3 ((Hybrid MBACO)).

Step 1. Input all the parameters, N_a, M_N, p_m, φ, ρ, and w, and initialize the pheromone trail intensity $\tau_{0i} = \tau_{1i} = \tau^0, i = 1, 2, \cdots, n$.

Step 2. Construct a colony of ants $x_1, \cdots, x_{N_a}$ via transition probability (5.33)–(5.34).

Step 3. *Run the online update (5.37) to each 0- and 1-pheromone intensity (τ_{0i} and τ_{1i}).*

Step 4. *Run mutation operation to each ant with the probability p_m.*

Step 5. *Apply the local search to improve each ant via (5.36) in the colony by calculating the fitness $\boldsymbol{F}(x)$ for each ant and its neighbors through computation schemes (Algorithm 5.1 or 5.2) for recourse together with the simplex algorithm, and then find the best ant x_{best} so far.*

Step 6. *Apply the offline update rule (5.38)–(5.39) to each 0- and 1-pheromone intensity.*

Step 7. *Repeat Step 2 to Step 5 for a given number of generations.*

Step 8. *Return the best ant x_{best} as the optimal solution of the two-stage FR-FLM-RFC, and $\boldsymbol{F}(x_{best})$ as the optimal value.*

5.5 Numerical Experiments

In this section, some numerical examples as well as some comparisons are provided to illustrate the solution performance of the hybrid MBACO algorithm for the proposed two-stage FR-FLM-RFC (7.1)–(7.3).

As the first example, we consider a firm which plans to open new facilities in ten potential sites; the capacities s_i, fixed costs c_i, and fuzzy random operating costs V_i of the sites $i, i = 1, 2, \cdots, 10$ are given in Table 5.1. We suppose that there are five customers whose fuzzy random demands $D_j, j = 1, 2, \cdots, 5$ are given in Table 5.2,

Table 5.1 Capacities, fixed costs, and variable costs of 10 sites

Facility site i	Capacity s_i	Fixed cost c_i	Variable cost V_i	Parameter Y_i
1	140	4	$(4+Y_1, 6+Y_1, 8+Y_1)$	$\mathscr{U}(1,2)$
2	120	12	$(7+Y_2, 9+Y_2, 10+Y_2)$	$\mathscr{U}(2,3)$
3	200	4	$(2+Y_3, 4+Y_3, 5+Y_3)$	$\mathscr{U}(1,2)$
4	190	9	$(4+Y_4, 4+Y_4, 7+Y_4)$	$\mathscr{U}(0,1)$
5	160	6	$(3+Y_5, 5+Y_5, 8+Y_5)$	$\mathscr{U}(1,2)$
6	180	10	$(4+Y_6, 6+Y_6, 7+Y_6)$	$\mathscr{U}(0,2)$
7	190	9	$(4+Y_7, 6+Y_7, 8+Y_7)$	$\mathscr{U}(2,4)$
8	130	10	$(8+Y_8, 10+Y_8, 12+Y_8)$	$\mathscr{U}(2,3)$
9	180	5	$(3+Y_9, 5+Y_9, 6+Y_9)$	$\mathscr{U}(3,4)$
10	170	6	$(4+Y_{10}, 6+Y_{10}, 7+Y_{10})$	$\mathscr{U}(1,2)$

Table 5.2 Fuzzy random demands of five clients

Customer j	Demand D_j	Parameter Z_j
1	$(16+Z_1, 18+Z_1, 20+Z_1)$	$\mathscr{U}(1,2)$
2	$(8+Z_2, 12+Z_2, 14+Z_2)$	$\mathscr{U}(1,3)$
3	$(13+Z_3, 16+Z_3, 18+Z_3)$	$\mathscr{U}(2,4)$
4	$(12+Z_4, 15+Z_4, 17+Z_4)$	$\mathscr{U}(2,3)$
5	$(10+Z_5, 12+Z_5, 15+Z_5)$	$\mathscr{U}(3,4)$

Table 5.3 The values of $r_j - t_{ij}$ in the case of ten sites and five clients

$r_j - t_{ij}$	$i=1$	2	3	4	5	6	7	8	9	10
$j=1$	6	3	5	4	5	6	8	4	6	4
2	5	7	5	7	4	6	5	4	5	3
3	4	5	3	6	5	6	4	6	4	6
4	7	4	8	6	5	7	4	5	4	8
5	6	8	3	6	5	8	6	4	5	6

where $\mathscr{U}(a,b)$ represents a random variable with uniform distribution on $[a,b]$. In addition, the unit price r_j charged to each customer and the unit transportation cost t_{ij} are listed in Table 5.3 in the form of $r_j - t_{ij}$.

From the above settings, we can formulate the location problem by using two-stage FR-FLM-RFC:

$$\left.\begin{array}{ll} \max & \mathscr{Q}(x) - 4x_1 - 12x_2 - 4x_3 - 9x_4 - 6x_5 - 10x_6 - 9x_7 - 10x_8 \\ & -5x_9 - 6x_{10} \\ \text{subject to } x_1, x_2, \ldots, x_{10} \in \{0,1\}, & \end{array}\right\} \quad (5.40)$$

where $\mathscr{Q}(x) = E[Q(x,\xi)]$, and

$$\left.\begin{array}{l} Q\Big(x, \xi(\omega,\gamma)\Big) = \max \sum_{i=1}^{10} \sum_{j=1}^{5} \left(r_j - t_{ij} - V_i(\omega,\gamma)\right) y_{ij}^{(\omega,\gamma)} \\ \qquad \text{subject to} \\ \qquad\qquad \sum_{i=1}^{10} y_{ij}^{(\omega,\gamma)} \le D_j(\omega,\gamma), j = 1,2,\cdots,5, \\ \qquad\qquad \sum_{j=1}^{5} y_{ij}^{(\omega,\gamma)} \le s_i x_i, i = 1,2,\cdots,10, \\ \qquad\qquad y_{ij}^{(\omega,\gamma)} \ge 0, i = 1,2,\cdots,10, j = 1,2,\cdots,5. \end{array}\right\} \quad (5.41)$$

We note that in problem (5.40)–(5.41), the demand-cost vector

$$\xi = (D_1, \cdots, D_5, V_1, V_2, \cdots, V_{10})$$

is a continuous fuzzy random vector. Therefore, fuzzy random simulation (Algorithm 2) should be utilized to compute the recourse function $\mathscr{Q}(x)$.

To solve this two-stage fuzzy random facility location problem, for any feasible solution x, we first generate 3,000 random sample points $\widehat{\omega}_i, i = 1,2,\cdots,3{,}000$, for the random simulation (such sample size is sufficient for the simulation of random expected value). For each ω_i generate 1,000 fuzzy sample points $\widehat{\zeta_{i,l}}, l = 1,2,\cdots,1000$. Based on the generated samples, the second-stage programming (5.41), which is a linear programming problem with $10 \times 5 = 50$ variables, is

Table 5.4 Results of hybrid MBACO algorithm with different parameters

	System Parameters						Results		
No.	N_a	M_N	p_m	φ	ρ	w	Optimal solution	Objective value	Error(%)
1	3	2	0.2	0.6	0.7	0.1	(1,0,1,1,0,1,1,0,0,0)	95.51	3.42
2	3	2	0.3	0.8	0.9	0.05	(0,0,1,1,0,1,1,0,0,0)	98.90	0.00
4	3	2	0.4	0.9	0.9	0.01	(0,0,1,1,0,1,1,0,0,0)	98.90	0.00
3	3	2	0.5	0.7	0.8	0.1	(0,0,1,1,0,1,0,0,0,0)	97.95	0.96
5	3	2	0.2	0.8	0.6	0.05	(0,0,1,1,0,1,0,0,0,0)	97.95	0.96
6	3	2	0.4	0.9	0.8	0.01	(0,0,1,1,0,1,1,0,0,0)	98.90	0.00
7	3	2	0.4	0.5	0.5	0.1	(0,0,1,1,0,1,0,0,0,0)	97.95	0.96
8	3	2	0.5	0.5	0.7	0.01	(0,0,1,1,0,0,0,0,0,0)	98.67	0.23
9	3	2	0.4	0.6	0.6	0.05	(0,0,1,1,0,1,0,0,1,0)	98.10	0.81
10	3	2	0.3	0.7	0.5	0.01	(0,0,1,1,0,1,0,0,0,0)	97.95	0.96

solved by the simplex algorithm for each pair $(\widehat{\omega}_i, \widehat{\zeta}_{i,l})$. Furthermore, we compute the values of $\mathscr{Q}(x)$ by the fuzzy random simulation (Algorithm 5.2). After that, the fuzzy random simulation together with the simplex algorithm is embedded into the MBACO to produce a hybrid algorithm (Algorithm 5.3) to search for the approximately optimal solutions.

Since in binary swarm optimization, the population size is usually set to be around the number of the dimension (see [63]), in order to demonstrate more clearly the performance of the algorithm, the population size here is set to be less than the dimension $n = 10$, that is, $(N_a, M_N) = (3,2)$ which implies that nine ants are used in the search. A run of the hybrid algorithm with $Gen = 100$ generations, we obtain the optimal location solutions with different parameters as listed in Table 5.4, where the *relative error* is given in the last column, which is defined by

$$Error = \frac{optimal\ value - objective\ value}{optimal\ value} \times 100\%. \tag{5.42}$$

It follows from Table 5.4 that the relative error does not exceed 3.42% when different values of parameters have been selected. In addition, the convergence of the objective value in iterations is shown in Fig. 5.6. The performance implies that the hybrid MBACO algorithm is robust to the parameter settings and effective to solve the FR-FLM-RFC.

There exist some recent evolutionary approaches to location problems with uncertainty in the literature, such as genetic algorithm (GA) (see [163]), binary particle swarm optimization (BPSO) (see [152]), and tabu search (TS) (see [168]). To further assess the performance of our proposed hybrid MBACO algorithm for two-stage FR-FLM-RFC, we compare the experimental results of the hybrid MBACO with those produced by other metaheuristic approaches of BPSO, binary GA (BGA), and binary TS (BTS), respectively. For the introduction on BPSO, please refer to Appendix C.

Table 5.5 The comparison results of different approaches in the case of ten sites and five clients

Approach	Optimal solution	Objective value
Hybrid MBACO	(0,0,1,1,0,1,1,0,0,0)	98.90
Hybrid BPSO	(0,0,1,1,0,1,0,0,0,0)	97.95
Hybrid BGA	(0,0,1,1,0,1,0,0,0,1)	95.22
Hybrid BTS	(0,0,1,1,0,1,0,0,0,0)	97.95

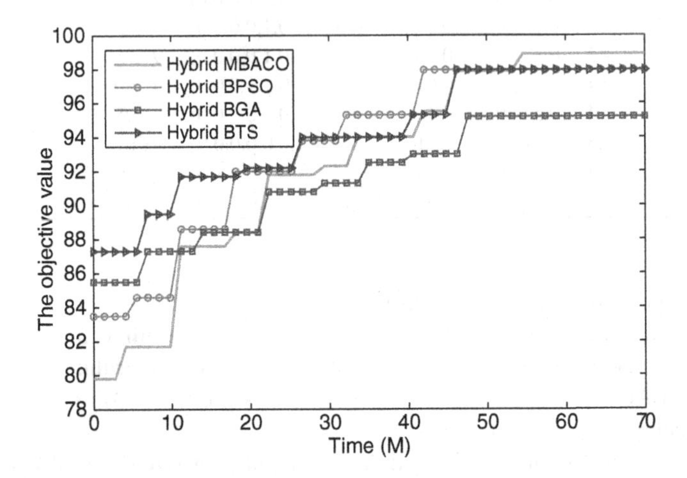

Fig. 5.6 The convergence of different approaches in the case of ten sites and five clients

In order to give a fair comparison, we keep population size in those metaheuristics the same as that of the MBACO, i.e., $P_{size} = 9$ in BPSO, $P_{size} = 9$ in BGA, and $N_{size} = 8$ in BTS, respectively. Furthermore, the best system parameters are selected from the different settings in a number of experiments, which are $c_1 = c_2 = 2$ in BPSO, $p_m = 0.3, p_c = 0.2$ in BGA, and $TT = \text{(int)rand}(8,10)$ in BTS, respectively. In the fuzzy random simulation (Algorithm 5.3), we generate 3,000 random sample points and 1,000 fuzzy sample points which is the same as in hybrid MBACO.

Running the fuzzy random simulation-based hybrid algorithms of BPSO, BGA, and BTS, respectively to the problem (5.40)–(5.41), within the same time duration of 70 minutes (M) (all the algorithms run more than 100 generations and have reached an iteration stability), we obtain the comparison results in Table 5.5 in which the second column lists the optimal solutions and objective values. Moreover, the convergence comparison expressed in the number of iterations is provided in Fig. 5.6.

In order to test the quality of the solutions obtained by the hybrid heuristics, we compute the objective values for all the possible combinations of the binary decision variables by using fuzzy random simulation, and obtain the optimal solution that is $x* = (0,0,1,1,0,1,1,0,0,0)$ with the objective value $z^* = 98.90$. From these comparison results, we see that the four approaches cost almost the same computation time in this example, and the hybrid MBACO algorithm finds the best solution among all the four different hybrid approaches, which equals exactly to the optimal solution obtained by exhaustive search.

Table 5.6 Capacities, fixed costs, and variable costs of 20 sites

Facility site i	Capacity s_i	Fixed cost c_i	Variable cost V_i	Parameter Y_i
1	150	18	$(6+Y_1, 8+Y_1, 10+Y_1)$	$\mathscr{U}(1,2)$
2	120	10	$(6+Y_2, 9+Y_2, 10+Y_2)$	$\mathscr{U}(2,3)$
3	180	14	$(2+Y_3, 4+Y_3, 5+Y_3)$	$\mathscr{U}(1,2)$
4	200	9	$(4+Y_4, 6+Y_4, 7+Y_4)$	$\mathscr{U}(0,1)$
5	240	16	$(3+Y_5, 5+Y_5, 8+Y_5)$	$\mathscr{U}(1,2)$
6	190	10	$(4+Y_6, 6+Y_6, 7+Y_6)$	$\mathscr{U}(0,2)$
7	130	12	$(4+Y_7, 6+Y_7, 8+Y_7)$	$\mathscr{U}(2,4)$
8	180	10	$(8+Y_8, 10+Y_8, 12+Y_8)$	$\mathscr{U}(2,3)$
9	200	8	$(3+Y_9, 5+Y_9, 6+Y_9)$	$\mathscr{U}(3,4)$
10	190	16	$(5+Y_{10}, 6+Y_{10}, 7+Y_{10})$	$\mathscr{U}(1,2)$
11	120	12	$(6+Y_{11}, 8+Y_{11}, 10+Y_{11})$	$\mathscr{U}(0,2)$
12	160	6	$(3+Y_{12}, 5+Y_{12}, 6+Y_{12})$	$\mathscr{U}(2,3)$
13	210	7	$(3+Y_{13}, 6+Y_{13}, 8+Y_{13})$	$\mathscr{U}(2,3)$
14	130	14	$(6+Y_{14}, 9+Y_{14}, 11+Y_{14})$	$\mathscr{U}(1,2)$
15	120	13	$(5+Y_{15}, 6+Y_{15}, 8+Y_{15})$	$\mathscr{U}(2,3)$
16	100	8	$(4+Y_{16}, 6+Y_{16}, 7+Y_{16})$	$\mathscr{U}(0,2)$
17	170	19	$(2+Y_{17}, 3+Y_{17}, 5+Y_{17})$	$\mathscr{U}(2,3)$
18	190	11	$(3+Y_{18}, 4+Y_{18}, 5+Y_{18})$	$\mathscr{U}(1,3)$
19	150	9	$(2+Y_{19}, 3+Y_{19}, 6+Y_{19})$	$\mathscr{U}(1,2)$
20	200	13	$(6+Y_{20}, 8+Y_{20}, 10+Y_{20})$	$\mathscr{U}(2,4)$

Next, by using hybrid MBACO algorithm, we try to solve a more complex fuzzy random location problem with 20 potential sites and eight clients. That means for each decision x and each realization $\xi(\omega,\gamma), (\omega,\gamma) \in \Omega \times \Gamma$, to determine the value of the objective function, we have to solve a large linear programming problems with $20 \times 8 = 160$ variables (such complexity already researches the limitation of the Visual C++ in the PC to handle linear programming problems). What's more, in this example, since the number of all the possibilities of x researches $2^{20} = 1048,576$, and to determine each possible value of objective function at x we have to call a fuzzy random simulation (Algorithm 5.2), the exhaustive search is not feasibly any longer for test in this situation.

The parameters within the location problem are provided in Tables 5.6, 5.7, and 5.8, and the location problem can be modeled as follows:

$$\left.\begin{array}{ll} \max & \mathscr{Q}(x) - 18x_1 - 10x_2 - 14x_3 - 9x_4 - 16x_5 - 10x_6 - 12x_7 \\ & \quad - 10x_8 - 8x_9 - 16x_{10} - 12x_{11} - 6x_{12} - 7x_{13} - 14x_{14} \\ & \quad - 13x_{15} - 8x_{16} - 19x_{17} - 11x_{18} - 9x_{19} - 13x_{20} \\ \text{subject to } & x_1, x_2, x_3, \ldots, x_{20} \in \{0,1\}, \end{array}\right\} \quad (5.43)$$

Table 5.7 Fuzzy random demands of eight clients

Customer j	Demand D_j	Parameter Z_j
1	$(26+Z_1,28+Z_1,30+Z_1)$	$\mathscr{U}(1,2)$
2	$(18+Z_2,22+Z_2,24+Z_2)$	$\mathscr{U}(1,3)$
3	$(23+Z_3,26+Z_3,28+Z_3)$	$\mathscr{U}(2,4)$
4	$(22+Z_4,25+Z_4,27+Z_4)$	$\mathscr{U}(1,2)$
5	$(20+Z_5,22+Z_5,25+Z_5)$	$\mathscr{U}(3,4)$
6	$(28+Z_6,30+Z_6,32+Z_6)$	$\mathscr{U}(1,3)$
7	$(20+Z_5,24+Z_5,25+Z_5)$	$\mathscr{U}(0,1)$
8	$(22+Z_6,26+Z_6,28+Z_6)$	$\mathscr{U}(1,2)$

Table 5.8 The values of $r_j - t_{ij}$ in the case of 20 sites and eight clients

$r_j - t_{ij}$	$j=1$	2	3	4	5	6	7	8
$i=1$	5	5	4	7	6	3	5	5
2	4	8	5	5	5	2	3	4
3	7	6	5	2	5	4	5	5
4	5	6	6	6	7	8	4	2
5	4	4	6	4	5	6	5	4
6	6	5	6	6	5	4	3	6
7	5	6	6	5	4	7	7	2
8	7	5	4	8	5	5	4	7
9	6	3	8	2	4	8	5	5
10	5	5	3	4	7	6	5	2
11	5	4	5	5	4	8	6	6
12	7	8	5	2	4	4	6	4
13	6	6	5	4	6	5	5	2
14	5	4	8	6	8	6	5	5
15	4	7	6	3	2	5	4	8
16	5	2	4	7	5	3	2	2
17	4	8	5	5	3	2	3	4
18	5	6	5	6	5	4	2	5
19	5	5	2	6	5	8	5	5
20	4	4	2	4	4	6	5	4

where $\mathscr{Q}(x) = E[Q(x,\xi)]$, and

$$\left.\begin{array}{l} Q\Big(x,\xi(\omega,\gamma)\Big) = \max \sum\limits_{i=1}^{20}\sum\limits_{j=1}^{8}\Big(r_j - t_{ij} - V_i(\omega,\gamma)\Big)y_{ij}^{(\omega,\gamma)} \\ \qquad\qquad \text{subject to} \\ \qquad\qquad\qquad \sum\limits_{i=1}^{20} y_{ij}^{(\omega,\gamma)} \le D_j(\omega,\gamma), j=1,2,\cdots,8, \\ \qquad\qquad\qquad \sum\limits_{j=1}^{8} y_{ij}^{(\omega,\gamma)} \le s_i x_i, i=1,2,\cdots,20, \\ \qquad\qquad\qquad y_{ij}^{(\omega,\gamma)} \ge 0, i=1,2,\cdots,20, j=1,2,\cdots,8. \end{array}\right\} \quad (5.44)$$

Table 5.9 The comparison results of different approaches in the case of 20 sites and eight clients

Approach	Optimal solution	Objective value
Hybrid MBACO	(0,0,0,1,0,0,0,0,0,1,0,0,1,0,0,0,0,1,0,1)	291.68
Hybrid BPSO	(1,0,1,0,0,0,0,0,1,0,0,1,0,0,0,0,1,0,1,0)	283.05
Hybrid BGA	(0,0,1,0,0,0,0,1,1,0,0,1,0,1,0,0,1,0,1,0)	282.08
Hybrid BTS	(0,0,1,0,0,0,0,0,1,0,0,0,0,0,0,0,1,1,1,0)	285.20

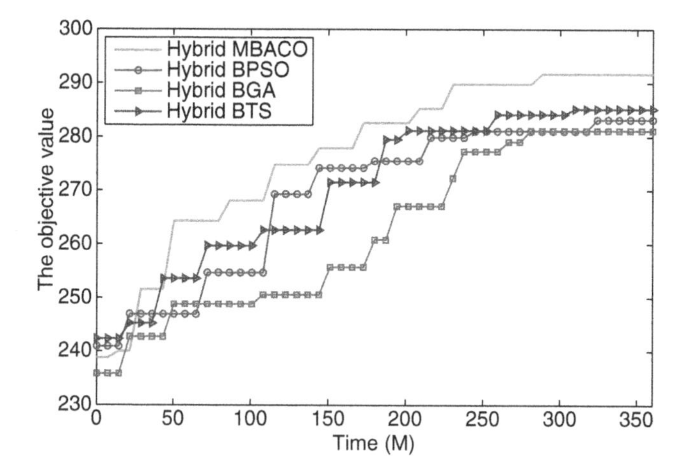

Fig. 5.7 The convergence of different approaches in the case of 20 sites and eight clients

Executing the different hybrid algorithms to problem (5.43)–(5.44), where the population size is set to be 20 (i.e., $(N_a, M_N) = (4,4)$ in MBACO), we obtain the results as listed in Table 5.9, and the convergence comparison in Fig. 5.7 (all the algorithms run more than 100 generations and have reached an iteration stability within the same time duration of 360 Ms). From these results, we can also see that again, hybrid MBACO algorithm finds a better solution than other three hybrid approaches.

Chapter 6
Two-Stage Fuzzy Stochastic Programming with Value-at-Risk: A Generic Model

In preceding Chaps. 4 and 5, we have discussed two concrete engineering optimization models (system reliability optimization and facility location selection) with fuzzy random parameters. It is not difficult to see that the fuzzy randomness exists pervasively in many real-life situations in terms of different manifestations. From a general point of view, this chapter is devoted to a generic risk optimization model in a fuzzy random environment, and an application in facility location selection is shown in the next chapter.

Risk goes hand in hand with uncertainty. Generally speaking, the more complex uncertainty the information exhibits, the more difficult it is to handle the risk involved. As a hybrid uncertainty embracing stochastic variability and fuzziness or imprecision, fuzzy randomness is a higher dimensional mode of uncertainty than the pure randomness, and it is no question more complicated and difficult to cope with the risk in such hybrid uncertainty than in random situations. In this chapter, we extend the notion of Value-at-Risk(VaR), a widely used risk metric in finance and originally defined by probability in randomness, to the scope of fuzzy randomness and, in a general sense, present a VaR-based two-stage risk optimization model with fuzzy random uncertainty.

In Sect. 6.1, we introduce the concepts of probabilistic VaR and credibilistic VaR metrics in the random and fuzzy settings, respectively. Then, we define a fuzzy random VaR (VaR with fuzzy random parameters) using chance measure and discuss its rationality.

In Sect. 6.2, making use of the criterion of fuzzy random VaR, we build a generic two-stage risk optimization model in a fuzzy random environment, named two-stage fuzzy stochastic programming with VaR (FSP-VaR). The two-stage structure in a general sense is explained, and some solution difficulties are indicated.

In Sect. 6.3, we discuss some theoretical properties of the two-stage FSP-VaR that include some sensitivity conditions of the fuzzy random VaR w.r.t. confidence level, value of fuzzy information in the two-stage decision-making process, and lower and upper bounds for the optimal solution.

In Sect. 6.4, we discuss how to compute the value of fuzzy random VaR function $\mathrm{VaR}_{1-\beta}(x)$ at any decision x. The discussion breaks into two cases: VaR function

S. Wang and J. Watada, *Fuzzy Stochastic Optimization: Theory, Models and Applications*,
DOI 10.1007/978-1-4419-9560-5_6,

with discrete fuzzy random parameters and VaR function with continuous fuzzy random parameters (general case). An approximate algorithm which combines fuzzy random simulation and bisection search is designed to estimate the fuzzy random VaR in the continuous case, and the convergence for the algorithm is discussed.

In Sect. 6.5, we discuss the solution approaches for the two-stage FSP-VaR by considering two different cases of integer decisions and continuous decisions. The former case can be tackled by similar approaches presented in Chaps. 4 and 5. As for the latter, the solution approach carries a hybrid mechanism which is based on a mutation-neighborhood-based particle swarm optimization (MN-PSO) with a mutation operator and a "Neighborhood-best-Global-best" update rule. In this hybrid approach, the simplex algorithm is utilized to determine the second-stage optimal return for each pair of decision and fuzzy random realization, the approximate algorithm is used to calculate the values of VaR, and the MPSO serves as an optimizer to cope with the continuous programming.

In Sect. 6.6, we conclude with a numerical example as well as some comparisons.

6.1 Fuzzy Random Value-at-Risk (VaR)

Value-at-Risk (VaR) as a risk criterion is introduced from the financial investment and is defined originally with probability measure on the random events. By fusing both the profit (or return) factor and the confidence level into one metric to characterize the risk value, VaR criterion exhibits its advantages in a wide rage of applications (see [30, 59]) and becomes a widely accepted measure of risk reported across market participants and industry segments. On the other hand, Wang, Watada, and Pedrycz [152] recently transplanted the idea of VaR in a fuzzy environment via credibility measure and discussed its applications. Before we move into the introduction of fuzzy random VaR, let us dabble a little here on probabilistic VaR and credibilistic VaR.

6.1.1 Probabilistic VaR and Credibilistic VaR

The probabilistic Value-at-Risk (VaR) of an investment is the likelihood of the greatest loss at some probability confidence level, which is defined as follows (see [30]):

Definition 6.1. Let $\mathscr{L}$ be the random loss variable in some investment. The probabilistic VaR of $\mathscr{L}$ with a confidence of $1-\beta$ is

$$\mathrm{VaR}_{\mathrm{Pr},1-\beta} = \sup\{\lambda \mid \Pr\{\mathscr{L} \geq \lambda\} \geq \beta\}, \tag{6.1}$$

where $\beta \in (0,1)$.

Here we just give the definition of probabilistic VaR; for more details on random VaR, refer to [30, 59, 65, 128].

Following the above definition for the probabilistic VaR, in the fuzzy environment, a fuzzy VaR of $\mathscr{L}$ with a confidence of $1-\beta$ can be expressed by using credibility measure in the form as below:

$$\mathrm{VaR}_{\mathrm{Cr},1-\beta} = \sup\{\lambda \mid \mathrm{Cr}\{\mathscr{L} \geq \lambda\} \geq \beta\}, \tag{6.2}$$

where $\mathscr{L}$ is a fuzzy loss variable in the investment and $\beta \in (0,1)$. More discussions on the credibilistic VaR can be found in [144, 152].

Since in the theory of fuzzy sets and possibility, Possibility (Pos) and Necessity (denoted by Nec) are a pair of well-known mutually dual measures, while credibility (Cr) is a self-dual measure. In order to further show the rationality of the credibilistic VaR defined in (6.2), the following remarks are necessary.

Remark 6.1. The possibility measure is not suitable to define VaR in fuzzy environment. In fact, on the one hand, if VaR at the confidence of $1-\beta(0<\beta<1)$ is defined by possibility as

$$\mathrm{VaR}_{\mathrm{Pos},1-\beta} = \sup\{\lambda \mid \mathrm{Pos}\{\mathscr{L} \geq \lambda\} \geq \beta\}, \tag{6.3}$$

then for any general continuous fuzzy loss variable $\mathscr{L}$, $\mathrm{Pos}\{\mathscr{L} \geq \lambda\}$ is a continuous function of λ; hence we get

$$\mathrm{Pos}\{\mathscr{L} \geq \mathrm{VaR}_{\mathrm{Pos},1-\beta}\} = \beta < 1.$$

We note from the definition of possibility that

$$1 = \mathrm{Pos}\{\mathscr{L} \in \Re\} = \mathrm{Pos}\{\mathscr{L} \geq \mathrm{VaR}_{\mathrm{Pos},1-\beta}\} \bigvee \mathrm{Pos}\{\mathscr{L} < \mathrm{VaR}_{\mathrm{Pos},1-\beta}\}, \tag{6.4}$$

where $\Re$ is the set of real numbers, therefore,

$$\mathrm{Pos}\{\mathscr{L} < \mathrm{VaR}_{\mathrm{Pos},1-\beta}\} = 1,$$

for any $(0<\beta<1)$. As a consequence, the $\mathrm{VaR}_{\mathrm{Pos},1-\beta}$ is always the largest loss at any $1-\beta$; the confidence level looses its original meaning.

On the other hand, if we assume that $\mathrm{VaR}_{\mathrm{Pos},1-\beta}$ satisfies the relationship

$$\mathrm{Pos}\{\mathscr{L} < \mathrm{VaR}_{\mathrm{Pos},1-\beta}\} = 1-\beta,$$

for any $\beta > 0$, it follows from (6.4) that

$$\mathrm{Pos}\{\mathscr{L} < \mathrm{VaR}_{\mathrm{Pos},1-\beta}\} = 1-\beta < 1 \Longrightarrow \mathrm{Pos}\{\mathscr{L} \geq \mathrm{VaR}_{\mathrm{Pos},1-\beta}\} = 1.$$

That implies the $\mathrm{VaR}_{\mathrm{Pos},1-\beta}$ could never be the largest loss at any confidence level; the possibility makes confidence interval unreliable. From the above two aspects, we can see that the VaR cannot be defined based only on the possibility measure.

Remark 6.2. Likewise, the necessity measure is not a viable choice for expressing fuzzy VaR. Following the same reasoning as presented in the case of possibility, on the one hand, if the fuzzy VaR at the confidence $1-\beta$ is defined by necessity as

$$\mathrm{VaR}_{\mathrm{Nec},1-\beta} = \sup\{\lambda \mid \mathrm{Nec}\{\mathscr{L} \geq \lambda\} \geq \beta\}, \tag{6.5}$$

we can get

$$\mathrm{Nec}\{\mathscr{L} \geq \mathrm{VaR}_{\mathrm{Nec},1-\beta}\} = \beta,$$

for any general continuous $\mathscr{L}$. Furthermore, by the definition of necessity, we have

$$0 = \mathrm{Nec}\{\mathscr{L} \in \emptyset\} = \mathrm{Nec}\{\mathscr{L} \geq \mathrm{VaR}_{\mathrm{Nec},1-\beta}\} \bigwedge \mathrm{Nec}\{\mathscr{L} < \mathrm{VaR}_{\mathrm{Nec},1-\beta}\}, \tag{6.6}$$

where $\emptyset$ is the empty set. Therefore, we have

$$\mathrm{Nec}\{\mathscr{L} < \mathrm{VaR}_{\mathrm{Nec},1-\beta}\} = 0,$$

for any $0 < \beta < 1$, which implies the $\mathrm{VaR}_{\mathrm{Nec},1-\beta}$ could never be the largest loss at any confidence level $1-\beta$. Again, the confidence level expressed in this way becomes meaningless.

On the other hand, if we start from

$$\mathrm{Nec}\{\mathscr{L} < \mathrm{VaR}_{\mathrm{Nec},1-\beta}\} = 1-\beta,$$

it follows from (6.6) that for any $\beta < 1$, we have

$$\mathrm{Nec}\{\mathscr{L} < \mathrm{VaR}_{\mathrm{Nec},1-\beta}\} = 1-\beta > 0 \Longrightarrow \mathrm{Nec}\{\mathscr{L} \geq \mathrm{VaR}_{\mathrm{Nec},1-\beta}\} = 0.$$

This implies the $\mathrm{VaR}_{\mathrm{Nec},1-\beta}$ seems always to be the largest loss for any confidence level, which is quite risky. As a consequence of above-mentioned aspects, it is not suitable either to characterize the VaR in terms of necessity.

6.1.2 Fuzzy Random VaR

In a fuzzy random environment, the Value-at-Risk can be defined by chance measure Ch as follows:

Definition 6.2. We let $\mathscr{L}$ be the loss variable with fuzzy random parameters in some investment. The fuzzy random Value-at-Risk for the investment with confidence $1-\beta$ (in the sense of mean chance) is given by (see Fig. 6.1):

$$\mathrm{VaR}_{\mathrm{Ch},1-\beta} = \sup\{\lambda \in \Re \mid \mathrm{Ch}\{\mathscr{L} \geq \lambda\} \geq \beta\}, \tag{6.7}$$

where $\beta \in (0,1)$ and Ch is the mean chance measure in (2.6).

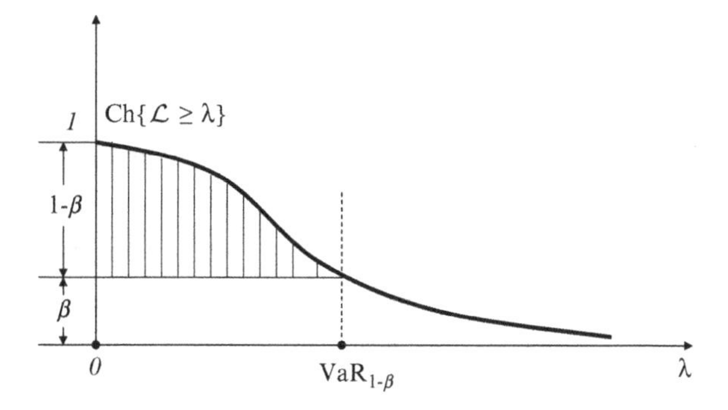

Fig. 6.1 Fuzzy random VaR

Remark 6.3. If the fuzzy random loss variable $\mathscr{L}$ in Definition 6.2 degenerates to a random loss variable, then we have

$$\begin{aligned}
\mathrm{VaR}_{\mathrm{Ch},1-\beta} &= \sup\{\lambda \in \Re \mid \mathrm{Ch}\{\mathscr{L} \geq \lambda\} \geq \beta\} \\
&= \sup\left\{\lambda \in \Re \mid \int_0^1 \Pr\{\mathrm{Cr}\{\mathscr{L}(\omega) \geq \lambda\} \geq r\}\,\mathrm{d}r \geq \beta\right\} \\
&= \sup\left\{\lambda \in \Re \mid \int_\Omega I_{\{\mathscr{L}(\omega)\geq\lambda\}}\Pr(\mathrm{d}\omega) \geq \beta\right\} \\
&= \sup\{\lambda \in \Re \mid \Pr\{\mathscr{L} \geq \lambda\} \geq \beta\},
\end{aligned}$$

where $I_{\{\cdot\}}$ is the indicator function of event $\{\cdot\}$. Hence, the Value-at-Risk in (6.7) becomes the right classical random Value-at-Risk in (6.1).

Remark 6.4. If the fuzzy random loss variable $\mathscr{L}$ in Definition 6.2 reduces to a fuzzy loss variable, then $\mathscr{L}(\omega) \equiv \mathscr{L}$ for any $\omega \in \Omega$. Thus, the Value-at-Risk in (6.7) becomes

$$\begin{aligned}
\mathrm{VaR}_{\mathrm{Ch},1-\beta} &= \sup\left\{\lambda \in \Re \mid \int_0^1 \Pr\{\mathrm{Cr}\{\mathscr{L}(\omega) \geq \lambda\} \geq r\}\,\mathrm{d}r \geq \beta\right\} \\
&= \sup\{\lambda \in \Re \mid \mathrm{Cr}\{\mathscr{L} \geq \lambda\} \geq \beta\},
\end{aligned}$$

which is just the fuzzy Value-at-Risk in (6.2).

In the following, without any confusion, we denote fuzzy random VaR by $\mathrm{VaR}_{1-\beta}$ without the subscript Ch for the sake of simplicity. The following theorem shows that the fuzzy random VaR defined as above coincides with the classic notion of Value-at-Risk.

Proposition 6.1. *Suppose* $\text{VaR}_{1-\beta}$ *is the fuzzy random Value-at-Risk on a fuzzy random loss variable* $\mathscr{L}$ *defined by (6.7), then the* $\text{VaR}_{1-\beta}$ *is the potentially largest loss to the confidence level* $1-\beta$.

Proof. We prove the theorem by contradiction. Suppose there exists a value $V' > \text{VaR}_{1-\beta}$ such that

$$\text{Ch}\{\mathscr{L} < V'\} = 1-\beta.$$

From the definition, we have

$$\begin{aligned}\text{Ch}\{\mathscr{L} \geq V'\} &= \int_{\Omega} \text{Cr}\left\{\mathscr{L}(\omega) \geq V'\right\} \text{Pr}(\text{d}\omega) \\ &= \int_{\Omega} 1 - \text{Cr}\left\{\mathscr{L}(\omega) < V'\right\} \text{Pr}(\text{d}\omega) \\ &= 1 - \int_{\Omega} \text{Cr}\left\{\mathscr{L}(\omega) < V'\right\} \text{Pr}(\text{d}\omega) \\ &= 1 - \text{Ch}\{\mathscr{L} < V'\} = \beta.\end{aligned}$$

Recalling that

$$\text{VaR}_{1-\beta} = \sup\{\lambda \mid \text{Ch}\{\mathscr{L} \geq \lambda\} \geq \beta\},$$

we have $V' \leq \text{VaR}_{1-\beta}$, which leads to a contradiction. As a consequence, $\text{VaR}_{1-\beta}$ is the largest loss at the confidence level $1-\beta$. The proof of the proposition is completed. □

Example 6.1. Suppose the loss $\mathscr{L}$ of an investment is a triangular fuzzy random variable $(X-100, X, X+100)$ ($), where X is a discrete random variable taking on values $X_1 = 50$ with probability 0.8 and $X_2 = 100$ with probability 0.2, respectively. Let us calculate the VaR at confidence 0.9.

From the assumptions, we know the loss $\mathscr{L}$ takes on the fuzzy values as follows: $\mathscr{L}(X_1) = (-50, 50, 150)$ with probability 0.8, and $\mathscr{L}(X_2) = (0, 100, 200)$ with probability 0.2. We can compute

$$\text{Cr}\{\mathscr{L}(X_1) \geq r\} = \begin{cases} 1, & r \leq -50 \\ \dfrac{150-r}{200}, & -50 < r \leq 150 \\ 0, & \text{otherwise,} \end{cases}$$

and

$$\text{Cr}\{\mathscr{L}(X_2) \geq r\} = \begin{cases} 1, & r \leq 0 \\ \dfrac{200-r}{200}, & 0 < r \leq 200 \\ 0, & \text{otherwise.} \end{cases}$$

By the definition, we obtain

$$\mathrm{Ch}\{\mathscr{L} \geq r\} = 0.8 \times \mathrm{Cr}\{\mathscr{L}(X_1) \geq r\} + 0.2 \times \mathrm{Cr}\{\mathscr{L}(X_2) \geq r\}$$

$$= \begin{cases} 1, & r \leq -50 \\ \dfrac{200-r}{250}, & -50 < r \leq 0 \\ \dfrac{160-r}{200}, & 0 < r \leq 150 \\ \dfrac{200-r}{1000}, & 150 < r \leq 200 \\ 0, & \text{otherwise.} \end{cases}$$

Therefore, by (6.7), the Value-at-Risk at a confidence level of 0.9 is

$$\mathrm{VaR}_{0.9} = \sup\left\{r \mid \frac{160-r}{200} \geq 0.1\right\} = 140.$$

6.2 Model Development

6.2.1 Mathematical Modeling

In the field of management and industrial engineering, there is a class of important and complicated problems, called two-stage problems, for example, the location problems we discussed in Chap. 5 is a typical two-stage problem. Many other examples in this sort of models are production planning, farming planning, material requirement planning, and so on. The two-stage programming (see [13,96,152,154]) is abstracted from those practical two-stage problems.

As we have elaborated in Chap. 5, in a two-stage programming, the two-stage structure is determined by the uncertain parameters and two categories of decisions of two stages. In this chapter, making use of the Value-at-Risk criterion with fuzzy random parameters, a generic two-stage programming model is built.

Let us start with the following loss-minimization problem with fuzzy random parameters:

$$\begin{aligned} \min_{x} \quad & \mathscr{L}(x,\xi) = f(x) - \max_{y} q^T(\xi)y \\ \text{subject to} \quad & x \in D, \\ & T(\xi)x + W(\xi)y = h(\xi), \\ & y \geq 0, \end{aligned} \tag{6.8}$$

where $\xi = (\xi_1, \xi_2, \cdots, \xi_m)$ is a vector of fuzzy random parameters defined from a probability space $(\Omega, \Sigma, \Pr)$ to a group of fuzzy variables on possibility space

$(\Gamma, \mathscr{A}, \text{Pos})$; x and y are two decision vectors which are $n_1 \times 1$ and $n_2 \times 1$, respectively; D is a feasible set of x, which can be a continuous or discrete region; and $f(\cdot)$ is a continuous function from $\Re^{n_1}$ to $\Re$. What's more, $q(\xi)$, $h(\xi)$, $T(\xi)$, and $W(\xi)$ are $n_2 \times 1, m_2 \times 1, m_2 \times n_1$, and $m_2 \times n_2$, respectively, which contain fuzzy random parameters and are represented through the following affine sums

$$\left.\begin{array}{l} q(\xi) = q^0 + \sum_{i=1}^{m} q^i \cdot \xi_i \\ h(\xi) = h^0 + \sum_{i=1}^{m} h^i \cdot \xi_i \\ T(\xi)) = T^0 + \sum_{i=1}^{m} T^i \cdot \xi_i \\ W(\xi) = W^0 + \sum_{i=1}^{m} W^i \cdot \xi_i \end{array}\right\} \tag{6.9}$$

where $q^i, h^i, T^i, W^i, i = 0, 1, \cdots, m$, are deterministic matrices of $n_2 \times 1, m_2 \times 1$, $m_2 \times n_1, m_2 \times n_2$, respectively.

Remodeling the problem (6.8) by putting the fuzzy random VaR criterion as the objective function, a new fuzzy stochastic optimization model can be formulated as

$$\left.\begin{array}{ll} \min\limits_{x} & \text{VaR}_{1-\beta}(x) \\ & = \sup\left\{\lambda \mid \text{Ch}\left\{f(x) - \max_y q^T(\xi)y \geq \lambda\right\} \geq \beta\right\} \\ \text{subject to } x \in D, & \\ & T(\xi)x + W(\xi)y = h(\xi), \\ & y \geq 0. \end{array}\right\} \tag{6.10}$$

Furthermore, in model (6.10), we divide the decisions into two groups. The decision x is called the first-stage decision, which has to be taken before the realizations $\xi(\omega, \gamma), (\omega, \gamma) \in \Omega \times \Gamma$, of the fuzzy random vector ξ. While the decision y is taken after the outcomes of ξ, which is named the second-stage decision. After the first-stage decision x is taken and an outcome $\xi(\omega, \gamma)$ is realized, the data

$$q(\xi(\omega, \gamma)), h(\xi(\omega, \gamma)), T(\xi(\omega, \gamma)), \text{ and } W(\xi(\omega, \gamma))$$

of the second-stage problem become known, and then the second-stage decision y is then selected so as to satisfy the fuzzy random constraints:

$$T(\xi)x + W(\xi)y = h(\xi).$$

Noting that the value of the second-decision y is determined as soon as x and ξ are known, in other words, y is completely determined by the selection of x and some realized value of ξ; hence, it is not a real decision to the two-stage model from a holistical point of view. Therefore, the only real decision is the first-stage decision, i.e., x.

The objective $\mathrm{VaR}_{1-\beta}(x)$ of model (6.10) represents the largest loss at confidence level $1-\beta$ among the scenarios corresponding to all the realizations of fuzzy random vector ξ. The loss

$$\mathscr{L}(x,\xi) = f(x) - \max_{y} q^T(\xi)y$$

here is made up of a deterministic term $f(x)$ which represents the fixed cost, and the second-stage objective $\max_y q^T(\xi)y$ which represents the (optimal) return at a given scenario (ω,γ). The latter is more difficult because, taken each realization $\xi(\omega,\gamma)$ of the fuzzy random vector ξ, y, or more precisely, $y^{(\omega,\gamma)}$ is determined by solving a linear programming. For the given realization $\xi(\omega,\gamma)$, we lay out the second-stage value function as

$$\begin{aligned}\mathscr{R}(x,\xi(\omega,\gamma)) = \max_{y} \Big\{ & q^T(\xi(\omega,\gamma))y^{(\omega,\gamma)} \mid T(\xi(\omega,\gamma))x + W(\xi(\omega,\gamma))y^{(\omega,\gamma)} \\ & = h(\xi(\omega,\gamma)), y^{(\omega,\gamma)} \geq 0 \Big\}. \end{aligned} \tag{6.11}$$

Noting that the optimal objective value of a linear programming is a Borel measurable extended real-valued function, for each given x, the second-stage value function $\mathscr{R}(x,\xi)$ is a fuzzy random variable; hence the loss variable $\mathscr{L}(x,\xi) = f(x) - \mathscr{R}(x,\xi)$ is also a fuzzy random variable.

Making use of (6.11), model (6.10) can be converted equivalently to the following form that is called *two-stage fuzzy random programming with Value-at-Risk* (FSP-VaR):

$$\left.\begin{array}{ll} \min\limits_{x} & \mathrm{VaR}_{1-\beta}(x) \\ \text{subject to } x \in D, & \end{array}\right\} \tag{6.12}$$

where

$$\begin{aligned}\mathrm{VaR}_{1-\beta}(x) &= \sup\{\lambda \mid \mathrm{Ch}\{f(x) - \mathscr{R}(x,\xi) \geq \lambda\} \geq \beta\} \\ &= \sup\left\{\lambda \;\Big|\; \int_{\Omega} \mathrm{Cr}\{f(x) - \mathscr{R}(x,\xi(\omega,\gamma)) \geq \lambda\}\,\Pr(\mathrm{d}r) \geq \beta\right\}, \end{aligned} \tag{6.13}$$

and the second-stage problem at each scenario (ω,γ) is

$$\left.\begin{array}{l} \mathscr{R}(x,\xi(\omega,\gamma)) = \max\limits_{y^{(\omega,\gamma)}} q^T(\xi(\omega,\gamma))y^{(\omega,\gamma)} \\ \qquad \text{subject to } T(\xi(\omega,\gamma))x + W(\xi(\omega,\gamma))y^{(\omega,\gamma)} = h(\xi(\omega,\gamma)), \\ \qquad\qquad y^{(\omega,\gamma)} \geq 0. \end{array}\right\} \tag{6.14}$$

Compared with (6.10), the form of (6.12)–(6.14) is able to represent more clearly the two-stage structure of the two-stage FSP-VaR.

6.2.2 *Difficulties*

With its own particular characteristics, the two-stage FSP-VaR carries similar difficulties to that we elaborated in Chap. 5. From (6.12) to (6.14), it is easy to realize that to solve the two-stage FSP-VaR, a key task is to calculate the objective value $\text{VaR}_{1-\beta}(x)$ at x. However, this process requires calculating the greatest loss value $\mathscr{L}(x,\xi(\omega,\gamma)) = f(x) - \mathscr{R}(x,\xi(\omega,\gamma))$ taken over all realizations $(\omega,\gamma) \in \Omega \times \Gamma$, where even for one realization, we need to solve a second-stage programming problem (6.14). What's more, in the real applications, the parameter vector ξ is usually a continuous fuzzy random vector with an infinite support. In other words, fuzzy random vector ξ takes on an infinite number of realizations. Hence, to obtain $\text{VaR}_{1-\beta}(x)$ for each x, we have to solve an infinite number of second-stage programming problems. So the value of $\text{VaR}_{1-\beta}(x)$ cannot be determined analytically. Thus, the conventional nonlinear mathematical programming methods are not applicable to the two-stage FSP-VaR (6.12)–(6.14).

In addition, as for the continuous decision case (search space D is a continuous region), an effective continuous search approach should be designed.

6.3 Model Analysis

6.3.1 *Sensitivity Analysis of Fuzzy Random VaR*

The following theorems present some sensitivity conditions of the fuzzy random $\text{VaR}_{1-\beta}$ w.r.t. the confidence level $1-\beta$.

Theorem 6.1. *Let ξ be a fuzzy random variable and $\text{VaR}_{1-\beta}$ the Value-at-Risk of ξ at confidence $1-\beta$. Then $\text{VaR}_{1-\beta}$ is left continuous of β.*

Proof. First we take a positive sequence $\{\beta_i\}$ such that $\beta_i \uparrow \beta$. Therefore, $\{\text{VaR}_{1-\beta_i}\}$ is a decreasing sequence. It suffices to prove

$$\lim_{i\to\infty} \text{VaR}_{1-\beta_i} = \text{VaR}_{1-\beta}.$$

We use a proof of contradiction. Suppose

$$\lim_{i\to\infty} \text{VaR}_{1-\beta_i} > \text{VaR}_{1-\beta}.$$

Taking

$$z = \frac{\lim_{i\to\infty} \text{VaR}_{1-\beta_i} + \text{VaR}_{1-\beta}}{2},$$

obviously, we have $\mathrm{VaR}_{1-\beta_i} > z > \mathrm{VaR}_{1-\beta}$ for each i. As a consequence,

$$\mathrm{Ch}\{\xi \geq z\} \geq \beta_i$$

for each i.

Letting $i \to \infty$, we have $\mathrm{Ch}\{\xi \geq z\} \geq \beta$; thus, $z \leq \mathrm{VaR}_{1-\beta}$, which leads to a contradiction. The proof of the theorem is completed. □

Theorem 6.2. *Let ξ be a fuzzy random variable and $\beta \in (0,1)$, and $\mathrm{VaR}_{1-\beta}$ the Value-at-Risk of ξ at confidence $1-\beta$. If there is at most one value x_β such that the equality*

$$\mathrm{Cr}\{\xi(\omega) \geq x_\beta\} = \beta$$

holds almost surely for $\omega \in \Omega$, then $\mathrm{VaR}_{1-\beta}$ is continuous at β.

Proof. By Theorem 6.1, $\mathrm{VaR}_{1-\beta}$ is a left continuous function w.r.t. $\beta \in (0,1)$. It suffices to prove $\mathrm{VaR}_{1-\beta}$ is right continuous at β.

Since there corresponds at most one value x_β for each given $\beta \in (0,1)$ such that

$$\mathrm{Cr}\{\xi(\omega) \geq x_\beta\} = \beta$$

almost surely, we obtain that there is at most one value x_β such that

$$\mathrm{Ch}\{\xi \geq x_\beta\} = \int_\Omega \mathrm{Cr}\{\xi(\omega) \geq x_\beta\}\,\mathrm{Pr}(\mathrm{d}\omega) = \beta.$$

It is easy to testify that $\mathrm{VaR}_{1-\beta}$ is decreasing w.r.t. β. Letting $\{\beta_i\}$ be any positive sequence such that $\beta_i \downarrow \beta$, therefore, $\{\mathrm{VaR}_{1-\beta_i}\}$ is a increasing sequence. To obtain the right continuity of $\mathrm{VaR}_{1-\beta}$, it suffices to prove that the limitation of $\{\mathrm{VaR}_{1-\beta_i}\}$ is equal to $\mathrm{VaR}_{1-\beta}$.

We use a proof of contradiction. Suppose

$$\lim_{i\to\infty} \mathrm{VaR}_{1-\beta_i} < \mathrm{VaR}_{1-\beta}.$$

Taking two real numbers z_1 and z_2 such that

$$\lim_{i\to\infty} \mathrm{VaR}_{1-\beta_i} < z_1 < z_2 < \mathrm{VaR}_{1-\beta},$$

we have $\mathrm{VaR}_{1-\beta_i} < z_1 < z_2 < \mathrm{VaR}_{1-\beta}$ for each i. Thus, the inequalities

$$\mathrm{Ch}\{\xi \geq z_1\} < \beta_i, \quad \mathrm{Ch}\{\xi \geq z_2\} < \beta_i$$

hold for each i. Since $\beta_i \downarrow \beta$, we obtain

$$\mathrm{Ch}\{\xi \geq z_1\} \leq \beta, \ \mathrm{Ch}\{\xi \geq z_2\} \leq \beta.$$

On one hand, if $\mathrm{Ch}\{\xi \geq z_i\} < \beta$, $i = 1$ or 2, then $z_i \geq \mathrm{VaR}_{1-\beta}$, $i = 1$ or 2, which is a contradiction with $z_1 < z_2 < \mathrm{VaR}_{1-\beta}$. On the other hand, if $\mathrm{Ch}\{\xi \geq z_i\} = \beta$, $i = 1, 2$, this also leads to a contradiction with the assumption. Therefore, we obtain

$$\lim_{i\to\infty} \mathrm{VaR}_{1-\beta_i} = \mathrm{VaR}_{1-\beta}.$$

The proof of the theorem is completed. □

Theorem 6.3. *Let ξ be a fuzzy random variable and $\beta \in (0,1)$, and $\mathrm{VaR}_{1-\beta}$ the Value-at-Risk of ξ at confidence $1-\beta$. Then $\mathrm{VaR}_{1-\beta}$ is continuous at $\beta \in (0,1)$ if and only if there is at most one value x_β such that $\mathrm{Ch}\{\xi \geq x_\beta\} = \beta$.*

Proof. Sufficiency: From the proof of Theorem 6.2, the sufficiency is valid.

Necessity: We use the proof of contradiction. Suppose there are two points t_1, t_2 such that $t_1 < t_2$ and

$$\mathrm{Ch}\{\xi \geq t_1\} = \mathrm{Ch}\{\xi \geq t_2\} = \beta;$$

then we get $\mathrm{Ch}\{\xi \geq t\} = \beta$ for any $t_1 \leq t \leq t_2$, and

$$\mathrm{VaR}_{1-\beta} = \sup\{x \mid \mathrm{Ch}\{\xi \geq x\} \geq \beta\} \geq t_2.$$

For any $\delta > 0$, if we take a β_0 with $0 < \beta_0 - \beta < \delta$, then for any x with

$$\mathrm{Ch}\{\xi \geq x\} \geq \beta_0 > \beta,$$

we have $x < t_1$, which follows from the definition of $\mathrm{VaR}_{1-\beta_0}$ that

$$\mathrm{VaR}_{1-\beta_0} = \sup\{x \mid \mathrm{Ch}\{\xi \geq x\} \geq \beta_0\} \leq t_1.$$

Thus, we obtain

$$\mathrm{VaR}_{1-\beta_0} \leq t_1 < t_2 \leq \mathrm{VaR}_{1-\beta},$$

which leads a contradiction to the continuity of $\mathrm{VaR}_{1-\beta}$ at β. The necessity is proved. □

6.3.2 Value of Fuzzy Information

As we have done in Sect. 5.2.1, we can also evaluate the value of fuzzy information in the decision process of the two-stage FSP-VaR, in terms of the difference between the fuzzy random solution (FRS) and the random solution (RS).

Recall that in the two-stage FSP-VaR, the Fuzzy Random Solution (FRS), which is based on (6.8), can be rewritten as

$$\begin{aligned} FRS_{1-\beta} &= \min_{x \in D} \sup\{\lambda \mid \mathrm{Ch}\{\mathscr{L}(x,\xi) \geq \lambda\} \geq \beta\} \\ &= \min_{x \in D} \mathrm{VaR}_{1-\beta}(x). \end{aligned} \tag{6.15}$$

Now, suppose the decision maker does not take fuzzy information into account when dealing with the decision making in a hybrid uncertain environment, a natural temptation is to replace the imprecise or fuzzy value $\xi(\omega)$ with some reasonable crisp one, denoted $\bar{\xi}(\omega)$ for each random scenario ω. Apparently, $\bar{\xi}$ is a random variable, and the original problem is converted into a stochastic programming problem, and an optimal solution $x^*\left(\bar{\xi}\right)$ can be obtained within such simplified situation:

$$x^*\left(\bar{\xi}\right) = \arg\min_{x \in D} \sup\left\{\lambda \mid \Pr\left\{\mathscr{L}\left(x,\bar{\xi}(\omega)\right) \geq \lambda\right\} \geq \beta\right\}. \tag{6.16}$$

After that, the solution $x^*\left(\bar{\xi}\right)$ will be input to the different scenarios of real situations with hybrid uncertainty of randomness and fuzziness. This random solution (RS) to fuzzy random problems can be written as follows:

$$\begin{aligned} RS_{1-\beta} &= \sup\left\{\lambda \mid \mathrm{Ch}\left\{\mathscr{L}\left(x^*\left(\bar{\xi}\right),\xi\right) \geq \lambda\right\} \geq \beta\right\} \\ &= \mathrm{VaR}_{1-\beta}\left(x^*\left(\bar{\xi}\right)\right), \end{aligned} \tag{6.17}$$

where $\beta \in (0,1)$ and $x^*\left(\bar{\xi}\right)$ is given by (6.16).

The value of fuzzy information (VFI) at confidence $1-\beta$ is then defined as

$$VFI_{1-\beta} = RS_{1-\beta} - FRS_{1-\beta}$$

for any $\beta \in (0,1)$.

Proposition 6.2. *Assume that $FRS_{1-\beta}$ and $RS_{1-\beta}$ are the fuzzy random solution and random solution defined in (6.15) and (6.17), respectively. We have*

$$VFI_{1-\beta} \geq 0,$$

for any $\beta \in (0,1)$.

Proof. Recalling that

$$RS_{1-\beta} = \mathrm{VaR}_{1-\beta}\left(x^*\left(\bar{\xi}\right)\right), \tag{6.18}$$

where $\beta \in (0,1)$, we note that $x^*\left(\bar{\xi}\right)$ is just one solution to the problem

$$\min_{x \in D} \mathrm{VaR}_{1-\beta}(x).$$

Hence, we have

$$RS_{1-\beta} \geq FRS_{1-\beta},$$

or equivalently,

$$VFI_{1-\beta} \geq 0,$$

for any $\beta \in (0,1)$. The proof is completed. □

6.3.3 Solution Bounds

Finally, considering a slightly different form of the original problem:

$$\begin{aligned}
\min_{x} \quad & \mathscr{L}(x,h(\xi)) = f(x) - \max_y q^T y \\
\text{subject to } & x \in D, \\
& Tx + Wy \geq h(\xi), \\
& y \geq 0,
\end{aligned} \tag{6.19}$$

where q, T, and W are fixed, we give a pair of bounds for the FRS in the following theorem.

Theorem 6.4. *Consider the two-stage FSP-VaR based on the original problem (6.19), suppose fuzzy random parameter $h(\xi)$ is bounded by two real-valued vectors $h_{\min}$ and $h_{\max}$, i.e., $h_{\min} \leq h(\xi) \leq h_{\max}$. If we denote $x_{\min}$ and $x_{\max}$ the optimal solutions to the problems $\mathscr{L}(x,h_{\min})$ and $\mathscr{L}(x,h_{\max})$ in (6.19), respectively, then we have*

$$\mathscr{L}(x_{\min},h_{\min}) \leq FRS_{1-\beta} \leq \mathscr{L}(x_{\max},h_{\max}),$$

for any $\beta \in (0,1)$.

Proof. Here, we only prove the inequality on the left-hand side; the one on the right-hand side can be obtained by the similar method. Let us write the feasible sets as

$$\mathscr{M}(x,h(\xi)) = \{y \geq 0 \mid Tx + Wy \geq h(\xi)\}$$

for any $x \in D$ where

$$D = \{x \mid x \geq 0, g_i(x) \leq 0, i = 1,2,\cdots,n\},$$

and realization $h(\xi(\omega,\gamma)), (\omega,\gamma) \in \Omega \times \Gamma$. Hence, we have

$$\mathscr{M}(x,h_{\min}) \supset \mathscr{M}(x,h(\xi(\omega,\gamma)))$$

for any decision $x \geq 0$ and $\omega \times \gamma \in \Omega \times \Gamma$. Therefore,

$$\mathscr{L}(x,h_{\min}) \leq \mathscr{L}(x,h(\xi(\omega,\gamma)) \tag{6.20}$$

for any $x \geq 0$ and $\omega \times \gamma \in \Omega \times \Gamma$.

Now we denote

$$\mathscr{S}(\star,\beta) = \sup\{\lambda \mid \mathrm{Ch}\{\star \geq \lambda\} \geq \beta\},$$

where $\beta \in (0,1)$. It follows from (6.20) that

$$\mathscr{S}\left(\mathscr{L}\left(x,h_{\min}\right),\beta\right) \leq \mathscr{S}\left(\mathscr{L}\left(x,h(\xi)\right),\beta\right) = \mathrm{VaR}_{1-\beta}(x) \tag{6.21}$$

for any decision x and $\beta \in (0,1)$. Furthermore, we note that

$$\begin{aligned}\mathscr{S}\left(\mathscr{L}\left(x,h_{\min}\right),\beta\right) &= \sup\left\{\lambda \mid \mathrm{Ch}\left\{\mathscr{L}\left(x,h_{\min}\right) \geq \lambda\right\} \geq \beta\right\} \\ &= \sup\left\{\lambda \mid I_{\{\mathscr{L}(x,h_{\min}) \geq \lambda\}} \geq \beta\right\} \\ &= \mathscr{L}\left(x,h_{\min}\right) \end{aligned} \tag{6.22}$$

for any $x \in D$ and $\beta \in (0,1)$, where $I_{\{\cdot\}}$ is the indicator function of event $\{\cdot\}$.

As a consequence of (6.21)–(6.22), we obtain

$$\mathscr{L}\left(x,h_{\min}\right) \leq \mathrm{VaR}_{1-\beta}(x)$$

for any $x \in D$ and $\beta \in (0,1)$. Furthermore, we obtain

$$\mathscr{L}\left(x,h_{\min}\right) \leq \min_{x} \mathrm{VaR}_{1-\beta}(x),$$

and finally,

$$\min_{x} \mathscr{L}\left(x,h_{\min}\right) \leq \min_{x} \mathrm{VaR}_{1-\beta}(x).$$

That is,

$$\mathscr{L}\left(x_{min},h_{\min}\right) \leq FRS_{1-\beta}$$

for any $\beta \in (0,1)$, which proves the theorem. □

6.4 Computing VaR

6.4.1 Discrete Case

Assuming that ξ in the two-stage FSP-VaR (6.12)–(6.14) is a discrete fuzzy random vector such that ω is a discrete random vector with a finite number of values ω_i with probability p_i, $i = 1,2,\cdots,N$, respectively, and for each i, $\xi(\omega_i)$ is a discrete fuzzy vector which takes on the following values:

$$\widehat{\xi}^{i1} = \left(\widehat{\xi}_1^{i1},\widehat{\xi}_2^{i1},\cdots,\widehat{\xi}_m^{i1}\right) \text{ with membership degree } \mu_{i1} > 0,$$

$$\widehat{\xi}^{i2} = \left(\widehat{\xi}_1^{i2}, \widehat{\xi}_2^{i2}, \cdots, \widehat{\xi}_m^{i2}\right) \quad \text{with membership degree } \mu_{i2} > 0,$$
$$\cdots \qquad \cdots$$
$$\widehat{\xi}^{iN_i} = \left(\widehat{\xi}_1^{iN_i}, \widehat{\xi}_2^{iN_i}, \cdots, \widehat{\xi}_m^{iN_i}\right) \quad \text{with membership degree } \mu_{iN_i} > 0,$$

and $\max_{j=1}^{N_i} \mu_{ij} = 1$. By the definition, the Value-at-Risk $\mathrm{VaR}_{1-\beta}(x)$ can be calculated as follows. Given each x, for $i = 1,2,\cdots,N;\ j = 1,2,\cdots,N_i$, we write $\mathscr{L}^{ij}(x) = f(x) - \mathscr{R}\left(x, \widehat{\xi}^{ij}\right)$, where

$$\begin{aligned}
\mathscr{R}\left(x, \widehat{\xi}^{ij}\right) = \max_{y}\ & q^T\left(\widehat{\xi}^{ij}\right) y \\
\text{subject to } & T\left(\widehat{\xi}^{ij}\right) x + W\left(\widehat{\xi}^{ij}\right) y = h\left(\widehat{\xi}^{ij}\right), \\
& y \ge 0.
\end{aligned} \tag{6.23}$$

Furthermore, given each pair (k,l), where $k = 1,2,\cdots,N; l = 1,2,\cdots,N_k$, we denote

$$\begin{aligned}
Q_i^{kl}(x) &= \mathrm{Cr}\left\{f(x) - \mathscr{R}(x, \xi(\omega_i)) \ge \mathscr{L}^{kl}(x)\right\} \\
&= \frac{1}{2}\left[\max_{j=1}^{N_i}\left\{\mu_{ij} \mid \mathscr{L}^{ij}(x) \ge \mathscr{L}^{kl}(x)\right\} + 1\right. \\
&\qquad \left. - \max_{j=1}^{N_i}\left\{\mu_{ij} \mid \mathscr{L}^{ij}(x) < \mathscr{L}^{kl}(x)\right\}\right]
\end{aligned} \tag{6.24}$$

for each $i, i = 1,2,\cdots,N$, and

$$\mathscr{Q}^{kl}(x) = \mathrm{Ch}\left\{f(x) - \mathscr{R}(x,\xi) \ge \mathscr{L}^{kl}(x)\right\} = \sum_{i=1}^{N} p_i Q_i^{kl}(x). \tag{6.25}$$

Thus,

$$\begin{aligned}
\mathrm{VaR}_{1-\beta}(x) &= \sup\{\lambda \mid \mathrm{Ch}\{f(x) - \mathscr{R}(x,\xi) \ge \lambda\} \ge \beta\} \\
&= \max_{k=1}^{N}\max_{l=1}^{N_k}\left\{\mathscr{L}^{kl}(x) \mid \mathscr{Q}^{kl}(x) \ge \beta\right\}.
\end{aligned} \tag{6.26}$$

As a consequence, the two-stage FSP-VaR (6.12)–(6.14) with discrete parameters can be formulated as

$$\begin{aligned}
\min_{x} \quad & \mathrm{VaR}_{1-\beta}(x) = \max_{k=1}^{N}\max_{l=1}^{N_k}\left\{\mathscr{L}^{kl}(x) \mid \mathscr{Q}^{kl}(x) \ge \beta\right\} \\
\text{subject to } & x \in D,
\end{aligned} \tag{6.27}$$

where

$$\mathscr{L}^{ij}(x) = f(x) - \mathscr{R}\left(x, \widehat{\xi}^{ij}\right),$$

$Q^{kl}(x)$ is calculated by (6.24)–(6.25), and $\mathscr{R}\left(x, \widehat{\xi}^{ij}\right)$ is determined by (6.23).

The following example illustrates the above calculation procedure of $\mathrm{VaR}_{1-\beta}(x)$ in the case of discrete fuzzy random parameters.

Example 6.2. Given the following two-stage FSP-VaR:

$$\begin{array}{ll}
\min\limits_{x} & \mathrm{VaR}_{0.7}(x) \\
 & = \sup\{\lambda \mid \mathrm{Ch}\{3x_1 + 5x_2 - \max\{3y_1 + 4y_2\} \geq \lambda\} \geq 0.3\} \\
\text{subject to} & x_1 - 1 \leq 0, \\
 & x_2 - 2 \leq 0, \\
 & y_1 + y_2 \leq \xi - x_1 - x_2, \\
 & x_1, x_2, y_1, y_2 \geq 0,
\end{array} \tag{6.28}$$

where ξ is a fuzzy random variable defined by

$$\xi(\omega) = \begin{cases} X_1, & \text{with probability } 1/3 \\ X_2, & \text{with probability } 2/3, \end{cases}$$

fuzzy variable X_1 takes on values $5, 6$, and 8 with membership degrees $1/3, 1$, and $3/5$, respectively, and X_2 assumes 2 and 3 with membership degrees $1/2$ and 1, respectively. Let us calculate $\mathrm{VaR}_{0.7}(1,1)$.

First of all, model (6.28) can be expressed equivalently by

$$\begin{array}{rl}
\min & \mathrm{VaR}_{0.7}(x) \\
\text{subject to} & 0 \leq x_1 \leq 1 \\
 & 0 \leq x_2 \leq 2,
\end{array} \tag{6.29}$$

where $\mathrm{VaR}_{0.7}(x) = \sup\{\lambda \mid \mathrm{Ch}\{3x_1 + 5x_2 - \mathscr{R}(x, \xi) \geq \lambda\} \geq 0.3\}$ and

$$\begin{array}{rl}
\mathscr{R}(x, \xi(\omega)(\gamma)) = \max & 3y_1 + 4y_2 \\
\text{subject to} & y_1 + y_2 \leq \xi(\omega)(\gamma) - x_1 - x_2, \\
 & y_1 \geq 0, \\
 & y_2 \geq 0.
\end{array} \tag{6.30}$$

By assumptions, fuzzy random variable ξ has the following realizations:

When $\xi(\omega) = X_1$ with probability $1/3$, $\widehat{\xi}^{11} = 5$ with membership degree $\mu_{11} = 1/3$, $\widehat{\xi}^{12} = 6$ with membership degree $\mu_{12} = 1$, and $\widehat{\xi}^{13} = 8$ with membership degree $\mu_{13} = 3/5$. When $\xi(\omega) = X_2$ with probability $2/3$, $\widehat{\xi}^{21} = 2$ with membership degree $\mu_{11} = 1/2$ and $\widehat{\xi}^{22} = 3$ with membership degree $\mu_{22} = 1$.

In order to calculate the Value-at-Risk

$$\mathrm{VaR}_{0.7}(1,1) = \sup\{\lambda \mid \mathrm{Ch}\{8 - \mathscr{R}(1,1,\xi) \geq \lambda\} \geq 0.3\},$$

we first compute $\mathscr{L}^{ij}(1,1) = 8 - \mathscr{R}(1,1,\widehat{\xi}^{ij})$ for each pair (i,j). Solving the second-stage programming problem (6.30) for each $\widehat{\xi}^{ij}$, we obtain

$$\mathscr{L}^{11}(1,1) = -4, \mathscr{L}^{12}(1,1) = -8, \mathscr{L}^{13}(1,1) = -16,$$

$$\mathscr{L}^{21}(1,1) = 8, \mathscr{L}^{22}(1,1) = 4.$$

Next, we calculate $Q_1^{ij}(1,1)$ and $Q_2^{ij}(1,1)$, for all i,j. Making use of (6.24), we have

$$\begin{aligned} Q_1^{11}(1,1) &= \frac{1}{2}\left[\max_{j=1}^{3}\left\{\mu_{1j} \mid \mathscr{L}^{1j}(x) \geq -4\right\} + 1 - \max_{j=1}^{3}\left\{\mu_{1j} \mid \mathscr{L}^{1j}(x) < -4\right\}\right] \\ &= \frac{1}{6}, \end{aligned}$$

similarly,

$$Q_1^{12}(1,1) = 7/10, Q_1^{13}(1,1) = 1, Q_1^{21}(1,1) = Q_1^{22}(1,1) = 0;$$

and

$$Q_2^{11}(1,1) = Q_2^{12}(1,1) = Q_2^{13}(1,1) = 1, Q_2^{21}(1,1) = 1/4, Q_2^{22}(1,1) = 1.$$

Furthermore, we compute $\mathscr{Q}^{ij}$ for each pair (i,j) by (6.25).

$$\mathscr{Q}^{11}(1,1) = \sum_{i=1}^{2} Q_i^{11}(1,1)p_i = \frac{1}{6} \times \frac{1}{3} + 1 \times \frac{2}{3} = \frac{13}{18},$$

and similarly, we obtain

$$Q^{12}(1,1) = 27/30, Q^{13}(1,1) = 1; Q^{21}(1,1) = 1/6, Q^{22}(1,1) = 2/3.$$

Finally, from (6.26), we obtain

$$\begin{aligned} \mathrm{VaR}_{0.7}(1,1) &= \max_{i=1}^{N}\max_{j=1}^{N_i}\left\{\mathscr{L}^{ij}(1,1) \mid \mathscr{Q}^{ij}(1,1) \geq 0.3\right\} \\ &= \mathscr{L}^{22}(1,1) = 4, \end{aligned}$$

where $N_1 = 3, N_2 = 2$.

6.4.2 Continuous Case

From Sect. 6.4.1, the VaR function $\mathrm{VaR}_{1-\beta}(x)$ with a discrete fuzzy random vector can be calculated directly through (6.23)–(6.26). In this section, for a continuous fuzzy random vector with an infinite support, we introduce an approximate algorithm to estimate the $\mathrm{VaR}_{1-\beta}(x)$. This approach combines fuzzy random simulation techniques, bisection search, and simplex algorithm.

Suppose $\xi = (\xi_1, \xi_2, \cdots, \xi_m)$ is a continuous fuzzy random vector. That is, the random parameter involved in ξ is a continuous random vector, and for any $\omega \in \Omega$, fuzzy vector $\xi(\omega)$ is also a continuous one. Let Ξ be the infinite support of ξ as

$$\Xi = \prod_{j=1}^{m} [a_j, b_j], \tag{6.31}$$

where $[a_j, b_j]$ is the support of $\xi_j, j = 1, 2, \cdots, m$. The approximate algorithm to $\mathrm{VaR}_{1-\beta}(x)$ with fuzzy random parameter ξ includes the following procedures **i–iv**.

(i) For the ξ with infinite realizations, we utilize the discretization method discussed in Sect. 4.3.3 to produce a sequence of fuzzy random vectors

$$\{\zeta_l\} = \{(\zeta_{l,1}, \zeta_{l,2}, \ldots, \zeta_{l,m})\}$$

which converges uniformly to original ξ, where for any $\omega \in \Omega$, $\zeta_l(\omega)$ is a discrete fuzzy vector. After producing the discretization $\{\zeta_l\}$, given a positive integer l, we denote

$$\mathrm{VaR}_{1-\beta}(x, \zeta_l) = \sup\{\lambda \mid \mathrm{Ch}\{f(x) - \mathscr{R}(x, \zeta_l) \geq \lambda\} \geq \beta\}.$$

In the following, we elaborate the approximate of $\mathrm{VaR}_{1-\beta}(x)$ through $\mathrm{VaR}_{1-\beta}(x, \zeta_l)$ for each feasible x.

(ii) For any $\omega \in \Omega$, the discrete fuzzy vector $\zeta_l(\omega)$ takes on a finite number of values as follows:

$$\widehat{\zeta}_l^k(\omega) = \left(\widehat{\zeta}_{l,1}^k(\omega), \widehat{\zeta}_{l,2}^k(\omega), \ldots, \widehat{\zeta}_{l,m}^k(\omega)\right),$$

with membership degree $\mu_{\omega,k}$, for $k = 1, 2, \ldots, N_\omega$. Therefore, for any $\lambda \in \Re$, we can calculate

$$\begin{aligned} Q_\omega(x, \lambda) &= \mathrm{Cr}\{f(x) - \mathscr{R}(x, \zeta_l(\omega)) \geq \lambda\} \\ &= \frac{1}{2}\left[\max_{j=1}^{N_\omega}\left\{\mu_{\omega,j} \mid \mathscr{L}_\omega^j(x) \geq \lambda\right\} + 1\right. \\ &\qquad \left. - \max_{j=1}^{N_\omega}\left\{\mu_{\omega,j} \mid \mathscr{L}_\omega^j(x) < \lambda\right\}\right], \end{aligned} \tag{6.32}$$

where $\mathscr{L}_{\omega}^{j}(x)=f(x)-\mathscr{R}\left(x,\widehat{\zeta}_l^{j}(\omega)\right)$ is obtained through solving the second-stage programming (6.14). Here, we utilize simplex algorithm to solve each second-stage programming.

(iii) Denote

$$Q(x,\lambda)=\mathrm{Ch}\{f(x)-\mathscr{R}(x,\zeta_l)\geq\lambda\}.$$

We generate N random samples $\omega_1,\omega_2,\cdots,\omega_N$ uniformly from the distribution of the random vector ω. Thus,

$$\mathscr{Q}(x,\lambda)=\int_{\Omega}Q_{\omega}(x,\lambda)\Pr(\mathrm{d}\omega)$$

can be computed by the following random simulation

$$\frac{Q_{\omega_1}(x,\lambda)+Q_{\omega_2}(x,\lambda)+\cdots+Q_{\omega_N}(x,\lambda)}{N}\to\mathscr{Q}(x,\lambda)\quad(N\to\infty),\tag{6.33}$$

with probability 1.

(iv) We note that $\mathscr{Q}(x,\lambda)$ is a nonincreasing real function of λ; therefore, for each feasible x,

$$\mathrm{VaR}_{1-\beta}(x,\zeta_l)=\sup\{\lambda\mid\mathscr{Q}(x,\lambda)\geq\beta\}$$

can be calculated by a bisection search.

Summarizing the above procedures **i-iv**, the approximate algorithm can be summarized as follows.

Algorithm 6.1 (Approximation to VaR).

Step 1. *Set $\lambda_L=e$ and $\lambda_R=E$, where e and E are a small and a large positive numbers, respectively.*
Step 2. *Let $\lambda_C=(\lambda_L+\lambda_R)/2$. Calculate $\mathscr{Q}(x,\lambda_C)$ by calling Algorithm 6.2.*
Step 3. *If $\mathscr{Q}(x,\lambda_C)\geq\beta$, then set $\lambda_L=\lambda_C$; otherwise, $\lambda_R=\lambda_C$.*
Step 4. *Repeat Steps 2–3 until $\lambda_R-\lambda_L<\varepsilon$, where $\varepsilon>0$ is a prescribed sufficiently small number.*
Step 5. *Return the value of* $\mathrm{VaR}_{1-\beta}(x,\zeta_l)=(\lambda_L+\lambda_R)/2$.

In Algorithm 6.1, the value of $\mathscr{Q}(x,\lambda)$ at each (x,λ) is determined by calling the following algorithm (fuzzy random simulation) which bases the procedures **i–iii**.

Algorithm 6.2 (Computing $\mathscr{Q}(x,\lambda)$).

Step 1. *Set $Q=0$ and $l=L$, where L is a sufficient large integer.*
Step 2. *Randomly generate a simple point $\widehat{\omega}$ from the distribution of the continuous random vector ω.*
Step 3. *Generate simples*

$$\widehat{\zeta}_l^{k}(\widehat{\omega})=\left(\widehat{\zeta}_{l,1}^{k}(\widehat{\omega}),\widehat{\zeta}_{l,2}^{k}(\widehat{\omega}),\ldots,\widehat{\zeta}_{l,m}^{k}(\widehat{\omega})\right)$$

through the discretization method from the support Ξ of ξ for $k=1,2,\cdots,N_{\widehat{\omega}}$.

Step 4. *Calculate* $\mathscr{L}^k_{\widehat{\omega}}(x)$ *by solving the second-stage programming (6.14) with the simplex algorithm for* $k = 1, 2, \cdots, N_{\widehat{\omega}}$.
Step 5. *Calculate* $Q_{\widehat{\omega}}(x, \lambda)$ *via (6.32).*
Step 6. $Q \leftarrow Q + Q_{\widehat{\omega}}(x, \lambda)$.
Step 7. *Repeat the Steps 2–6 N times.*
Step 8. *Return the value of* $\mathscr{Q}(x, \lambda) = Q/N$.

6.4.3 Convergence

The following theorem shows the convergence of the approximate algorithm (Algorithm 6.1) to Value-at-Risk $\mathrm{VaR}_{1-\beta}(x)$.

Theorem 6.5. *Consider the two-stage FSP-VaR (6.12)–(6.14). Suppose the matrix* W *is fixed, and* $q(\xi) = q(\xi_1), (T(\xi), h(\xi)) = (T(\xi_2), h(\xi_2))$, *where the fuzzy random vector* $\xi = (\xi_1, \xi_2)$ *is continuous with compact support (6.31), and* $\{\zeta_l\}$ *and* $\{\eta_k\}$ *are the discretizations of* ξ_1 *and* ξ_2, *respectively, then given any feasible decision x, we have*

$$\lim_{l \to \infty} \lim_{k \to \infty} \mathrm{VaR}_{1-\beta}(x, \zeta_l, \eta_k) = \mathrm{VaR}_{1-\beta}(x) \tag{6.34}$$

for almost every $\beta \in (0, 1)$.

Proof. By the assumption that $q(\xi) = q(\xi_1)$,

$$(T(\xi), h(\xi)) = (T(\xi_2), h(\xi_2)),$$

the two-stage FSP-VaR (6.12)–(6.14) can be expressed as

$$\mathrm{VaR}_{1-\beta}(x) = \sup\{\lambda \mid \mathrm{Ch}\{f(x) - \mathscr{R}(x, \xi_1, \xi_2) \geq \lambda\} \geq \beta\}, \tag{6.35}$$

where

$$\begin{aligned} \mathscr{R}(x, \xi_1, \xi_2) = \max_{y} \ & q^T(\xi_1) y \\ \text{subject to } & T(\xi_2)x + W(\xi_2)y = h(\xi_2), \\ & y \geq 0. \end{aligned} \tag{6.36}$$

Since matrix W is fixed, by the theory of parametric programming (see [13, Chap. 3, Theorem 5]), the second-stage value function $\mathscr{R}(x, \xi_1, \xi_2)$ is convex with respect (w.r.t.) to ξ_1 and is concave w.r.t. ξ_2, respectively. Thus,

$$\mathscr{L}(x, \xi_1, \xi_2) = f(x) - \mathscr{R}(x, \xi_1, \xi_2)$$

is continuous w.r.t. ξ_1 and ξ_2, respectively. Noting that the support Ξ in (6.31) of $\xi = (\xi_1, \xi_2)$ is compact, hence, $\mathscr{L}(x, \xi_1, \xi_2)$ is uniformly continuous w.r.t. ξ_1 and ξ_2, respectively.

Furthermore, we recall that the discretizations $\{\zeta_l\}$ and $\{\eta_k\}$ converge uniformly to ξ_1 and ξ_2, respectively. Therefore, from the convergence results on fuzzy random programming with recourse (see [96, Theorem 2]), given a fixed η_k, for any $\varepsilon \geq 0$, we have

$$\lim_{l\to\infty} \mathrm{Ch}\{|\mathscr{L}(x,\zeta_l,\eta_k) - \mathscr{L}(x,\xi_1,\eta_k)| \geq \varepsilon\} = 0.$$

Since convergence in chance implies convergence in distribution for fuzzy random variables, we obtain

$$\lim_{l\to\infty} \mathrm{Ch}\{\mathscr{L}(x,\zeta_l,\eta_k) \geq \lambda\} = \mathrm{Ch}\{\mathscr{L}(x,\xi_1,\eta_k) \geq \lambda\}, \tag{6.37}$$

provided $\lambda \in \mathfrak{R}$ is a continuity point of function $\mathrm{Ch}\{\mathscr{L}(x,\xi_1,\eta_k) \geq \lambda\}$.

We first prove

$$\lim_{l\to\infty} \mathrm{VaR}_{1-\beta}(x,\zeta_l,\eta_k) = \mathrm{VaR}_{1-\beta}(x,\eta_k) \tag{6.38}$$

for any fixed η_k and $\beta \in E_1$, where

$$\mathrm{VaR}_{1-\beta}(x,\eta_k) = \sup\{\lambda \mid \mathrm{Ch}\{\mathscr{L}(x,\xi_1,\eta_k) \geq \lambda\} \geq \beta\}$$

and $E_1 \subset (0,1)$ is the set of continuity points of $\mathrm{VaR}_{1-\beta}(x,\eta_k)$ w.r.t. β.

Suppose $\beta \in E_1$, from the continuity condition (Theorem 6.3) for fuzzy random VaR

$$\mathrm{VaR}_{1-\beta}(x) = \sup\{\lambda \mid \mathrm{Ch}\{\mathscr{L}(x,\xi_1,\eta_k) \geq \lambda\} \geq \beta\},$$

there is at most one value λ such that

$$\mathrm{Ch}\{\mathscr{L}(x,\xi_1,\eta_k) \geq \lambda\} = \beta.$$

On the one hand, for any $\lambda < \mathrm{VaR}_\beta(x,\eta_k)$ where λ is a continuity point of $\mathrm{Ch}\{\mathscr{L}(x,\xi_1,\eta_k) \geq \lambda\}$, we have $\mathrm{Ch}\{\mathscr{L}(x,\xi_1,\eta_k) \geq \lambda\} > \beta$. From (6.37), there is some positive integer N such that

$$\mathrm{Ch}\{\mathscr{L}(x,\zeta_l,\eta_k) \geq \lambda\} > \beta$$

for $l \geq N$. Therefore, $\mathrm{VaR}_{1-\beta}(x,\zeta_l,\eta_k) \geq \lambda$ for all $l \geq N$, provided $\lambda < \mathrm{VaR}_{1-\beta}(x,\eta_k)$ is a continuity point of $\mathrm{Ch}\{\mathscr{L}(x,\xi_1,\eta_k) \geq \lambda\}$. Producing an increasing sequence $\{\lambda_n\} \uparrow \mathrm{VaR}_{1-\beta}(x,\eta_k)$, we obtain

$$\liminf_{l\to\infty} \mathrm{VaR}_{1-\beta}(x,\zeta_l,\eta_k) \geq \mathrm{VaR}_{1-\beta}(x,\eta_k). \tag{6.39}$$

On the other hand, similarly, we can prove

$$\limsup_{l\to\infty} \mathrm{VaR}_{1-\beta}(x,\zeta_l,\eta_k) \leq \mathrm{VaR}_{1-\beta}(x,\eta_k). \tag{6.40}$$

Combining (6.39) and (6.40) proves the validity of (6.38).

Similarly as in the proof of (6.38), we can obtain

$$\lim_{k\to\infty} \mathrm{VaR}_{1-\beta}(x,\eta_k) = \mathrm{VaR}_{1-\beta}(x), \tag{6.41}$$

for any $\beta \in E_2$, where E_2 is the continuity set of $\mathrm{VaR}_{1-\beta}(x)$ w.r.t. β. Therefore, we have

$$\lim_{l\to\infty}\lim_{k\to\infty} \mathrm{VaR}_{1-\beta}(x,\zeta_l,\eta_k) = \mathrm{VaR}_{1-\beta}(x), \tag{6.42}$$

for any $\beta \in E_1 \cap E_2$.

We note that $\mathrm{VaR}_{1-\beta}(x,\eta_k)$ and $\mathrm{VaR}_{1-\beta}(x)$ are nonincreasing functions w.r.t. $\beta \in (0,1)$; hence, their sets of discontinuity points, $(0,1)\setminus E_1$ and $(0,1)\setminus E_2$, both are at most countable. This fact implies that the Lebesgue measure of set $(0,1)\setminus E_1 \cap E_2$ is zero. Thus, the result (6.42) is valid for almost every $\beta \in (0,1)$. □

Corollary 6.1. *Consider the two-stage FSP-VaR (6.12)–(6.14). Suppose the matrix W is fixed, and $q(\xi)$ or $(T(\xi),h(\xi))$ is fixed, where the vector ξ of fuzzy random parameters is a continuous fuzzy random vector with compact support (6.31), and $\{\zeta_l\}$ is the discretization of ξ, then for any given feasible decision x, we have*

$$\lim_{l\to\infty} \mathrm{VaR}_{1-\beta}\,(x,\zeta_l) = \mathrm{VaR}_{1-\beta}(x), \tag{6.43}$$

provided $\beta \in (0,1)$ is a continuity point of $\mathrm{VaR}_\beta(x)$.

6.5 Solution Approaches

In what follows, the solution approaches of the two-stage FSP-VaR (6.12)–(6.14) are discussed for two different cases: integer decisions and continuous decisions.

6.5.1 Integer-Decision Case

We note that if the feasible set D of x becomes a binary vector set $\{0,1\}^{n_1}$, the two-stage FSP-VaR becomes

$$\begin{array}{ll} \min\limits_{x} & \mathrm{VaR}_{1-\beta}(x) \\ \text{subject to } x \in \{0,1\}^{n_1}, & \end{array} \tag{6.44}$$

where the objective value $\mathrm{VaR}_{1-\beta}(x)$ is computed through (6.13), and the second-stage value function $\mathscr{R}(x,\xi(\omega,\gamma))$ is determined by solving the subprogramming problem (6.14) for each realization pair (ω,γ). In the hybrid MBACO algorithm

(Algorithm 5.3), if we replace the fuzzy random simulation to expected value (Algorithm 5.2 in Chap. 5) with the approximate algorithm (Algorithm 6.1) to VaR, then a solution approach can be formed to solve problem (6.44). Hence, the above problem (6.44) is also able to be solved directly by a hybrid vehicle which consists of the approximate algorithm (Algorithm 6.1) and the MBACO mechanism (Chap. 5).

On the other hand, when the feasible set D becomes a collection of integers $\mathbb{Z}$, then the two-stage FSP-VaR can be rewritten as

$$\begin{array}{ll} \min\limits_{x} & \mathrm{VaR}_{1-\beta}(x) \\ \text{subject to } x \in \mathbb{Z}^{n_1}, & \end{array} \tag{6.45}$$

where $\mathrm{VaR}_{1-\beta}(x)$ and $\mathscr{R}(x, \xi(\omega, \gamma))$ are also determined by (6.13) and (6.14), respectively. It is not difficult to see that combining the approximate algorithm (Algorithm 6.1) and the metaheuristic mechanism of GA (Chap. 4) also yields to a solution approach to the above two-stage fuzzy random integer programming problem (6.45).

To the above two forms of binary and integer decisions, the solution approaches to the two-stage FSP-VaR (6.12)–(6.14) are similar to that of the reliability model and the location model discussed in Chaps. 4 and 5, respectively. Hence, we omit the detailed discussion on them and focus on the solution for the case of continuous decisions in the next section.

6.5.2 *A Hybrid MN-PSO Algorithm for Continuous-Decision Case*

This section is devoted to the solution approach to two-stage FSP-VaR (6.12)–(6.14) with continuous decisions. In this case, the feasible set D can be expressed as

$$D = \{x \mid g_i(x) \leq 0, i = 1, 2, \cdots, n_1; x \geq 0\},$$

where $g_i(\cdot), i = 1, 2, \cdots, n_1$ are continuous function from $\Re^{n_1}$ to $\Re$.

In the Sect. 6.4, an approximate algorithm (Algorithm 6.1) has been introduced to determine the value of $\mathrm{VaR}_{1-\beta}(x)$ for each feasible decision x. In order to solve the two-stage FSP-VaR (6.12)–(6.14), we design a hybrid mutation-neighborhood-based particle swarm optimization (MN-PSO) algorithm in which the approximate algorithm deals with the VaR computation while the MN-PSO handles the continuous optimization (see Fig. 6.2).

Particle swarm optimization (PSO), a nature-inspired evolutionary computation algorithm, was originally developed by Kennedy and Eberhart [62] in 1995. This evolutionary computation technique uses collaboration among a population of simple search agents (called particles) to find optima in solution spaces; it has been shown effective in optimizing difficult multidimensional problems in variety

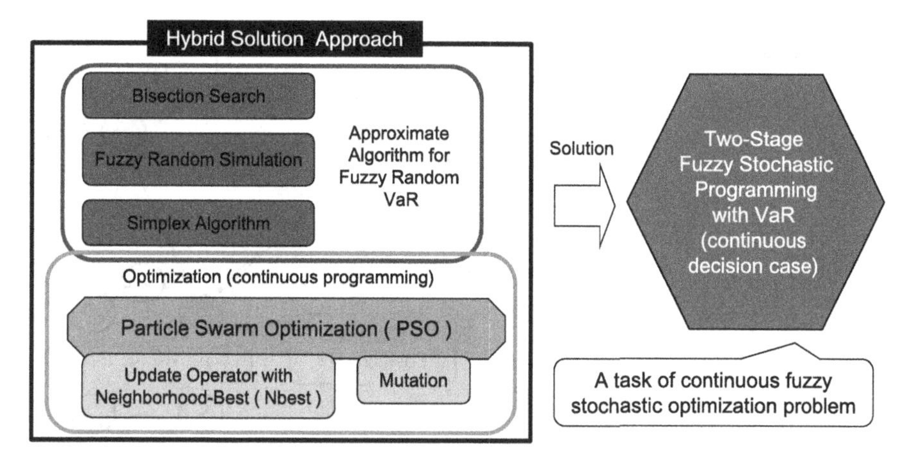

Fig. 6.2 The structure of hybrid approach for two-stage FSP-VaR

of fields (see [57, 64, 111, 119]). In PSO, a potential solution is represented as a particle x_i and a direction v_{id} in which the particle will travel. Each particle maintains a record of the position of its previous personal best (Pbest) performance in a vector called p_{id}. The variable g is the index of the particle with global best (Gbest) performance so far in the population. An iteration comprises evaluation of each particle, then stochastic adjustment of v_{id} in the direction of particle x_i's Pbest position p_{id} and the Gbest position p_{gd} of any particle in the population via the following formulas (see also Fig. 6.3):

$$v_{id} = \mathscr{W} * v_{id} + c_1 * \text{rand}() * (p_{id} - x_i) + c_2 * \text{rand}() * \left(p_{gd} - x_i\right) \tag{6.46}$$

$$x_i = x_i + v_{id}, \tag{6.47}$$

where c_1 and c_2 are learning rates generated in the interval [0,4], rand() is a uniform random number in the interval [0,1], and $\mathscr{W}$ is the inertia weight.

In order to solve the two-stage FSP-VaR more effectively with the mechanism of PSO, we incorporate several modifications as below:

(a) Most PSO algorithms employ the type of Pbest-Gbest-based update formula (the "Pbest" and "Gbest" denote personal-best and global-best particles, respectively). However, in a realistic population-based optimization mechanism, an individual should be influenced not only by the personal-best position of itself and the global-best position of the population, but also by that of its neighbors (such neighborhood-related individual development mode is obvious in any real-life society). Thus, to further improve the global search in the optimization, it is desirable to consider the individual's neighborhood. From this point of view, we introduce an Nbest-Gbest-based update rule (the "Nbest" denotes the neighborhood-best particles) by adjusting the velocity in the directions of the personal-best particles in the neighborhood and the global-best particle.

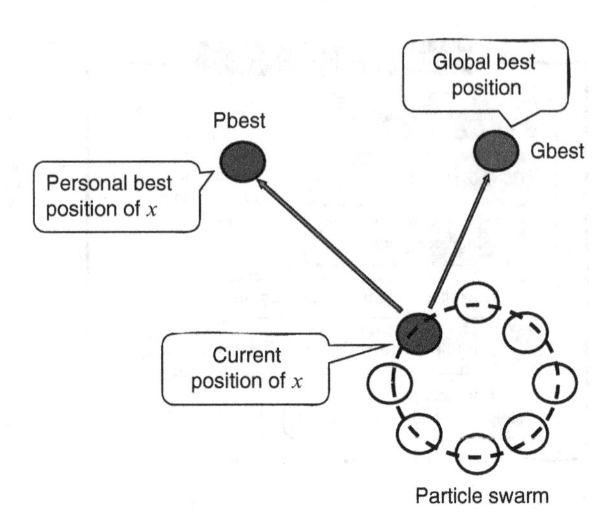

Fig. 6.3 Pbest-Gbest-based update process

(b) Furthermore, a mutation operator is applied to further extend the search space of modified PSO so as to decrease the probability of getting trapped in a local optimum.
(c) The approximate algorithm is embedded into the MN-PSO to evaluate the particles.

In the following, we introduce the hybrid MN-PSO algorithm in detail.

6.5.2.1 Solution Representation and Initialization

A positive real number vector $x = (x_1, x_2, \cdots, x_n)$ is used as a particle to represent a solution of the two-stage FSP-VaR (6.12)–(6.14).

We randomly generate the particle $x = (x_1, x_2, \cdots, x_n)$ by checking the feasibility of x as follows:

$$g_i(x) \leq 0, i = 1, 2, \cdots, n, x \geq 0. \tag{6.48}$$

Repeat the above process P_{size} times; we get initial particles $x_1, x_2, \cdots, x_{P_{size}}$.

6.5.2.2 Evaluation by Approximate Algorithm

Denote $\mathbf{Fit}(\cdot)$ the fitness function, and let the fitness of each particle x be the minus of the Value-at-Risk, i.e.,

$$\mathbf{Fit}(x) = -\mathrm{VaR}_{1-\beta}(x).$$

Hence, the particles of smaller objective values are evaluated with higher fitness. For each particle x, the fitness $\mathbf{Fit}(x)$ is calculated by the approximate algorithm (Algorithm 6.1).

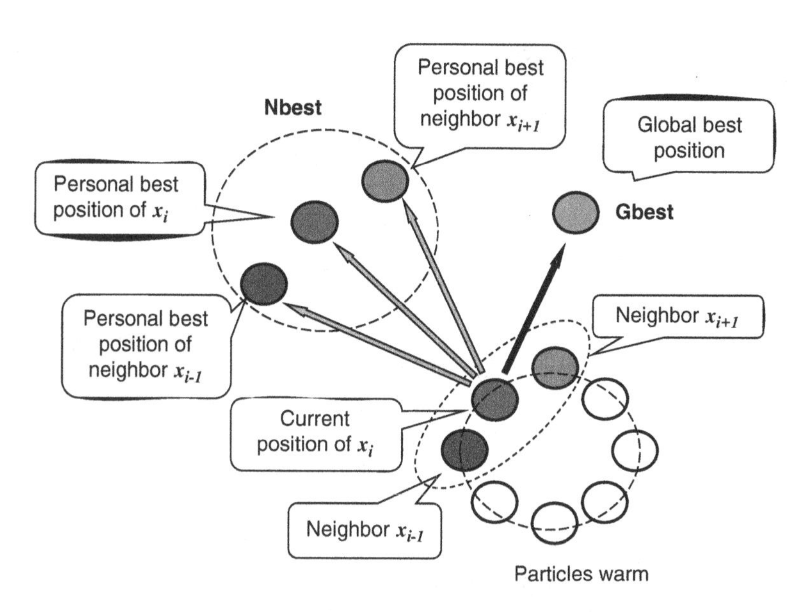

Fig. 6.4 Nbest-Gbest-based update process

6.5.2.3 Nbest-Gbest-Based Update

In the update process, we first determine the Gbest particle p_{gd} (with the highest fitness) of the population and for each particle x_i, find the p_{id} with the best previous performance, where $i = 1, 2, \cdots, P_{size}$. Then, we compute the velocity vector v_{id} and update x_i for $i = 1, 2, \cdots, P_{size}$ through following Nbest-Gbest-based update formula (see Fig. 6.4):

$$v_{id} = \mathscr{W} * v_{id} + c_1 * d_N(x_i) + c_2 * \text{rand}() * (p_{gd} - x_i), \tag{6.49}$$

$$x_i = x_i + v_{id}. \tag{6.50}$$

In the above formula, $d_N(x_i), i = 1, 2, \cdots, P_{size}$ are the average distances from x_i to the best positions in its neighbors, which are defined as

$$d_N(x_1) = \sum_{k=1}^{2} \text{rand}() * \frac{(p_{k,d} - x_1)}{2}, \tag{6.51}$$

$$d_N(x_i) = \sum_{k=i-1}^{i+1} \text{rand}() * \frac{(p_{k,d} - x_i)}{3}, i = 2, 3, \cdots, P_{size} - 1, \tag{6.52}$$

$$d_N(x_{P_{size}}) = \sum_{k=P_{size}-1}^{P_{size}} \text{rand}() * \frac{(p_{k,d} - x_{P_{size}})}{2}, \tag{6.53}$$

where rand() is a uniform random number in the interval [0,1] and c_1 and c_2 are learning rates.

To well adjust the convergence of the particles, here we employ the time-varying learning rates (see [125]) as follows:

$$c_1 = 2 * \frac{G_{max} - G_n}{G_{max}} + 1,$$

and

$$c_2 = 2 * \frac{G_n}{G_{max}} + 1,$$

where G_{max} and G_n are the indexes of the maximum and current generations, respectively. $\mathscr{W}$ is the inertia weight which is set by the following expression [22]:

$$\mathscr{W} = \frac{2}{|2 - \phi - \sqrt{\phi^2 - 4\phi}|},$$

where $\phi = c_1 + c_2$.

If the updated x_i is feasible (by checking (6.48)), then we keep it as a new particle of the next generation. Otherwise, we reupdate through (6.49)–(6.50) until a feasible new particle is obtained. Repeat the above process P_{size} times; we yield a new generation of particles $x'_1, x'_2, \cdots, x'_{P_{size}}$.

6.5.2.4 Mutation

We predetermine a parameter $P_m \in (0,1)$ as the probability of mutation. For each particle $x_i = (x_{i1}, x_{i2}, \cdots, x_{in}), i = 1, 2, \cdots, P_{size}$, if rand() $< P_m$, then we generated a number N_m between 1 and n, and replace the particle

$$x_i = (x_{i1}, \cdots, x_{iN_m}, x_{iN_m+1} \cdots, x_{in})$$

by a new one

$$x_i = (x''_{i1}, \cdots, x''_{iN_m}, x_{iN_m+1} \cdots, x_{in}),$$

where each $x''_{ij} \in \Re, j = 1, 2, \cdots, N_m$. The above process is repeated until the new particle satisfies the constraint (6.48).

6.5.2.5 Hybrid Algorithm Procedure

The hybrid MN-PSO algorithm for solving model (6.12)–(6.14) is summarized as follows; see also flowchart in Fig. 6.5.

Algorithm 6.3. (Hybrid MN-PSO Algorithm)

Step 1. Initialize a population of particles $x_1, \cdots, x_{P_{size}}$ by checking (6.48).
Step 2. Calculate the fitness **Fit**(x) *for all particles through the approximate algorithm (Algorithm 6.1), and evaluate each particle according to the fitness.*

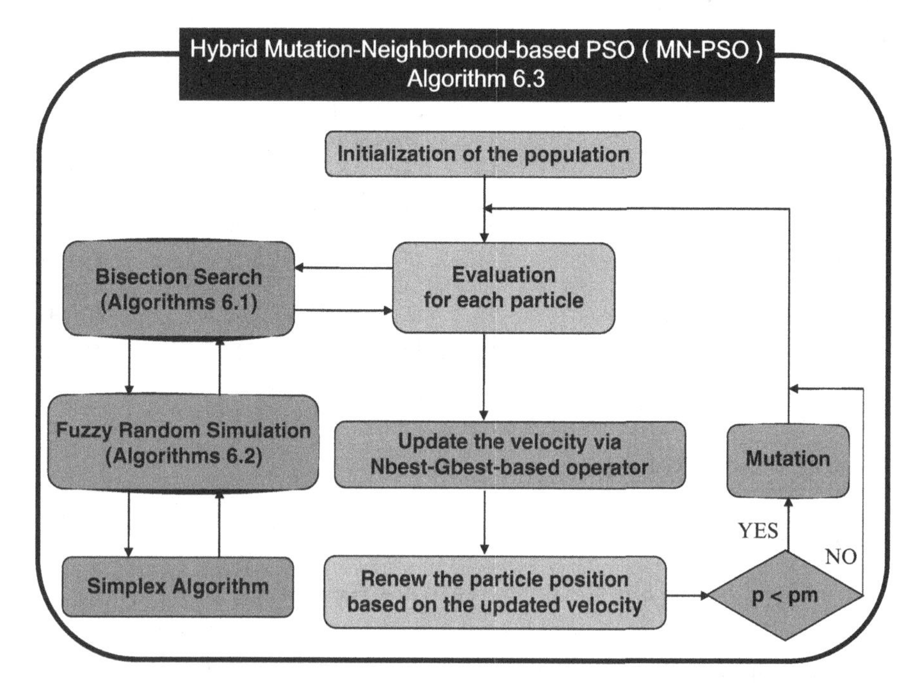

Fig. 6.5 Flowchart of the hybrid MN-PSO algorithm

Step 3. *Determine the $d_N(x_i)$ for each particle x_i, and the p_{gd} for the population.*
Step 4. *Update all the particles by formulas (6.49) and (6.50).*
Step 5. *Run mutation operator to each particle with probability P_m.*
Step 6. *Repeat Step 2 to Step 5 for a given number of generations.*
Step 7. *Return the particle p_{gd} as the optimal solution to the two-stage FSP-VaR (6.12)–(6.14), and*

$$-\mathbf{Fit}(p_{\mathbf{gd}}) = \mathrm{VaR}_{\mathbf{1}-\beta}(p_{\mathbf{gd}})$$

the corresponding optimal value.

6.6 Numerical Experiments

Let use consider the following two-stage FSP-VaR at a confidence of 0.8:

$$\left.\begin{array}{ll} \min & \mathrm{VaR}_{0.8}(x) \\ \text{subject to} & (x_1-1)^2+(x_2-2)^2+(x_3-3)^2 \\ & +(x_4-4)^2+(x_5-5)^2+(x_6-6)^2 \leq 100, \\ & x_i \geq 0, i=1,2,\cdots,6, \end{array}\right\} \tag{6.54}$$

Table 6.1 Distributions of fuzzy random parameters

i	Fuzzy random variable ξ_i	X_i
1	$\mathcal{N}_{FR}(X_1,2)$	$\mathcal{U}(2,3)$
2	$\mathcal{N}_{FR}(X_2,4)$	$\mathcal{U}(4,5)$
3	$\mathcal{N}_{FR}(X_3,6)$	$\mathcal{U}(2,3)$
4	(X_4-2,X_4+1,X_4+3)	$\mathcal{U}(4,6)$
5	(X_5+1,X_5+3,X_5+4)	$\mathcal{U}(3,4)$
6	(X_6+3,X_6+5,X_6+6)	$\mathcal{U}(2,3)$
7	(X_7+20,X_7+21,X_7+22)	$\mathcal{U}(1,2)$
8	(X_8+21,X_8+22,X_8+23)	$\mathcal{U}(4,5)$
9	(X_9+24,X_9+25,X_9+26)	$\mathcal{U}(3,4)$
10	$(X_{10}+28,X_{10}+29,X_{10}+30)$	$\mathcal{U}(3,4)$

where the Value-at-Risk objective is

$$\begin{aligned}&\mathrm{VaR}_{0.8}(x)\\&=\sup\left\{\lambda\mid \mathrm{Ch}\left\{x_1^4x_2+x_2^3x_4+x_3^2x_4+x_4^3x_5+x_5x_6-\mathscr{R}(x,\xi)\geq\lambda\right\}\geq 0.2\right\},\end{aligned}$$

and the second-stage programming at each scenario $(\omega,\gamma)\in\Omega\times\Gamma$ is

$$\left.\begin{aligned}&\mathscr{R}(x,\xi(\omega,\gamma))\\&\qquad=\max\ \xi_1(\omega,\gamma)y_1+\xi_2(\omega,\gamma)y_2+\xi_3(\omega,\gamma)y_3+\xi_4(\omega,\gamma)y_4+y_5\\&\qquad\text{subject to } y_1+y_2-3y_4-y_5\\&\qquad\qquad=2\xi_5(\omega,\gamma)+3\xi_6(\omega,\gamma)+x_1+2x_2+x_4,\\&\qquad\qquad 8y_1-10y_2+6y_3+5y_4-13y_5\\&\qquad\qquad=\xi_7(\omega,\gamma)+4x_1+4x_3-x_5,\\&\qquad\qquad y_1+8y_2+14y_3-y_4+12y_5+3y_6\\&\qquad\qquad=\xi_8(\omega,\gamma)-x_3+3x_6,\\&\qquad\qquad 10y_3+y_4-3y_6\\&\qquad\qquad=\xi_9(\omega,\gamma)+2\xi_{10}(\omega,\gamma)-x_1+x_2+6x_3,\\&\qquad\qquad y_k\geq 0,k=1,\cdots,10,\end{aligned}\right\}\quad(6.55)$$

here $\xi_i,i=1,2,\cdots,10$ are fuzzy random variables with distributions given in Table 6.1, respectively.

To solve the problem (6.54)–(6.55), we first compute $\mathrm{VaR}_{0.8}(x)$ through the approximate algorithm (Algorithm 6.1). For each feasible input $x=(x_1,x_2,x_3,x_4,x_5,x_6)$, we generate randomly 3,000 simples $\widehat{\omega}_i=(\widehat{X}_{1i},\widehat{X}_{2i},\cdots,\widehat{X}_{10i})$,

Table 6.2 Results of hybrid MN-PSO algorithm with different parameters

	Parameters		Results		
No.	P_{size}	P_m	Optimal solution	Objective value	Error(%)
1	20	0.2	(0.100, 0.206, 4.215, 0.562, 4.375, 14.382)	880.94	4.10
2	20	0.3	(0.418, 0.157, 7.592, 0.514, 2.320, 13.372)	871.59	2.99
3	20	0.4	(0.714, 0.115, 2.759, 0.292, 3.207, 14.332)	880.02	3.99
4	20	0.5	(1.202, 0.089, 1.860, 0.778, 2.309, 14.357)	879.42	3.91
5	30	0.2	(0.035, 0.253, 4.893, 0.778, 1.727, 14.314)	861.21	1.76
6	30	0.3	(0.246, 0.124, 4.106, 0.420, 2.999, 14.137)	846.29	0.00
7	30	0.4	(0.036, 0.341, 3.619, 0.509, 4.789, 14.423)	857.73	1.35
8	30	0.5	(0.306, 0.184, 3.003, 0.136, 1.936, 13.470)	862.43	1.91

$i = 1,2,\cdots,3,000$, from the distribution of random vector $\omega = (X_1, X_2, \cdots, X_{10})$. Then, for each $\widehat{\omega}_i$, we generate 1,000 sample points $\widehat{\zeta}^j(\widehat{\omega}_i), j = 1,2,\cdots,1,000$ via discretization method. Each

$$\mathscr{L}^j_{\widehat{\omega}_i}(x) = x_1^4 x_2 + x_2^3 x_4 + x_3^2 x_4 + x_4^3 x_5 + x_5 x_6 - \mathscr{R}\left(x, \widehat{\zeta}^j(\widehat{\omega}_i)\right)$$

is determined by solving the second-stage programming (6.55) with the simplex algorithm. Furthermore, the value of

$$\mathrm{Ch}\left\{x_1^4 x_2 + x_2^3 x_4 + x_3^2 x_4 + x_4^3 x_5 + x_5 x_6 - \mathscr{R}(x, \xi) \geq \lambda\right\}$$

for each λ is estimated through the random simulation (6.32)–(6.33). Finally, we obtain the objective value $\mathrm{VaR}_{0.8}(x)$ by the bisection method.

Embedding the approximate algorithm into the MN-PSO algorithm, we run the hybrid MN-PSO algorithm (Algorithm 6.3) with 200 generations and obtain the optimal solutions with different parameters shown in Table 6.2, where the *relative error* is given in the last column, which is defined in the same way as (5.42) in Chap. 5. It follows from Table 6.2 that the relative error does not exceed 4.10% when different parameters are selected. The performance implies that the hybrid MN-PSO algorithm is robust to the parameter settings and effective to solve the two-stage FSP-VaR. In addition, the trend line of the VaR with respect to the confidence level $1-\beta$ is provided in Fig. 6.6.

To further assess the performance of our proposed hybrid MN-PSO algorithm for two-stage FSP-VaR, we compare the experimental results of the hybrid MN-PSO with those produced by other hybrid approaches of (regular) particle swarm optimization (PSO)(see [44, 97]) and genetic algorithm (GA) (see [50, 91, 166]).

The hybrid MN-PSO, PSO, and GA algorithms with population size of 30 are run to the problem (6.54)–(6.55) with the same time duration of 50 minutes (in this duration, all the algorithms have run more than 200 generations and reached an iteration stability within). The best parameter settings are picked out for PSO and GA in the experiments with different system parameters: in PSO, $c_1 = c_2 = 2$ is the best for PSO, and the crossover and mutation rates $(P_c, P_m) = (0.4, 0.2)$ are the best for GA.

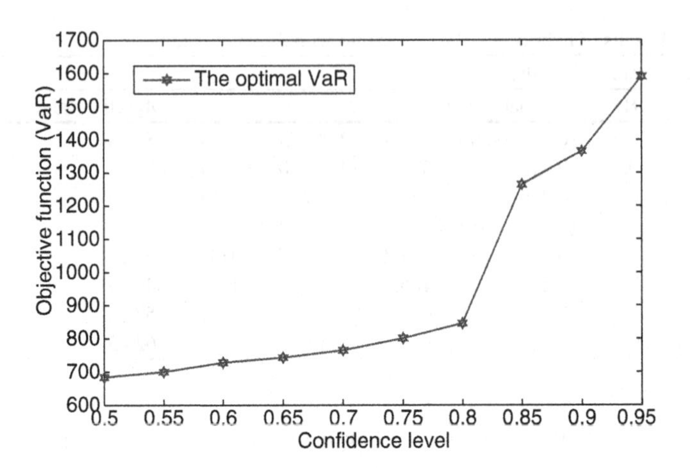

Fig. 6.6 The trend line of the VaR with respect to $1 - \beta$

Table 6.3 The comparison results of different approaches

Approach	Optimal solution	Objective value
Hybrid MN-PSO	(0.246, 0.124, 4.106, 0.420, 2.999, 14.137)	846.29
Hybrid PSO	(0.221, 0.100, 3.240, 0.766, 4.026, 15.000)	868.75
Hybrid GA	(0.852, 0.174, 7.977, 0.442, 3.980, 11.822)	878.23

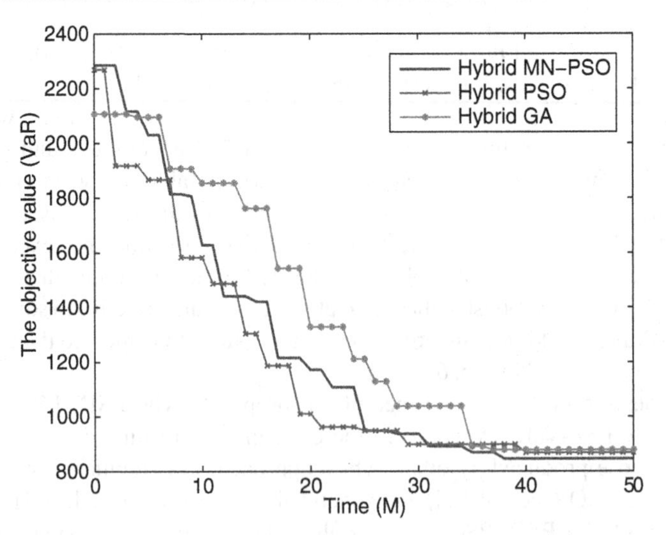

Fig. 6.7 Convergence comparison of different approaches

In Table 6.3, we compare the optimal solutions and objective values of hybrid approaches of MN-PSO, PSO, and GA. In addition, the convergence comparison in iteration cycle is provided in Fig. 6.7. From these results, we can see that the hybrid MN-PSO finds the best solution among the three approaches within the same time duration.

Chapter 7
VaR-Based Fuzzy Random Facility Location Model with Variable Capacity

In this chapter, we revisit the facility location problem. Applying the two-stage fuzzy stochastic programming with VaR (FSP-VaR) discussed in Chap. 6 to the context of facility location selection with variable capacity, we present another two-stage facility location model in the fuzzy random environment which owns a quite different structure from the location model of Chap. 5.

In Sect. 7.1, we build the model which is named two-stage VaR-based fuzzy random facility location model with variable capacity (VaR-FRFLM-VC). Different from the previously discussed recourse-based location model, the VaR-FRFLM-VC has an objective of VaR minimization and a fair of mixed binary-continuous decisions. It is the latter that makes the model structure (two-stage fuzzy stochastic mixed 0-1 programming) more difficult than the previous models.

In Sect. 7.2, we introduce a hybrid modified particle swarm optimization (MPSO) approach to the two-stage VaR-FRFLM-VC. This hybrid solution vehicle fuses a phenotype-genotype-based binary PSO mechanism that handles the search for the best binary location decision, the Nbest-Pbest-based continuous PSO mechanism (Chap. 6) that copes with searching for the best continuous capacity decision and updating genotype-location particles, mutation operators for both continuous and binary particles, and the approximate algorithm for VaR (Chap. 6) to compute the objective (VaR).

In Sect. 7.3, we conclude with numerical experiments as well as the comparisons with other approaches.

7.1 Problem Formulation

7.1.1 Mathematical Modeling

First of all, let us give a complete notation list for this facility location selection model with variable capacity though some of them have been already introduced in Chap. 5.

S. Wang and J. Watada, *Fuzzy Stochastic Optimization: Theory, Models and Applications*,
DOI 10.1007/978-1-4419-9560-5_7,

Indices and constants

i	index of facilities, $1 \leq i \leq n$
j	index of clients, $1 \leq j \leq m$
r_j	unit price charged to client j
c_i	fixed cost for opening and operating facility i
W_i	maximum capacity of each facility i
t_{ij}	unit transportation cost from i to j
$1-\beta$	confidence level of the Value-at-Risk

Fuzzy random parameters

D_j	fuzzy random demand of client j
V_i	fuzzy random unit variable operating cost of facility i
ξ	fuzzy random demand-cost vector $\xi = (D_1, \cdots, D_m, V_1, \cdots, V_n)$

Decision variables

x_i	location decision which is a binary variable
x	location decision vector which is $x = (x_1, x_2, \cdots, x_n)$
s_i	capacity decision of facility i
s	capacity decision vector which is $s = (s_1, s_2, \cdots, s_n)$
$y_{ij}^{(\omega,\gamma)}$	quantity supplied to client j from facility i at scenario (ω, γ).

The assumptions here are the same as that in Chap. 5, except for the first item on capacity.

Assumptions

1. The capacity for each facility is variable and then is also a decision to determine.
2. Each customer's demand cannot be overserved, but it is possible that the demand is not fully served.
3. The total supply from one facility to all clients cannot exceed the capacity of the facility.
4. Fuzzy random demand-cost vector $\xi = (D_1, \cdots, D_m, V_1, \cdots, V_n)$ is defined from a probability space $(\Omega, \Sigma, \Pr)$ to a collection of fuzzy vectors on possibility space $(\Gamma, \mathscr{A}, \mathrm{Pos})$.

Now, we can introduce the fuzzy random VaR criterion to the location problem and formulate a two-stage *VaR-based fuzzy random facility location model with variable capacity* (VaR-FRFLM-VC) as follows, where the VaR is at a confidence level $1-\beta$.

Model

$$\left.\begin{array}{ll}\min & \mathrm{VaR}_{1-\beta}(x,s)\\ \text{subject to} & x_i\in\{0,1\},i=1,2,\cdots,n,\\ & 0\le s_i\le W_ix_i,i=1,2,\cdots,n,\end{array}\right\}\tag{7.1}$$

where

$$\mathrm{VaR}_{1-\beta}(x,s)=\sup\left\{\lambda\mid \mathrm{Ch}\left\{\sum_{i=1}^{n}c_ix_i-\mathscr{R}(x,s,\xi)\ge\lambda\right\}\ge\beta\right\},\tag{7.2}$$

and the second-stage problem for each scenario (ω,γ) is

$$\left.\begin{array}{l}\mathscr{R}\Big(x,s,\xi(\omega,\gamma)\Big)=\max\sum_{i=1}^{n}\sum_{j=1}^{m}\big(r_j-V_i(\omega,\gamma)-t_{ij}\big)y_{ij}^{(\omega,\gamma)}\\ \qquad\text{subject to}\\ \qquad\qquad\sum_{i=1}^{n}y_{ij}^{(\omega,\gamma)}\le D_j(\omega,\gamma),j=1,2,\cdots,m,\\ \qquad\qquad\sum_{j=1}^{m}y_{ij}^{(\omega,\gamma)}\le s_ix_i,i=1,2,\cdots,n,\\ \qquad\qquad y_{ij}^{(\omega,\gamma)}\ge 0,i=1,2,\cdots,n,j=1,2,\cdots,m.\end{array}\right\}\tag{7.3}$$

The objective of this VaR-FRFLM-VC (7.1)–(7.3) is to minimize the VaR of the investment by determining the optimal locations as well as the capacities of the new facilities to open. Here in (7.2), $\sum_{i=1}^{n}c_ix_i$ represents the fixed cost at each location decision $x=(x_1,x_2,\cdots,x_n)$, and $\mathscr{R}(x,s,\xi)$ is the variable return based on the fuzzy random parameter ξ. Herewith, the $\mathrm{VaR}_{1-\beta}(x,s)$ in (7.2) represents the largest loss at the confidence $1-\beta$.

In contrast to the recourse-based location model (FR-FLM-RFC) of Chap. 5, where the first-stage decision (real decision) is singly the location x, in the VaR-FRFLM-VC, a location-capacity decision fair (x,s) is the first-stage decision (real decision), which should be made before the realizations $D_j(\omega,\gamma)$ and $V_i(\omega,\gamma)$ of the fuzzy random demand D_j and cost V_i are observed, respectively, where the scenario $(\omega,\gamma)\in\Omega\times\Gamma$.

Furthermore, we note that the objective function is

$$\begin{aligned}&\mathrm{VaR}_{1-\beta}(x,s)\\ &=\sup\left\{\lambda\mid \mathrm{Ch}\left\{\sum_{i=1}^{n}c_ix_i-\mathscr{R}(x,s,\xi)\ge\lambda\right\}\ge\beta\right\}\\ &=\sup\left\{\lambda\mid\int_{\Omega}\mathrm{Cr}\left\{\sum_{i=1}^{n}c_ix_i-\mathscr{R}\Big(x,s,\xi(\omega,\gamma)\Big)\ge\lambda\right\}\mathrm{Pr}(\mathrm{d}\omega)\ge\beta\right\},\end{aligned}\tag{7.4}$$

for each decision (x,s). Hence, in order to determine the value of objective $\mathrm{VaR}_{1-\beta}(x,s)$, we have to solve N second-stage problems (7.3) to determine the return

$$\mathscr{R}(x,s,\xi(\omega,\gamma)),$$

where N is the number of all the scenarios $(\omega,\gamma) \in \Omega \times \Gamma$ which could be infinite if the fuzzy random vector is a continuous one.

7.1.2 *Difficulties*

From the form (7.1)–(7.3), it is easy to see that the two-stage VaR-FRFLM-VC is a task of two-stage mixed 0-1 fuzzy stochastic programming problem. In addition to the difficulties that the previous models of two-stage FR-FLM-RFC and two-stage FSP-VaR have borne (no analytical expression for objective function, an infinite number of second-stage programming problems to solve to determine the objective value at each decision in a continuous fuzzy random parameter case), a key issue is how to cope with the optimal search with the mixed continuous-binary types of decision when solving the two-stage VaR-FRFLM-VC. This will be discussed in the next section.

7.2 A Hybrid Modified PSO Solution Approach

Recall that the standard particle swarm optimization (PSO) is a real-coded algorithm which only fits the continuous optimization problems, while the binary particle swarm optimization(BPSO) is an algorithm tailor-made for solving optimization problems featuring binary decision variables. A brief description for BPSO can be found in Appendix C, while for more on this subject, see [63, 77, 151, 152].

Integrating the continuous and binary types of PSO, approximation algorithm to fuzzy random VaR, and mutation operators, a hybrid modified PSO approach (see its structure in Fig. 7.1) is designed for the two-stage VaR-FRFLM-VC (7.1)–(7.3). Several modifications are useful characteristics that enhance the performance of the hybrid approach:

(a) A phenotype-genotype mechanism (see [77, 152]) in the BPSO is employed not only to enhance the searching ability of the binary particles but more importantly, phenotype-genotype mechanism makes it possible for the binary particles to update simultaneously and consistently with the continuous type of particles.
(b) The Nbest-Gbest-based update rule (Chap. 6) is utilized to improve the global search performance of the continuous type of particles (capacity and genotype-location particles) in the optimization process.

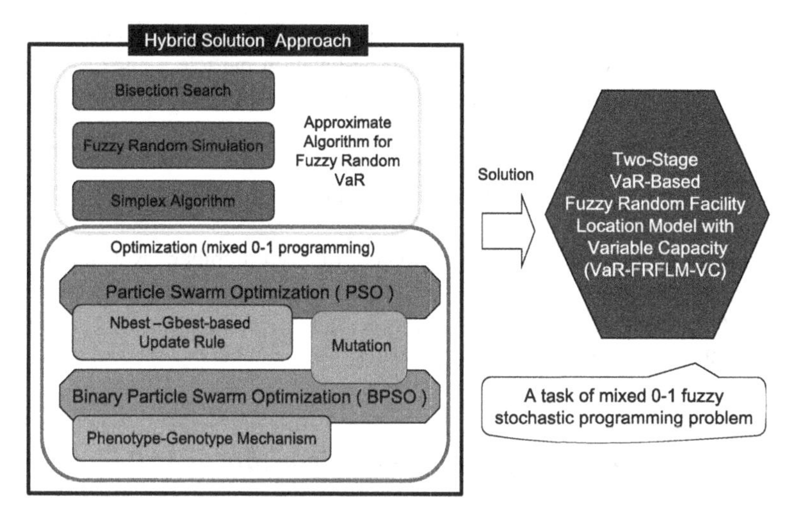

Fig. 7.1 The structure of hybrid approach for two-stage VaR-FRFLM-VC

(c) Two mutation operators are applied to the binary location particles and capacity particles to decrease the probability of their getting trapped in a local optimum so as to further enhance the search ability of the hybrid algorithm.

The detailed procedures of this hybrid MPSO algorithm arc givcn as follows.

7.2.1 Solution Representation

A mixed continuous-binary valued vector

$$(x,s) = (\langle x_1, s_1\rangle, \langle x_2, s_2\rangle, \cdots, \langle x_n, s_n\rangle)$$

is used as a particle pair to represent a solution (location-capacity) of the two-stage VaR-FRFLM-VC (7.1)–(7.3), where $x_i \in \{0,1\}, 0 \le s_i \le W_i x_i, i = 1,2,\cdots,n$.

7.2.2 Initialization

First of all, we randomly generate the initial binary phenotype location particle

$$x_p = (x_{p,1}, x_{p,2}, \cdots, x_{p,n})$$

as follows:

$$\begin{aligned}&\texttt{for}(i=1; i<=n; i++)\\&\quad \texttt{if}(\text{rand}()>0.5)\ \texttt{then}\ x_{p,i}=1;\ \texttt{else}\ x_{p,i}=0,\end{aligned} \tag{7.5}$$

where rand() is a random number coming from the uniform distribution over the interval [0,1], and initialize the genotype-location particle $x_g = x_p$. Then, we generate a capacity particle $s = (s_1, s_2, \cdots, s_n)$ by the following method:

$$\texttt{for}(i=1; i<=n; i++)\\ \quad \texttt{if}(x_{p,i}=1)\ \texttt{then}\ s_i = \text{rand}(0, W_i);\ \texttt{else}\ s_i = 0, \tag{7.6}$$

where rand(a,b) is a uniformly distributed random number over the interval [a,b]. Repeat the above process P_{size} times; we get P_{size} initial binary phenotype and genotype-location particles $x_{p,1}, x_{p,2}, \cdots, x_{p,P_{size}}; x_{g,1}, x_{g,2}, \cdots, x_{g,P_{size}}$, and P_{size} capacity particles $s_1, s_2, \cdots, s_{P_{size}}$, respectively.

7.2.3 Evaluation by Approximation Algorithm to VaR

In Chap. 6, an approximate algorithm for fuzzy random VaR and its convergence have been fully discussed. Here in the VaR-FRFLM-VC (7.1)–(7.3), we employ the approximate algorithm to estimate the objective value

$$\text{VaR}_{1-\beta}(x,s) = \sup\left\{\lambda \mid \text{Ch}\left\{\sum_{i=1}^{n} c_i x_i - \mathscr{R}(x,s,\xi) \geq \lambda\right\} \geq \beta\right\} \tag{7.7}$$

for each (x,s).

Denote $\mathbf{Fit}(\cdot)$ the fitness function, and let the fitness of each decision (x,s) be the minus of the Value-at-Risk, i.e.,

$$\mathbf{Fit}(x,s) = -\text{VaR}_{1-\beta}(x,s).$$

Therefore, the particles of smaller objective values are evaluated with higher fitness. For each (x,s), the fitness value $\mathbf{Fit}(x,s)$ is calculated by the approximate algorithm (Algorithm 6.1).

7.2.4 Update of Genotype-Location and Capacity Particles

In the update process, we first need to determine the global best particle pair (x_{Gbest}, s_{Gbest}) (with the highest fitness), where the x_{Gbest} is the best phenotype-location particle so far, and for each $(x_{p,k}, s_k)$, find the $(x_{Pbest,k}, s_{Pbest,k})$ with the highest fitness so far, where $k = 1, 2, \cdots, P_{size}$. Then, for each k, we determine the velocity vector pair $(v_{x,k}, v_{s,k})$ through the Nbest-Gbest-based update formula:

$$v_{x,k} = \mathscr{W} * v_{x,k} + c_1 * d_N(x_{p,k}) + c_2 * \text{rand}() * \left(x_{Gbest} - x_{p,k}\right), \tag{7.8}$$

$$v_{s,k} = \mathscr{W} * v_{s,k} + c_1 * d_N(s_k) + c_2 * \text{rand}() * \left(s_{Gbest} - s_k\right), \tag{7.9}$$

where $d_N(x_{p,k}), k = 1, 2, \cdots, P_{size}$ are given by

$$d_N(x_{p,1}) = \sum_{j=1}^{2} \text{rand}() * \left(\frac{x_{Pbest,j} - x_{p,1}}{2} \right), \tag{7.10}$$

$$d_N(x_{p,k}) = \sum_{j=k-1}^{k+1} \text{rand}() * \left(\frac{x_{Pbest,j} - x_{p,k}}{3} \right), k = 2, 3, \cdots, P_{size} - 1, \tag{7.11}$$

$$d_N\left(x_{p,P_{size}}\right) = \sum_{j=P_{size}-1}^{P_{size}} \text{rand}() * \left(\frac{x_{Pbest,j} - x_{p,P_{size}}}{2} \right), \tag{7.12}$$

respectively, and the $d_N(s_k), k = 1, 2, \cdots, P_{size}$ can be expressed similarly. Here, the learning rates c_1, c_2, and inertia weight $\mathscr{W}$ are set in the same way as that in Chap. 5:

$$c_1 = 2 * \frac{G_{max} - G_n}{G_{max}} + 1, \tag{7.13}$$

$$c_2 = 2 * \frac{G_n}{G_{max}} + 1, \tag{7.14}$$

$$\mathscr{W} = \frac{2}{|2 - \phi - \sqrt{\phi^2 - 4\phi}|}, \tag{7.15}$$

where G_{max} and G_n are the indexes of the maximum and current generations, respectively, and $\phi = c_1 + c_2$.

Next, each genotype-location particle $x_{g,k}$ and capacity particle s_k are updated by the following operations:

$$x_{g,k} = x_{g,k} + v_{x,k} \tag{7.16}$$

$$s_k = s_k + v_{s,k}, \tag{7.17}$$

respectively.

7.2.5 *Update of Phenotype-Location Particles and Reupdate of Capacity Particles*

All the phenotype-location particles $x_{p,k}, k = 1, 2, \cdots, P_{size}$ are updated according to the following rule:

$$\begin{aligned} &\texttt{for}(i = 1; i <= n; i++) \\ &\quad \texttt{if}(\text{rand}() < S(x_{g,ki}))\ \texttt{then}\ x_{p,ki} = 1; \texttt{else}\ x_{p,ki} = 0, \end{aligned} \tag{7.18}$$

where $x_{g,ki}$ and $x_{p,ki}$ are the components of the vectors $x_{g,k}$ and $x_{p,k}$, respectively, and $S(\cdot)$ is a sigmoid function with $S(x) = 1/1 + e^{-x}$. Furthermore, we reupdate the capacity particles s_k with the following constraint:

$$
\begin{array}{l}
\texttt{for}(i = 1; i <= n; i++) \\
\{ \\
\qquad \texttt{if}(x_{p,ki} = 0) \texttt{ then } s_{ki} = 0; \\
\quad \texttt{else} \\
\qquad\quad if(s_{ki} = 0) \texttt{ then } s_{ki} = \text{rand}(0, W_i), \\
\}
\end{array} \tag{7.19}
$$

where s_{ki} is a component of capacity particle s_k, for $k = 1, 2, \cdots, P_{size}$.

Making use of formulas (7.8)–(7.19), we yield a new generation of phenotype-location and capacity particle pairs

$$
\left(x'_{p,1}, s'_1\right), \left(x'_{p,2}, s'_2\right), \cdots, \left(x'_{p,P_{size}}, s'_{P_{size}}\right).
$$

7.2.6 *Mutation*

We predetermine two parameters $P_{m,L}, P_{m,C} \in (0, 1)$ representing the probability of mutation for the location and capacity particles, respectively. The following mutation operation is applied to all velocity vectors of location particles after the update process (7.16) of the genotype-location particles:

$$
\begin{array}{l}
\texttt{for}(k = 1; k <= P_{size}; k++) \\
\qquad \texttt{if}(\text{rand}() < P_{m,L}) \texttt{ then } v_{x,k} = -v_{x,k}.
\end{array} \tag{7.20}
$$

On the other hand, the mutation of capacity particles is implemented following the update operation (7.19). For each capacity particle $s_k = (s_{k1}, s_{k2}, \cdots, s_{kn}), k = 1, 2, \cdots, P_{size}$, if $\text{rand}() < P_{m,C}$, then we generate a number N_m between 1 and n, and mutate the capacity particle as follows:

$$
\begin{array}{l}
\texttt{for}(i = 1; i <= N_m; i++) \\
\qquad \texttt{if}(s_{ki} > 0) \texttt{ then } s_{ki} = \text{rand}(0, W_i).
\end{array} \tag{7.21}
$$

The mutation (7.21) can be shown in Fig. 7.2.

7.2.7 *Hybrid Algorithm Procedure*

The hybrid MPSO algorithm to VaR-FRFLM-VC (7.1)–(7.3) is summarized as follows (see also flowchart in Fig. 7.3).

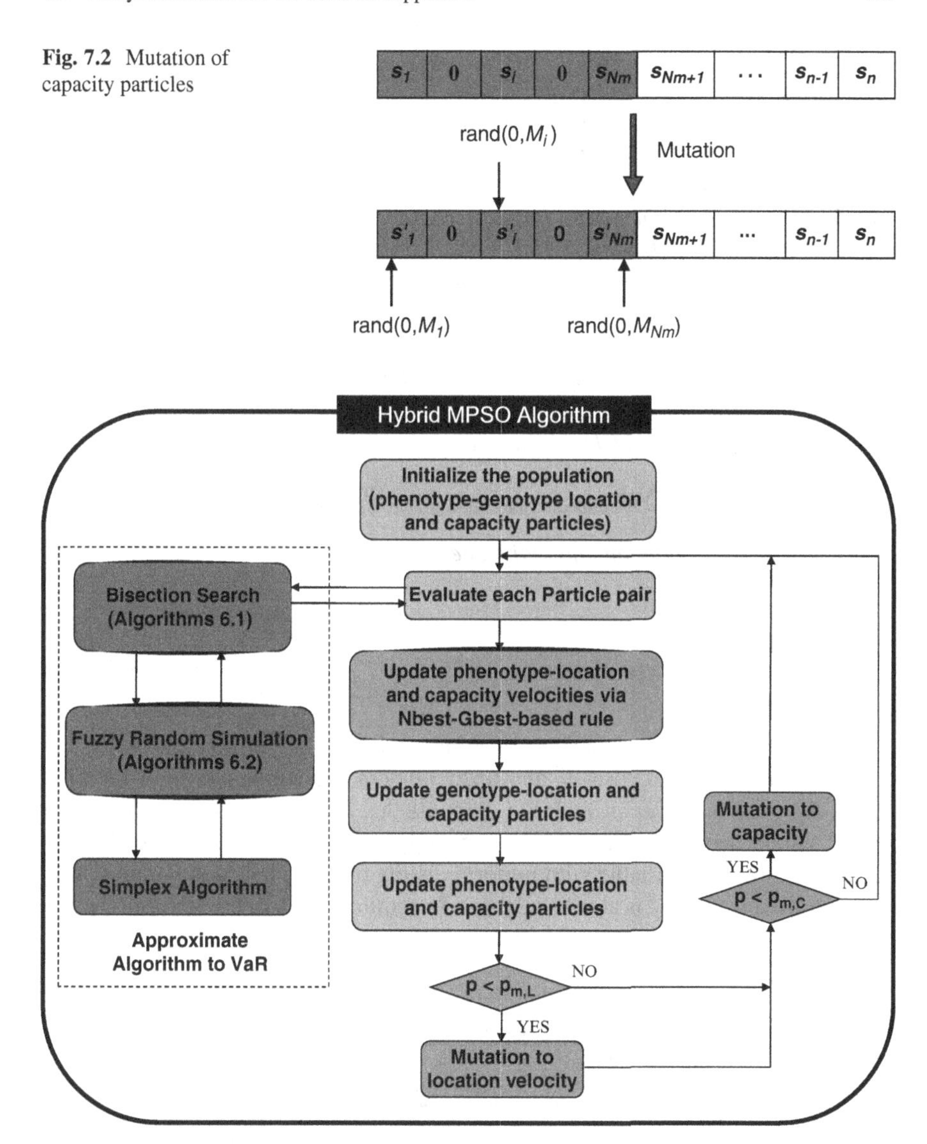

Fig. 7.2 Mutation of capacity particles

Fig. 7.3 Flowchart of the hybrid MPSO algorithm

Algorithm 7.1 (Hybrid MPSO Algorithm).

Step 1. *Initialize a population of phenotype-genotype location particles* $x_{p,k}, x_{g,k}$, *and capacity particles* s_k, *for* $k = 1, 2, \cdots, P_{size}$, *by using (7.5)–(7.6).*

Step 2. *Calculate the fitness* $\mathbf{Fit}(x_p, s)$ *for all particles through the approximate algorithm to VaR (Algorithm 6.1), and evaluate each particle pair according to the fitness;*

Step 3. *Find the global best particles x_{Gbest} (phenotype) and s_{Gbest} for the population, and determine the $d_N(x_p)$ and $d_N(s)$ for each phenotype location particle x_p and capacity particle s;*

Step 4. *Update all the genotype location and capacity particles by formulas (7.8)–(7.17);*

Step 5. *Update each phenotype location particle by (7.18), and re-update each capacity particle with (7.19).*

Step 6. *Run mutation operators (7.20) and (7.21) to each location velocity with probability $P_{m,L}$ and each capacity particle with probability $P_{m,C}$, respectively.*

Step 7. *Repeat Step 2 to Step 6 for a given number of generations;*

Step 8. *Return the particle pair (x_{Gbest}, s_{Gbest}) as the optimal solution to the VaR-FRFLM-VC (7.1)–(7.3), and*

$$\mathrm{VaR}_{1-\beta}(x_{Gbest}, s_{Gbest}) = -\mathbf{Fit}(x_{Gbest}, s_{Gbest})$$

as the corresponding optimal value.

7.3 Numerical Experiments

We consider a firm which plans to open new facilities in ten potential sites; the capacity limits W_i, fixed costs c_i, and fuzzy random operating costs V_i of the sites $i, i = 1, 2, \cdots, 10$ are given in Table 7.1. Suppose that there are five customers whose fuzzy random demands $D_j, j = 1, 2, \cdots, 5$ are given in Table 7.2, where $\mathscr{U}(a,b)$ represents a random variable with uniform distribution on $[a,b]$, and the unit price r_j charged to customer j is also listed there. In addition, the unit transportation costs $t_{ij}, i = 1, 2, \cdots, 10; j = 1, 2, \cdots, 5$ are given in Table 7.3.

Table 7.1 Capacity limits, fixed costs, and variable costs

Site i	Capacity limit W_i	Fixed cost c_i	Variable cost V_i	Parameter Y_i
1	250	8	$(7+Y_1, 9+Y_1, 10+Y_1)$	$\mathscr{U}(1,2)$
2	220	15	$(6+Y_2, 8+Y_2, 10+Y_2)$	$\mathscr{U}(2,3)$
3	300	16	$(8+Y_3, 10+Y_3, 11+Y_3)$	$\mathscr{U}(1,2)$
4	290	12	$(12+Y_4, 13+Y_4, 15+Y_4)$	$\mathscr{U}(0,1)$
5	260	6	$(13+Y_5, 15+Y_5, 16+Y_5)$	$\mathscr{U}(1,2)$
6	250	12	$(8+Y_6, 9+Y_6, 10+Y_6)$	$\mathscr{U}(0,2)$
7	320	17	$(6+Y_7, 7+Y_7, 8+Y_7)$	$\mathscr{U}(2,4)$
8	330	8	$(8+Y_8, 10+Y_8, 12+Y_8)$	$\mathscr{U}(2,3)$
9	280	9	$(13+Y_9, 15+Y_9, 16+Y_9)$	$\mathscr{U}(3,4)$
10	370	12	$(10+Y_{10}, 11+Y_{10}, 12+Y_{10})$	$\mathscr{U}(1,2)$

Table 7.2 Fuzzy random demands

Customer j	r_j	Demand D_j	Parameter Z_j
1	24	$(20+Z_1, 22+Z_1, 23+Z_1)$	$\mathscr{U}(1,2)$
2	22	$(18+Z_2, 20+Z_2, 21+Z_2)$	$\mathscr{U}(1,3)$
3	28	$(16+Z_3, 18+Z_3, 19+Z_3)$	$\mathscr{U}(2,4)$
4	26	$(22+Z_4, 23+Z_4, 24+Z_4)$	$\mathscr{U}(2,3)$
5	19	$(20+Z_5, 22+Z_5, 23+Z_5)$	$\mathscr{U}(3,4)$

Table 7.3 The values of unit transportation costs t_{ij}

t_{ij}	$j=1$	2	3	4	5
$i=1$	16	17	24	19	13
2	21	15	20	22	10
3	19	17	25	18	16
4	18	14	22	15	13
5	14	18	23	21	14
6	18	16	22	17	11
7	16	17	24	22	13
8	20	18	22	21	15
9	18	17	20	22	14
10	20	14	22	16	13

Within the above settings, we can formulate the location problem by using two-stage VaR-FRFLM-VC as

$$\left.\begin{array}{ll}\min & \mathrm{VaR}_{1-\beta}(x,s)\\ \text{subject to} & x_i\in\{0,1\}, i=1,2,\cdots,10,\\ & 0\le s_i\le M_i x_i, i=1,2,\cdots,10,\end{array}\right\}\tag{7.22}$$

where

$$\mathrm{VaR}_{1-\beta}(x,s)=\sup\left\{\lambda \mid \mathrm{Ch}\left\{\sum_{i=1}^{10} c_i x_i-\mathscr{R}(x,s,\xi)\ge\lambda\right\}\ge\beta\right\},\tag{7.23}$$

and

$$\left.\begin{array}{l}\mathscr{R}\Big(x,s,\xi(\omega,\gamma)\Big)=\max \displaystyle\sum_{i=1}^{10}\sum_{j=1}^{5}\left(r_j-t_{ij}-V_i(\omega,\gamma)\right)y_{ij}^{(\omega,\gamma)}\\ \qquad\text{subject to}\\ \qquad\qquad \displaystyle\sum_{i=1}^{10} y_{ij}^{(\omega,\gamma)}\le D_j(\omega,\gamma), j=1,2,\cdots,5,\\ \qquad\qquad \displaystyle\sum_{j=1}^{5} y_{ij}^{(\omega,\gamma)}\le s_i x_i, i=1,2,\cdots,10,\\ \qquad\qquad y_{ij}^{(\omega,\gamma)}\ge 0, i=1,2,\cdots,10, j=1,2,\cdots,5.\end{array}\right\}\tag{7.24}$$

In problem (7.22)–(7.24), the demand-cost vector

$$\xi = (D_1, \cdots, D_5, V_1, V_2, \cdots, V_{10})$$

is a continuous fuzzy random vector; the approximate algorithm (Algorithm 6.1) is therefore used to determine the value of objective function $\mathrm{VaR}_{1-\beta}(x,s)$. In the approximation, for any feasible solution (x,s), we first generate 5,000 random sample points $\widehat{\omega}_i, i = 1,2,\cdots,5{,}000$, for the random simulation (6.33). For each ω_i, we generate 1,000 fuzzy sample points $\widehat{\zeta}_l^j, j = 1,2,\ldots,1{,}000$ via discretization method (fuzzy simulation).

Based on the generated fuzzy samples, for each λ and random sample $\widehat{\omega}_i$,

$$Q_{\widehat{\omega}_i}(x,s,\lambda) = \mathrm{Cr}\left\{\sum_{i=1}^{10} c_i x_i - \mathscr{R}(x,s,\zeta_l(\widehat{\omega}_i)) \geq \lambda\right\}$$

is calculated by (6.32), where $\mathscr{R}(x,s,\zeta_l(\widehat{\omega}_i))$ (second-stage programming (7.24)) is determined by the simplex algorithm.

Furthermore, for each given λ,

$$\mathscr{Q}(x,s,\lambda) = \int_{\Omega} Q_\omega(x,s,\lambda)\,\Pr(\mathrm{d}\omega)$$

is computed by random simulation (6.33). Finally, we determine the value of $\mathrm{VaR}_{1-\beta}(x,s)$ by an iteration of varying the value of λ in a bisection search.

The hybrid MPSO algorithm (Algorithm 7.1) which contains the above approximation is run to solve the problem (7.22)–(7.24). In the hybrid MPSO algorithm, we set the population size $P_{size} = 20$, and run the algorithm with 200 generations for different confidence levels of 0.9, 0.85, and 0.8. The optimal solutions with different parameters are listed in Table 7.4, where the *relative error* is given in the last column.

It follows from Table 7.4 that the relative error does not exceed $1.10\%, 1.43\%$, and 1.90% for the different confidence levels $1-\beta = 0.9, 1-\beta = 0.85$, and $1-\beta = 0.8$, respectively, when different parameters are selected. In addition, the convergence of the best objective value at difference confidence levels is shown in Fig. 7.4. The performance implies the hybrid MBACO algorithm is robust to the parameter settings when dealing with the VaR-FRFLM-VC.

To further evaluate the performance of our proposed hybrid MPSO algorithm for the two-stage VaR-FRFLM-VC, we design some other hybrid approaches: a hybrid PSO by combining the approximate algorithm to VaR with the regular continuous-binary PSO mechanism and a hybrid GA by integrating the approximate algorithm and a GA with mixed continuous-binary variables. Then we compare the experimental results of the three different approaches.

Here, the population size is set as 20, and the maximum generation is set as 200. In addition, the best mutation and crossover rates (P_m, P_C) are selected for hybrid

Table 7.4 Results of hybrid MBACO algorithm with different parameters

	System parameters			Results		
No.	$1-\beta$	$P_{m,L}$	$P_{m,C}$	Optimal solution	Objective	Error(%)
1	0.90	0.2	0.4	$(\langle 0,0\rangle, \langle 1,220.0\rangle, \langle 0,0\rangle, \langle 1,67.4\rangle, \langle 1,185.0\rangle, \langle 0,0\rangle, \langle 0,0\rangle, \langle 0,0\rangle, \langle 1,200.3\rangle, \langle 0,0\rangle)$	−285.8	0.14
2	0.90	0.3	0.3	$(\langle 0,0\rangle, \langle 1,97.7\rangle, \langle 0,0\rangle, \langle 1,119.1\rangle, \langle 1,142.3\rangle, \langle 0,0\rangle, \langle 0,0\rangle, \langle 0,0\rangle, \langle 1,51.3\rangle, \langle 0,0\rangle)$	−284.0	0.77
3	0.90	0.4	0.2	$(\langle 0,0\rangle, \langle 1,127.0\rangle, \langle 0,0\rangle, \langle 1,204.0\rangle, \langle 1,182.4\rangle, \langle 0,0\rangle, \langle 0,0\rangle, \langle 0,0\rangle, \langle 1,234.3\rangle, \langle 0,0\rangle)$	−283.0	1.10
4	0.90	0.2	0.3	$(\langle 0,0\rangle, \langle 1,148.8\rangle, \langle 0,0\rangle, \langle 1,95.6\rangle, \langle 1,121.2\rangle, \langle 0,0\rangle, \langle 0,0\rangle, \langle 0,0\rangle, \langle 1,49.5\rangle, \langle 0,0\rangle)$	−286.2	0.00
5	0.90	0.3	0.4	$(\langle 0,0\rangle, \langle 1,218.7\rangle, \langle 0,0\rangle, \langle 1,148.0\rangle, \langle 1,188.2\rangle, \langle 0,0\rangle, \langle 0,0\rangle, \langle 0,0\rangle, \langle 1,159.7\rangle, \langle 0,0\rangle)$	−284.6	0.56
6	0.85	0.2	0.4	$(\langle 0,0\rangle, \langle 1,92.8\rangle, \langle 0,0\rangle, \langle 1,100.9\rangle, \langle 1,260.0\rangle, \langle 0,0\rangle, \langle 0,0\rangle, \langle 0,0\rangle, \langle 1,247.8\rangle, \langle 0,0\rangle)$	−294.6	0.44
7	0.85	0.3	0.3	$(\langle 0,0\rangle, \langle 1,91.7\rangle, \langle 0,0\rangle, \langle 1,46.7\rangle, \langle 1,180.2\rangle, \langle 0,0\rangle, \langle 0,0\rangle, \langle 0,0\rangle, \langle 1,147.4\rangle, \langle 0,0\rangle)$	−292.5	1.15
8	0.85	0.4	0.2	$(\langle 0,0\rangle, \langle 1,111.7\rangle, \langle 0,0\rangle, \langle 1,208.8\rangle, \langle 1,239.7\rangle, \langle 0,0\rangle, \langle 0,0\rangle, \langle 0,0\rangle, \langle 1,177.6\rangle, \langle 0,0\rangle)$	−295.9	0.00
9	0.85	0.2	0.3	$(\langle 0,0\rangle, \langle 1,138.4\rangle, \langle 0,0\rangle, \langle 1,175.3\rangle, \langle 1,54.6\rangle, \langle 0,0\rangle, \langle 0,0\rangle, \langle 0,0\rangle, \langle 1,91.8\rangle, \langle 0,0\rangle)$	−291.7	1.43
10	0.85	0.3	0.4	$(\langle 0,0\rangle, \langle 1,187.4\rangle, \langle 0,0\rangle, \langle 1,147.5\rangle, \langle 1,138.9\rangle, \langle 0,0\rangle, \langle 0,0\rangle, \langle 0,0\rangle, \langle 1,186.8\rangle, \langle 0,0\rangle)$	−292.6	1.14
11	0.80	0.2	0.4	$(\langle 0,0\rangle, \langle 1,172.7\rangle, \langle 0,0\rangle, \langle 1,144.0\rangle, \langle 1,199.3\rangle, \langle 0,0\rangle, \langle 0,0\rangle, \langle 0,0\rangle, \langle 1,105.4\rangle, \langle 0,0\rangle)$	−310.2	0.00
12	0.80	0.3	0.3	$(\langle 0,0\rangle, \langle 1,215.7\rangle, \langle 0,0\rangle, \langle 1,167.1\rangle, \langle 1,173.9\rangle, \langle 0,0\rangle, \langle 0,0\rangle, \langle 0,0\rangle, \langle 1,54.6\rangle, \langle 0,0\rangle)$	−308.8	0.45
13	0.80	0.4	0.2	$(\langle 0,0\rangle, \langle 1,49.3\rangle, \langle 0,0\rangle, \langle 1,266.0\rangle, \langle 1,250.7\rangle, \langle 0,0\rangle, \langle 0,0\rangle, \langle 0,0\rangle, \langle 1,241.5\rangle, \langle 0,0\rangle)$	−304.0	1.90
14	0.80	0.2	0.3	$(\langle 0,0\rangle, \langle 1,160.3\rangle, \langle 0,0\rangle, \langle 1,183.1\rangle, \langle 1,260.2\rangle, \langle 0,0\rangle, \langle 0,0\rangle, \langle 0,0\rangle, \langle 1,260.0\rangle, \langle 0,0\rangle)$	−309.0	0.39
15	0.80	0.3	0.4	$(\langle 0,0\rangle, \langle 1,212.3\rangle, \langle 0,0\rangle, \langle 1,287.3\rangle, \langle 1,145.6\rangle, \langle 0,0\rangle, \langle 0,0\rangle, \langle 0,0\rangle, \langle 1,160.2\rangle, \langle 0,0\rangle)$	−306.3	1.26

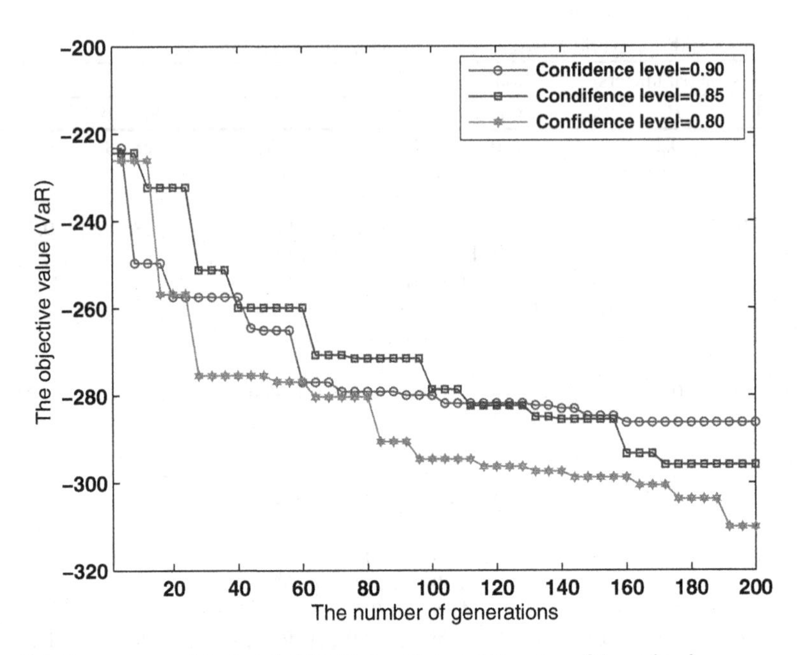

Fig. 7.4 Convergence of the hybrid MPSO algorithm at different confidence levels

Table 7.5 The comparison results of different approaches

Approach	Optimal solution	Objective value
$1-\beta=0.9$		
Hybrid MPSO	$(\langle 0,0\rangle, \langle 1,148.8\rangle, \langle 0,0\rangle, \langle 1,95.6\rangle, \langle 1,121.2\rangle, \langle 0,0\rangle, \langle 0,0\rangle, \langle 0,0\rangle, \langle 1,49.5\rangle, \langle 0,0\rangle)$	−286.2
Hybrid PSO	$(\langle 0,0\rangle, \langle 1,160.3\rangle, \langle 0,0\rangle, \langle 1,100.2\rangle, \langle 1,180.7\rangle, \langle 0,0\rangle, \langle 0,0\rangle, \langle 0,0\rangle, \langle 1,279.8\rangle, \langle 0,0\rangle)$	−280.1
Hybrid GA	$(\langle 0,0\rangle, \langle 1,173.5\rangle, \langle 0,0\rangle, \langle 1,199.5\rangle, \langle 1,111.3\rangle, \langle 0,0\rangle, \langle 0,0\rangle, \langle 0,0\rangle, \langle 1,145.2\rangle, \langle 1,102.2\rangle)$	−276.4
$1-\beta=0.85$		
Hybrid MPSO	$(\langle 0,0\rangle, \langle 1,138.4\rangle, \langle 0,0\rangle, \langle 1,175.3\rangle, \langle 1,54.6\rangle, \langle 0,0\rangle, \langle 0,0\rangle, \langle 0,0\rangle, \langle 1,91.8\rangle, \langle 0,0\rangle)$	−295.9
Hybrid PSO	$(\langle 0,0\rangle, \langle 1,125.1\rangle, \langle 0,0\rangle, \langle 1,200.0\rangle, \langle 1,180.0\rangle, \langle 0,0\rangle, \langle 0,0\rangle, \langle 0,0\rangle, \langle 1,160.2\rangle, \langle 0,0\rangle)$	−285.1
Hybrid GA	$(\langle 0,0\rangle, \langle 1,163.5\rangle, \langle 0,0\rangle, \langle 1,186.6\rangle, \langle 1,86.5\rangle, \langle 0,0\rangle, \langle 0,0\rangle, \langle 0,0\rangle, \langle 1,255.8\rangle, \langle 1,74.5\rangle)$	−283.6
$1-\beta=0.8$		
Hybrid MPSO	$(\langle 0,0\rangle, \langle 1,172.7\rangle, \langle 0,0\rangle, \langle 1,144.0\rangle, \langle 1,199.3\rangle, \langle 0,0\rangle, \langle 0,0\rangle, \langle 0,0\rangle, \langle 1,105.4\rangle, \langle 0,0\rangle)$	−310.2
Hybrid PSO	$(\langle 0,0\rangle, \langle 1,160.1\rangle, \langle 0,0\rangle, \langle 1,183.3\rangle, \langle 1,259.7\rangle, \langle 0,0\rangle, \langle 0,0\rangle, \langle 0,0\rangle, \langle 1,160.0\rangle, \langle 0,0\rangle)$	−299.4
Hybrid GA	$(\langle 0,0\rangle, \langle 1,129.1\rangle, \langle 0,0\rangle, \langle 1,232.7\rangle, \langle 1,171.7\rangle, \langle 0,0\rangle, \langle 0,0\rangle, \langle 0,0\rangle, \langle 1,127.5\rangle, \langle 0,0\rangle)$	−299.6

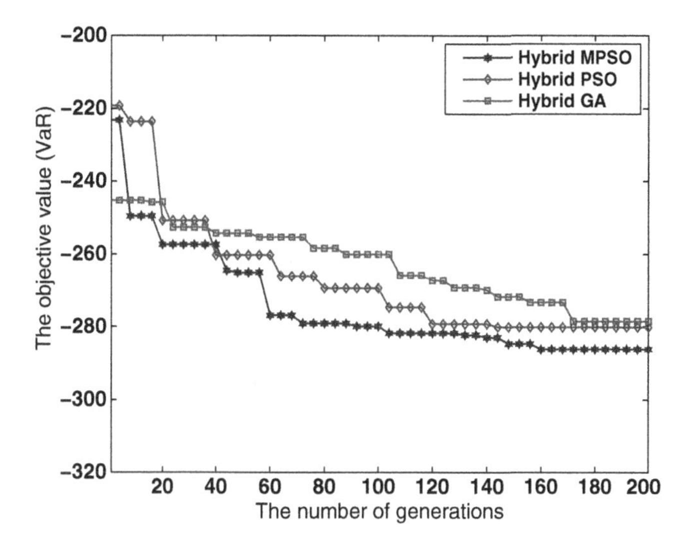

Fig. 7.5 The convergence comparison of different approaches when $1-\beta=0.90$

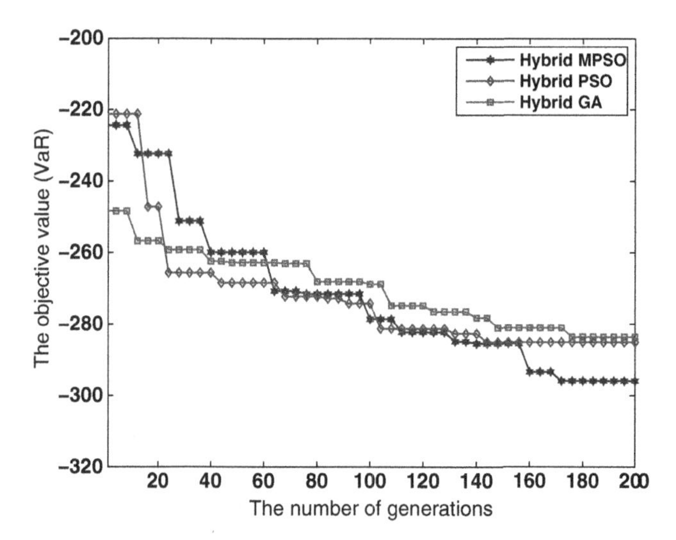

Fig. 7.6 The convergence comparison of different approaches when $1-\beta=0.85$

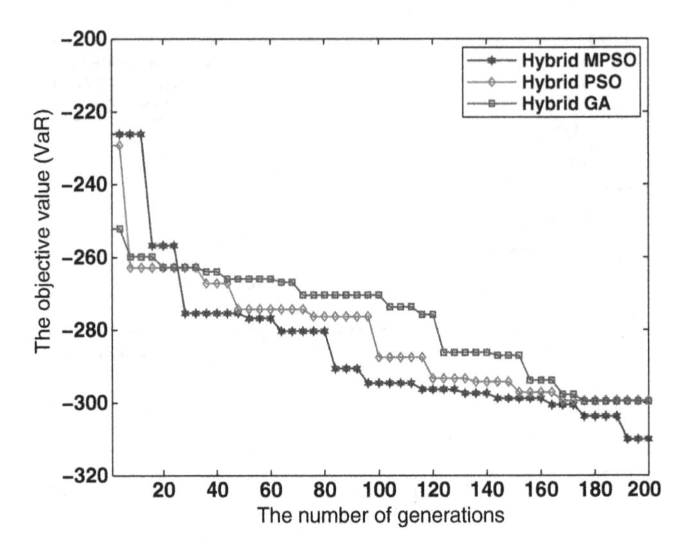

Fig. 7.7 The convergence comparison of different approaches when $1-\beta = 0.80$

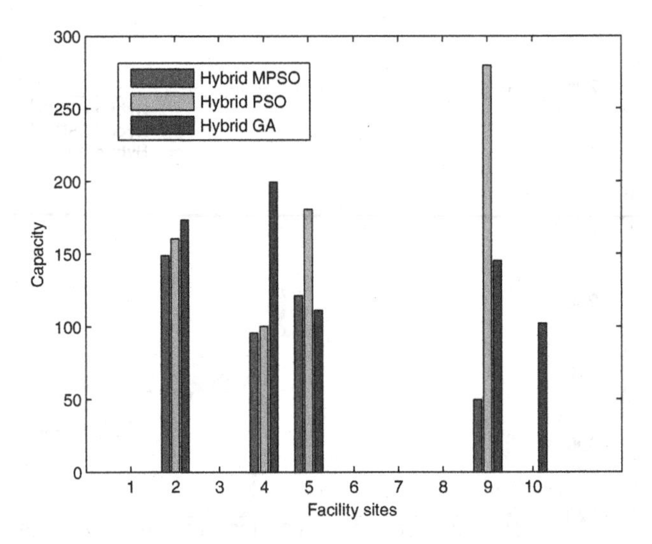

Fig. 7.8 Location-capacity comparison of different approaches when $1-\beta = 0.90$

GA after a number of experiments: $(P_m, P_C) = (0.3, 0.3)$ for confidence level of 0.9, $(P_m, P_C) = (0.4, 0.2)$ for confidence level of 0.85, and $(P_m, P_C) = (0.4, 0.2)$ for confidence level of 0.80. On the other hand, in the PSO approaches, the learning rates are set as the same values as in (7.13)–(7.14).

Applying the different approaches to the problem (7.22)–(7.24), we obtain the comparison results in Table 7.5. Furthermore, the convergence comparison is provided in Figs. 7.5–7.7, from which we can see that the performance of the hybrid

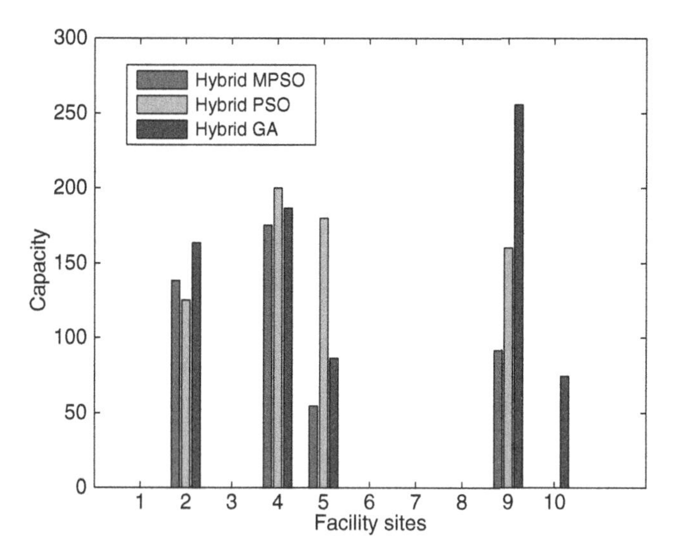

Fig. 7.9 Location-capacity comparison of different approaches when $1-\beta = 0.85$

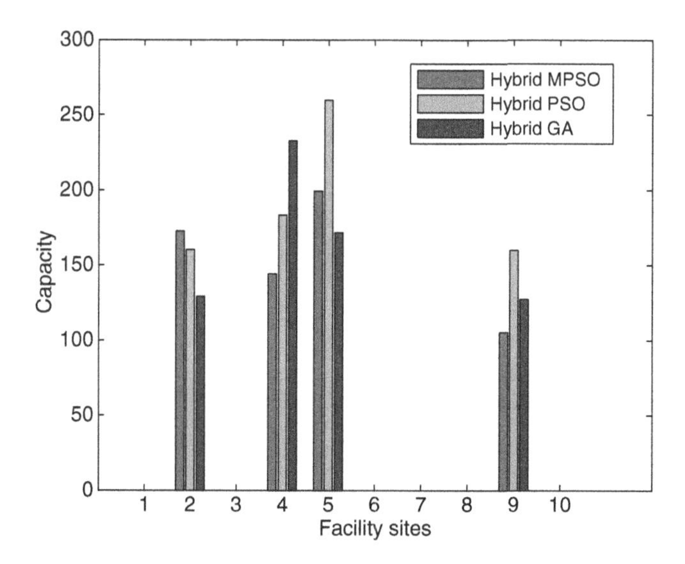

Fig. 7.10 Location-capacity comparison of different approaches when $1-\beta = 0.80$

MPSO and PSO is much better (they descend much faster) than the hybrid GA in the early iterations, and the hybrid MPSO outperforms the other two algorithms throughout all the generations. In addition, the location-capacity comparison is provided in Figs. 7.8–7.10. From all the above comparison results, we can see the hybrid MPSO algorithm outperforms the other approaches when dealing with the two-stage VaR-FRFLM-VC.

Part III
Real-Life Applications

Chapter 8
Case Study I: Dam Control System Design

In this chapter, as a real-life application of the fuzzy random system reliability models in Chap. 4, we present a case study on a dam control system design problem.

In Sect. 8.1, we introduce the background of this real-life problem.

In Sect. 8.2, we discuss a key issue in the whole problem-solving process: how to determine the lifetime distribution for each component. It shows that in a real-life situation, since the component lifetime has a nature of long lifetime and being vulnerable to ambient environment, the manufacturer is prone to provide the lifetime data in terms of estimation other than a precise prediction which is not easy to obtain. Hence, to determine a realistic and reliable lifetime distribution for each component, we analyze the lifetime by considering not only the testing-based statistic data but also the estimation of the product specialists, and fuse both into the lifetime distribution. Furthermore, the formed distributions are modified into more realistic forms by dam experts considering the high humidity surrounding the system which further shortens the component lifetimes. The determined lifetime distributions are intrinsically fuzzy random variables.

In Sect. 8.3, we apply the fuzzy random redundancy allocation models (FR-RAMs I and II, Chap. 4) to solve the dam control system design problem. In addition, a comparison is presented between the solutions considering and ignoring the fuzziness in the lifetime distributions.

8.1 Problem Background

We consider a dam control system which is located across the Carrizal river of Villahermosa City in the southeast of Mexico. This control system is used to manage the water flow into the river under water discharging conditions: diary water consumption and avenues. The system is principally composed of a control system, a hydraulic system, and three radial gates in which the control system measures, monitors, and regulates the water level and consumptions as well as positions of the gates.

S. Wang and J. Watada, *Fuzzy Stochastic Optimization: Theory, Models and Applications*, DOI 10.1007/978-1-4419-9560-5_8, © Springer Science+Business Media New York 2012

Table 8.1 The components mounted in each control subsystem

i	Name	Function
1	CJ1W-PA202	Power supply
2	CJ1M-CPU11	CPU
3	S8VS-24024	Power supply
4	CJ1W-PH41U	Analog input module
5	CJ1W-DA041	Analog output module
6	CJ1W-ID211	Digital input module
7	CJ1W-OD212	Digital output module
8	CJ1W-OD212	Digital output module
9	CJ1W-PRT21	PROFIBUS communication module

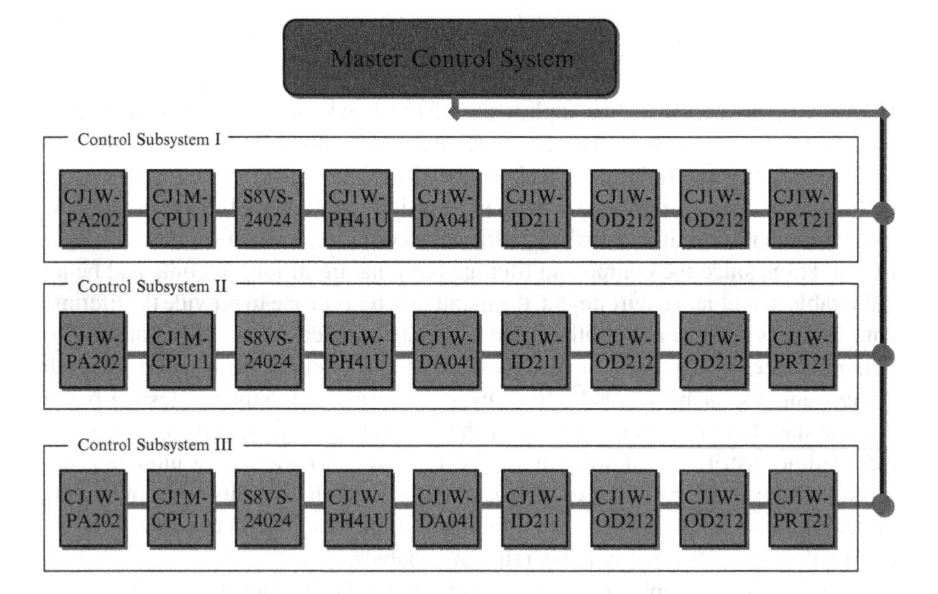

Fig. 8.1 The functional structure of the dam control system

The system has a PC and a master PLC (programmable logic controller) with three identical slave PLCs that control the three gates, respectively. Each slave PLC is a control subsystem which contains components, as shown in Table 8.1. Each component is critical and indispensable to subsystem, and the malfunction of any component may interrupt the data communication and hence bog down the control subsection. The functional structure of the dam control system is as in Fig. 8.1.

The automation company that installed and programmed the control system is now considering to reinstall the dam control system by establishing a redundancy system in order to further enhance its reliability. Noting that the three control subsystems are identical, therefore our problem is to design an optimal redundancy allocation to the subsystem.

8.2 Determining Lifetime Distributions

Since the components are equipped with aluminum electrolytic capacitors inside whose lifetimes are very sensitive to the ambient environment (temperature, humidity, etc.), the component lifetime varies accordingly depending on the temperature in the working environment. Due to this nature, it is very difficult (almost impossible) to obtain a precise prediction for the lifetimes of those devices. The following Table 8.2 shows the lifetime information of components at the temperature of 40°C which are some estimations provided by the Product Preventive Maintenance Report.

The internal temperature of the control system has been measured which varies from 38°C to 42°C. According to the Product Preventive Maintenance Report, making use of the relation between the lifetime and the temperature (Arrhenius equation, see [74]), the lifetime values can be calculated with respect to different temperatures in [38°C, 42°C].

Thus, the components mounted in each control subsystem are categorized into three groups: power supply I (CJ1W-PA202), power supply II (S8VS-24024), and CPU-communication-I/O units (CJ1M-CPU11, CJ1W-PRT21, CJ1W-PH41U, CJ1W-DA041, CJ1W-ID211, and CJ1W-OD212), in accordance with their lifetimes. Particularly, Figs. 8.2–8.4 give the change trend in lifetime of power supplies and the CPU-communication-I/O units depending on the changes in the ambient temperature. From the lifetime distribution structures in these figures, it is easy to see that the lifetime distributions are intrinsically distributions of fuzzy random variables: the temperature changes stochastically, and in each random scenario (temperature), the lifetime has an imprecise value which is a fuzzy number.

In addition to the temperature, as explained by the product manual, some other factors such as humidity and sunlight also account for the acceleration of lifetime reduce. Since the control system is set inside the dam to which the ambient humidity is much higher than the normal level, hence, the experts in charge of maintaining the system proposed to further modify the lifetime distributions of the components by taking into account the high-humidity factor. The following Figs. 8.5–8.7 provide the lifetime modifications by the dam experts for the power supplies and CPU-communication-I/O units, respectively, at the temperature of 40°C.

Table 8.2 Lifetime data for each component at 40°C

i	Name	Lifetime (years)	Temperature (°C)
1	CJ1W-PA202	8 min., 10 max.	40
2	CJ1M-CPU11	Approximately 10	40
3	S8VS-24024	10 years min., 12 max.	40
4	CJ1W-PH41U	Approximately 10	40
5	CJ1W-DA041	Approximately 10	40
6	CJ1W-ID211	Approximately 10	40
7	CJ1W-OD212	Approximately 10	40
8	CJ1W-OD212	Approximately 10	40
9	CJ1W-PRT21	Approximately 10	40

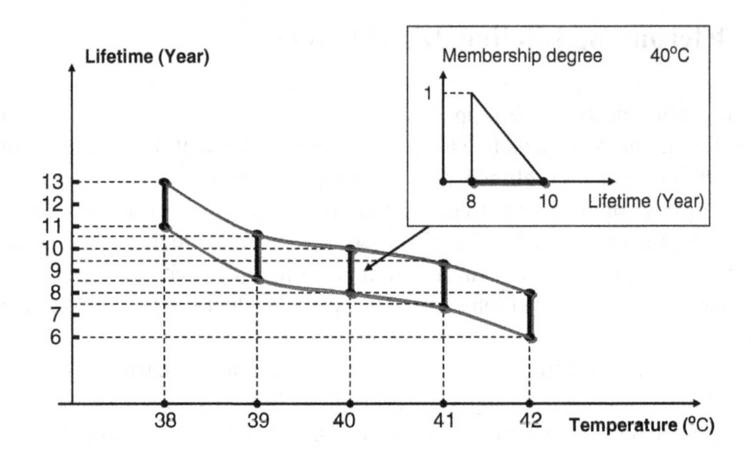

Fig. 8.2 The lifetime changing trend of CJ1W-PA202 depending on temperature

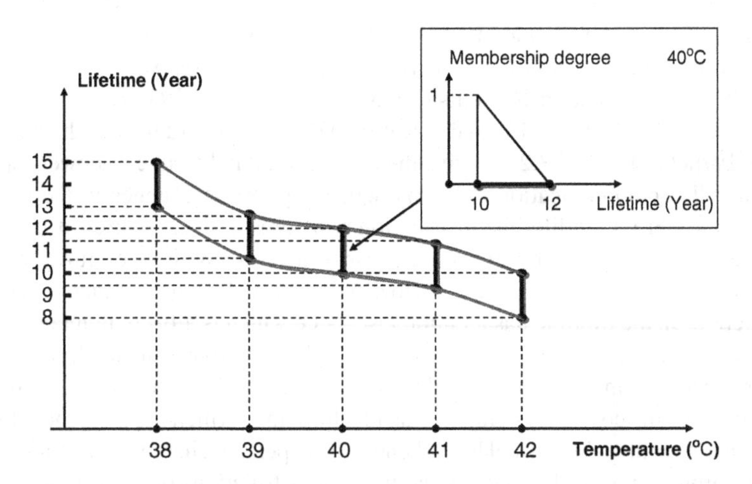

Fig. 8.3 The lifetime changing trend of S8VS-24024 depending on temperature

After the modification by incorporating dam expert knowledge, we treat the temperature $\mathscr{T}$ as a random variable with a uniform distribution $\mathscr{T} \sim \mathscr{U}(38,42)$, and the final lifetimes of all the components along the temperature interval [38°C, 42°C] are given in terms of following three types of fuzzy probabilistic distributions:

1. For power supply CJ1W-PA202 (lifetime ξ_1):

$$\xi_1 = \begin{cases} (105 - 2.5\mathscr{T}, 107 - 2.5\mathscr{T})_{\triangleright}, & \mathscr{T} \in [38,39] \\ (27 - 0.5\mathscr{T}, 29 - 0.5\mathscr{T})_{\triangleright}, & \mathscr{T} \in [39,41] \\ (68 - 1.5\mathscr{T}, 69 - 1.5\mathscr{T})_{\triangleright}, & \mathscr{T} \in [41,42], \end{cases} \tag{8.1}$$

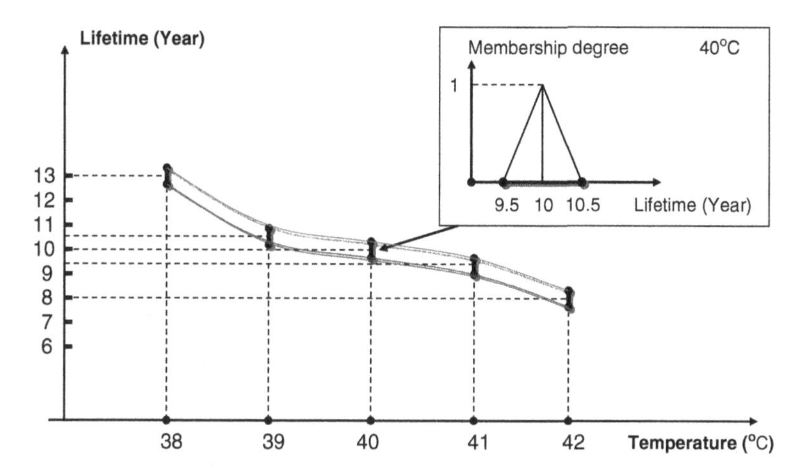

Fig. 8.4 The lifetime changing trend of CPU-communication-I/O units depending on temperature

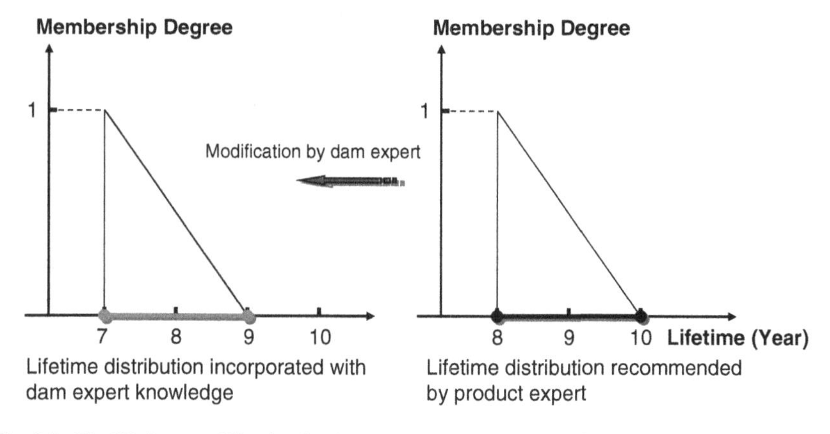

Fig. 8.5 The lifetime modification by dam experts for CJ1W-PA202 at 40°C

where $(a,b)_{\triangleright} = (a,a,b)$ is a semitriangular fuzzy number with the following membership degree:

$$\mu_{(a,b)_{\triangleright}}(x) = \frac{b-x}{b-a}, \quad x \in [a,b].$$

2. For power supply S8VS-24024 (lifetime ξ_3):

$$\xi_3 = \begin{cases} (107 - 2.5\mathscr{T}, 109 - 2.5\mathscr{T})_{\triangleright}, & \mathscr{T} \in [38,39] \\ (29 - 0.5\mathscr{T}, 31 - 0.5\mathscr{T})_{\triangleright}, & \mathscr{T} \in [39,41] \\ (70 - 1.5\mathscr{T}, 72 - 1.5\mathscr{T})_{\triangleright}, & \mathscr{T} \in [41,42]. \end{cases} \tag{8.2}$$

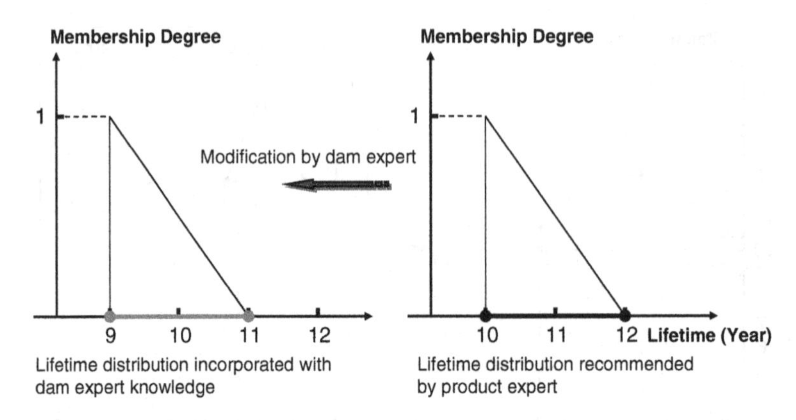

Fig. 8.6 The lifetime modification by dam experts for S8VS-24024 at 40°C

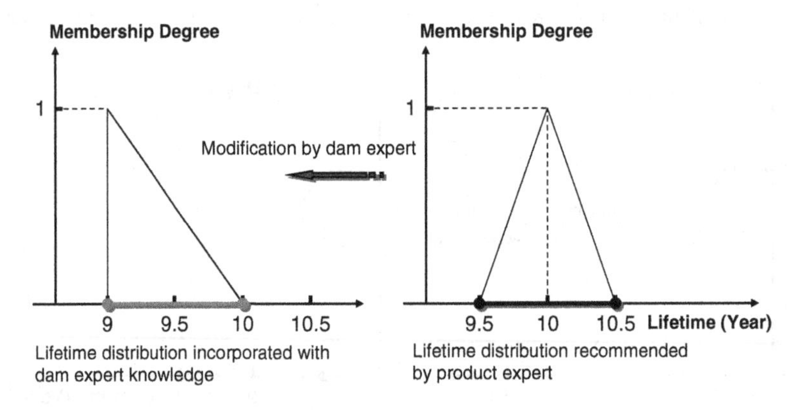

Fig. 8.7 The lifetime modification by dam experts for CPU-communication-I/O units at 40°C

3. For CPU-communication-I/O units (lifetime $\xi_i, i = 2,4,\cdots,9$):

$$\xi_i = \begin{cases} (107 - 2.5\mathscr{T}, 108 - 2.5\mathscr{T})_{\triangleright}, & \mathscr{T} \in [38,39] \\ (29 - 0.5\mathscr{T}, 30 - 0.5\mathscr{T})_{\triangleright}, & \mathscr{T} \in [39,41] \\ (70 - 1.5\mathscr{T}, 71 - 1.5\mathscr{T})_{\triangleright}, & \mathscr{T} \in [41,42]. \end{cases} \tag{8.3}$$

8.3 Modeling and Solution

Based on the above description, the fuzzy random variable is a tailor-made mathematical tool to characterize the lifetime distributions which carry twofold uncertainty information, i.e., stochastic variability (temperature fluctuation) and expert knowledge (lifetime estimation and further modification).

Table 8.3 The price of each component

i	Component	Price
1	CJ1W-PA202	127.2 $
2	CJ1M-CPU11	355 $
3	S8VS-24024	343 $
4	CJ1W-PH41U	1143.9 $
5	CJ1W-DA041	681.5 $
6	CJ1W-ID211	169.1 $
7	CJ1W-OD212	179.1 $
8	CJ1W-OD212	179.1 $
9	CJ1W-PRT21	1009.4 $

Furthermore, recall the identical structures of the three control subsystems which work independently to control the three different gates owing equal importance to the control system; they can be treated separately as three identical series system. Since a series system is essentially a special case of the parallel-series system, the system reliability optimization models discussed in Chap. 4 are able to be applied here to handle this problem.

Other system parameters are provided by the automation company as follows: prices of each component are listed in Table 8.3, maximum capital $c^0 = 13{,}000$ \$, and lower and upper bounds for the component number $l = 9, u = 40$. We use decision variable x_i to represent the number of redundant elements for each component i for $i = 1, 2, \cdots, 9$, and denote $x = (x_1, x_2, \cdots, x_9)$ as the decision vector. Maximizing the reliability by FR-RAM I, the dam control system design problem is formulated as

$$\left.\begin{array}{l}\max R_{t^0}(x) = \displaystyle\int_{\Omega} \mathrm{Cr}\left\{\bigwedge_{i=1}^{9}\left(\sum_{j=1}^{x_i}\xi_{ij}(\mathscr{T})\right) \geq t^0\right\}\Pr(\mathrm{d}\mathscr{T}) \\ \text{subject to} \\ \qquad 127.2x_1 + 355x_2 + 343x_3 + 1143.9x_4 + 681.5x_5 + 169.1x_6 \\ \qquad\quad +179.1(x_7 + x_8) + 1009.4x_9 \leq 13{,}000, \\ \qquad 9 \leq \displaystyle\sum_{i=1}^{9} x_i \leq 40, \\ \qquad x_i \in \mathbb{N}, \text{for } i = 1, \cdots, 9.\end{array}\right\} \qquad (8.4)$$

Alternatively, if we consider minimizing the cost as the objective, the problem can be presented as a cost minimization model (FR-RAM II) meeting a reliability r^0 as follows:

$$\left.\begin{array}{l}\min\ 127.2x_1+355x_2+343x_3+1143.9x_4+681.5x_5+169.1x_6\\ \qquad +179.1(x_7+x_8)+1009.4x_9\\ \text{subject to}\\ \qquad R_{t^0}(x)=\int_\Omega \mathrm{Cr}\left\{\bigwedge_{i=1}^{9}\left(\sum_{j=1}^{x_i}\xi_{ij}(\mathscr{T})\right)\geq t^0\right\}\Pr(\mathrm{d}\mathscr{T})\geq r^0\\ \qquad 9\leq\sum_{i=1}^{9}x_i\leq 40,\\ \qquad x_i\in\mathbb{N}, \text{for } i=1,\cdots,9.\end{array}\right\}\tag{8.5}$$

Noting that each lifetime $\xi_i(\mathscr{T})$ at any random realization $\mathscr{T}$ is a semitriangular fuzzy number, it has a convex distribution. Hence, making use of Theorem 4.1 in Chap. 4, the reliability $R_{t^0}(x)$ for each $x=(x_1,x_2,\cdots,x_9)$ can be calculated following the procedure given below.

First, we can write the subsystem lifetime as

$$\begin{aligned}\bigwedge_{i=1}^{9}\left(\sum_{j=1}^{x_i}\xi_{ij}(\mathscr{T})\right) &= x_1(105-2.5\mathscr{T},\,107-2.5\mathscr{T})_\triangleright\\ &\bigwedge x_3(107-2.5\mathscr{T},\,109-2.5\mathscr{T})_\triangleright\\ &\bigwedge\left(\min_{2\leq i\leq 9, i\neq 3}\{x_i\}\right)\cdot(107-2.5\mathscr{T},\,108-2.5\mathscr{T})_\triangleright,\end{aligned}\tag{8.6}$$

when $\mathscr{T}\in[38,39]$,

$$\begin{aligned}\bigwedge_{i=1}^{9}\left(\sum_{j=1}^{x_i}\xi_{ij}(\mathscr{T})\right) &= x_1(27-0.5\mathscr{T},\,29-0.5\mathscr{T})_\triangleright\\ &\bigwedge x_3(29-0.5\mathscr{T},\,31-0.5\mathscr{T})_\triangleright\\ &\bigwedge\left(\min_{2\leq i\leq 9, i\neq 3}\{x_i\}\right)\cdot(29-0.5\mathscr{T},\,30-0.5\mathscr{T})_\triangleright,\end{aligned}\tag{8.7}$$

when $\mathscr{T}\in[39,41]$, and

$$\begin{aligned}\bigwedge_{i=1}^{9}\left(\sum_{j=1}^{x_i}\xi_{ij}(\mathscr{T})\right) &= x_1(68-1.5\mathscr{T},\,70-1.5\mathscr{T})_\triangleright\\ &\bigwedge x_3(70-1.5\mathscr{T},\,72-1.5\mathscr{T})_\triangleright\\ &\bigwedge\left(\min_{2\leq i\leq 9, i\neq 3}\{x_i\}\right)\cdot(70-1.5\mathscr{T},\,71-1.5\mathscr{T})_\triangleright,\end{aligned}\tag{8.8}$$

when $\mathscr{T}\in[41,42]$, respectively.

Then, we denote $\mu_{[38,39]}(t)$ as the membership function of

$$\bigwedge_{i=1}^{9}\left(\sum_{j=1}^{x_i}\xi_{ij}(\mathscr{T})\right)$$

for $\mathscr{T}\in[38,39]$, and $\mu_{I,[38,39]}(t)$, $\mu_{II,[38,39]}(t)$, and $\mu_{III,[38,39]}(t)$ as the membership functions of

$$x_1(105-2.5\mathscr{T},\,107-2.5\mathscr{T})_{\triangleright},$$
$$x_3(107-2.5\mathscr{T},\,109-2.5\mathscr{T})_{\triangleright},$$

and

$$\left(\min_{2\le i\le 9, i\ne 3}\{x_i\}\right)\cdot(107-2.5\mathscr{T},\,108-2.5\mathscr{T})_{\triangleright},$$

respectively, and with the corresponding peak points (modes)

$$m_{I,[38,39]}=x_1(105-2.5\mathscr{T}),$$
$$m_{II,[38,39]}=x_3(107-2.5\mathscr{T}),$$

and

$$m_{III,[38,39]}=\left(\min_{2\le i\le 9, i\ne 3}\{x_i\}\right)\cdot(107-2.5\mathscr{T}),$$

respectively, when $\mathscr{T}\in[38,39]$.

Next, from (4.5) to (4.7) in Chap. 4, the membership function $\mu_{[38,39]}(t)$ for $\mathscr{T}\in[38,39]$ can be determined by

$$\mu_{[38,39]}(t)=\begin{cases}0, & t<m_{\min}\\ \mu_{\min}(t), & m_{\min}\le t<m_{\mathrm{mid}}\\ \min\left\{\mu_{\min}(t),\mu_{\mathrm{mid}}(t)\right\}, & m_{\mathrm{mid}}\le t<m_{\max}\\ \min\left\{\mu_{\min}(t),\mu_{\mathrm{mid}}(t),\mu_{\max}(t)\right\}, & t\ge m_{\max},\end{cases}\tag{8.9}$$

where $\mu_{\min},\mu_{\max}$, and μ_{mid} are designated to be the corresponding membership functions from $\mu_{\min},\mu_{\mathrm{mid}}$, and $\mu_{\max}$ in accordance with their peak points

$$m_{\min}=\min\left\{m_{I,[38,39]},m_{II,[38,39]},m_{III,[38,39]}\right\},$$
$$m_{\max}=\max\left\{m_{I,[38,39]},m_{II,[38,39]},m_{III,[38,39]}\right\},$$

and

$$m_{\mathrm{mid}}=\left\{m_{I,[38,39]},m_{II,[38,39]},m_{III,[38,39]}\right\}\setminus\left\{m_{\min},m_{\max}\right\}.$$

Table 8.4 Solution results for problem (8.4)

t^0 (Year)	Optimal solution	Reliability	Cost
20	(4,3,3,3,3,3,3,3,3)	0.983	12689.1 $
21	(5,3,3,3,3,3,3,3,3)	0.981	12816.3 $
22	(5,3,3,3,3,3,3,3,3)	0.948	12816.3 $
23	(4,3,3,3,3,3,3,3,3)	0.902	12689.1 $
24	(4,3,3,3,3,3,3,3,3)	0.856	12689.1 $
25	(4,3,3,3,3,3,3,3,3)	0.806	12689.1 $

Likewise, the membership functions $\mu_{[39,41]}(t)$ and $\mu_{[41,42]}(t)$ for the cases that $\mathscr{T} \in [39,41]$ and $\mathscr{T} \in [41,42]$, respectively, can be also determined.

Furthermore, for any $\mathscr{T} \in [38,42]$, we have

$$\mathrm{Cr}\left\{\bigwedge_{i=1}^{9}\left(\sum_{j=1}^{x_i}\xi_{ij}(\mathscr{T})\right) \geq 25\right\} = \begin{cases} \dfrac{1-\mu_{\mathscr{T}}(25)}{2}, & 25 \leq m_{\min}^{\mathscr{T}} \\ \dfrac{\mu_{\mathscr{T}}(25)}{2}, & \text{otherwise,} \end{cases} \tag{8.10}$$

where

$$\mu_{\mathscr{T}} = \begin{cases} \mu_{[38,39]}, & \mathscr{T} \in [38,39] \\ \mu_{[39,41]}, & \mathscr{T} \in [39,41] \\ \mu_{[41,42]}, & \mathscr{T} \in [41,42], \end{cases}$$

and $m_{\min}^{\mathscr{T}}$ is the corresponding $m_{\min}$ when random $\mathscr{T}$ falls into the different intervals of $[38,39]$, $[39,41]$, and $[41,42]$.

Finally, the reliability $R_{t^0}(x)$ at any redundancy allocation x can be computed by using stochastic simulation (refer to Algorithm 4.1 in Chap. 4).

Now, we solve problems (8.4) and (8.5) without much difficulty and obtain the following solutions in Tables 8.4 and 8.5.

As a further discussion, let us analyze the value loss in reliability and in cost if we do not consider the fuzzy information which does exist stubbornly inside the distribution, but rather, use some sort of simplified precise lifetime distributions in the modeling process.

From the description in Sect. 8.2, the imprecise or fuzzy information can be categorized into two parts: the fuzzy or vague value of lifetime estimated by the product experts, and the lifetime modifications by dam experts. Hence, the discussion goes along with the following two cases.

Case A. If we do not consider any fuzzy information in the first place, which means we take into account neither the fuzzy estimation by the product experts nor the lifetime modification by the dam experts, then, the real imprecise lifetime distribution of each component will be simplified to a precise one by assigning certain "reasonable" crisp values to replace the fuzzy realizations.

Apparently, no other values among all the possible realizations in each stochastic scenario could be more "reasonable" than the one with membership degree of 1, so

Table 8.5 Solution results for problem (8.5)

t^0 (Year)	r^0	Optimal solution	Cost	Reliability
20	0.90	(4,3,3,3,3,3,3,3,3)	12689.1 \$	0.983
21	0.90	(4,3,3,3,3,3,3,3,3)	12689.1 \$	0.960
22	0.90	(4,3,3,3,3,3,3,3,3)	12689.1 \$	0.936
23	0.90	(4,3,3,3,3,3,3,3,3)	12689.1 \$	0.902
20	0.85	(4,3,3,3,3,3,3,3,3)	12689.1 \$	0.983
21	0.85	(4,3,3,3,3,3,3,3,3)	12689.1 \$	0.960
22	0.85	(4,3,3,3,3,3,3,3,3)	12689.1 \$	0.936
23	0.85	(4,3,3,3,3,3,3,3,3)	12689.1 \$	0.902
24	0.85	(4,3,3,3,3,3,3,3,3)	12689.1 \$	0.856
20	0.80	(4,3,3,3,3,3,3,3,3)	12689.1 \$	0.983
21	0.80	(4,3,3,3,3,3,3,3,3)	12689.1 \$	0.960
22	0.80	(4,3,3,3,3,3,3,3,3)	12689.1 \$	0.936
23	0.80	(4,3,3,3,3,3,3,3,3)	12689.1 \$	0.902
24	0.80	(4,3,3,3,3,3,3,3,3)	12689.1 \$	0.856
25	0.80	(4,3,3,3,3,3,3,3,3)	12689.1 \$	0.806

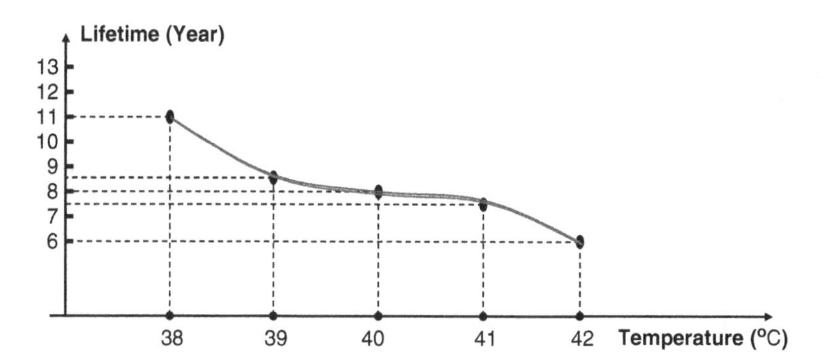

Fig. 8.8 Simplified lifetime distribution for power supply (CJ1W-PA202)

they (peak points) are utilized to form the new lifetime distribution. For instance, the simplified lifetime distribution of power supply (CJ1W-PA202) is made as in Fig. 8.8, and the simplified lifetime distributions can be also generated in a similar way for S8VS-24024 and the CPU-communication-I/O units.

As a consequence, the original fuzzy random lifetime distributions ξ_i for $i = 1,2,\cdots,9$ are simplified to be functions of the random variable $\mathscr{T}$ which are also random variables, and now the system reliability is rewritten as

$$R_{t^0}(x) = \Pr\left\{ \bigwedge_{i=1}^{9} \left(\sum_{j=1}^{x_i} \xi_{ij}(\mathscr{T}) \right) \geq t^0 \right\}, \tag{8.11}$$

which is the classic reliability defined by probability.

Table 8.6 Solution comparison for reliability maximization problem in Case A

t^0(Year)	FRS	RS
21	(5,3,3,3,3,3,3,3,3) Reliability 0.981	(4,3,3,3,3,3,3,3,3) Reliability 0.948
22	(5,3,3,3,3,3,3,3,3) Reliability 0.948	(4,3,3,3,3,3,3,3,3) Reliability 0.934

Table 8.7 Solution comparison for cost minimization problem in Case A

t^0(Year)	r^0	FRS	RS
20	0.85	(4,3,3,3,3,3,3,3,3) Optimal solution	(3,3,3,3,3,3,3,3,3) Infeasible solution
20	0.80	(4,3,3,3,3,3,3,3,3) Optimal solution	(3,3,3,3,3,3,3,3,3) Infeasible solution
21	0.80	(4,3,3,3,3,3,3,3,3) Optimal solution	(3,3,3,3,3,3,3,3,3) Infeasible solution

Replacing the reliability function with (8.11), the problems (8.4) and (8.5) are solved by calculating the reliability and cost using the allocations that are obtained via stochastic programming approaches to the simplified problems. The comparisons between the original fuzzy random solution (FRS) and the simplified random solution (RS) for the cases of reliability maximization and cost minimization are provided in Tables 8.6 and 8.7, respectively.

The solution comparisons show that in the reliability maximization problem, if we apply the simplified random solution (RS) obtained by a stochastic programming approach to the real situation, the "optimal" reliability obtained is just 0.948 for threshold lifetime of 21 years, and 0.934 for 22 years threshold lifetime, while the real fuzzy random solution (FRS) gives corresponding reliability values of 0.981 and 0.948 which are higher than the former results. So the reliability loss here are 0.033 and 0.014. On the other hand, as to the cost minimization problem, when the lifetime threshold and reliability target pair (t^0, r^0) is set as (20, 0.85), (20, 0.80), and (21, 0.80), all the RSs become infeasible solutions in the real fuzzy random environment letting alone being the optimal solutions.

Distinctly, the above comparisons show that the ignorance and simplification of the substantial imprecision or fuzziness changes the solution structure in the optimization process and therefore causes the latent risks to the decision (value loss).

Case B. Now, if we simplify the lifetime distributions from the modification of the dam experts, then all the imprecise lifetime distributions (8.1)–(8.3) will be replaced by the precise probability lifetime distributions that are formed using the crisp values with membership degree 1. Similarly as the discussions in Case A, the comparisons are provided in Tables 8.8 and 8.9 for reliability maximization and cost minimization problems, respectively.

From Tables 8.8 and 8.9, it is easy to see that in the reliability maximization problem for lifetime threshold $t^0 = 23$, the RS leads to a higher cost of 12,816.3 $ compared with the original fuzzy random solution (FRS) which just costs 12,689.1 $ while holding the same reliability as RS. So the capital loss is 127.2 $. Moreover,

Table 8.8 Solution comparison for reliability maximization problem in Case B

t^0(Year)	FRS	RS
23	(4,3,3,3,3,3,3,3,3) Reliability 0.902 Cost 12689.1 $	(5,3,3,3,3,3,3,3,3) Reliability 0.902 Cost 12816.3 $

Table 8.9 Solution comparison for cost minimization problem in Case B

t^0(Year)	r^0	FRS	RS
23	0.90	(4,3,3,3,3,3,3,3,3) Optimal solution	— No solution
24	0.85	(4,3,3,3,3,3,3,3,3) Optimal solution	— No solution
25	0.80	(4,3,3,3,3,3,3,3,3) Optimal solution	— No solution

to the cost minimization problem, ignoring all the possible lifetime values in the original imprecise lifetime distribution except for the one with membership degree 1, the simplified model finds no solution when the lifetime threshold and reliability target pair (t^0, r^0) takes on (23, 0.90), (24, 0.85), and (25, 0.80). This is a disaster for the decision makers, which means that they have to lower the reliability target or, alternatively, relax the capital limit to obtain the solutions. In either way, they lose value.

Chapter 9
Case Study II: Location Selection for Frozen Food Plants

This chapter presents another real-life case study which is about a location selection problem for frozen food plants. It serves as applications of the two-stage fuzzy random facility location model with recourse and fixed capacity (FR-FLM-RFC, Chap. 5) and the two-stage fuzzy stochastic programming with VaR (FSP-VaR, Chap. 6) in location selection with fixed capacity.

In Sect. 9.1, we briefly introduce the background of this frozen food plant location problem.

In Sect. 9.2, we discuss the distribution identification process for the demands (uncertain parameters) of clients (restaurants). First, we do questionnaires so as to acquire the basic demand quantity from the clients; they are essentially fuzzy numbers (estimation). Then, we transform each fuzzy data into an easily processable form and analyze statistically the transformed fuzzy data with Software $\mathbb{R}$. Finally, we identify the distributions (fuzzy random variables) of demand for each university.

In Sect. 9.3, we apply the two-stage location model of FR-FLM-RFC discussed in Chap. 5 to this frozen food plant location problem and find the optimal solution with the maximum expected profit. Furthermore, in expected value objective, the fuzzy random solution obtained is compared with a simplified random solution.

In Sect. 9.4, we remodel the problem with the two-stage model of FSP-VaR discussed in Chap. 6, and obtain another optimal solution with the minimum VaR. Also, the solution is compared with a simplified random solution in VaR objective.

9.1 Problem Background

We deal with a location selection problem for frozen food plants in the region of Taipei. The plants to be established will serve as meat suppliers for five chain restaurants newly opened in the campuses of five universities, respectively, in Taipei City: Shih Chien University (Taipei), National Chengchi University, Soochow University, Chinese Culture University, and National Taipei University of Education (see Fig. 9.1).

S. Wang and J. Watada, *Fuzzy Stochastic Optimization: Theory, Models and Applications*, DOI 10.1007/978-1-4419-9560-5_9, © Springer Science+Business Media New York 2012

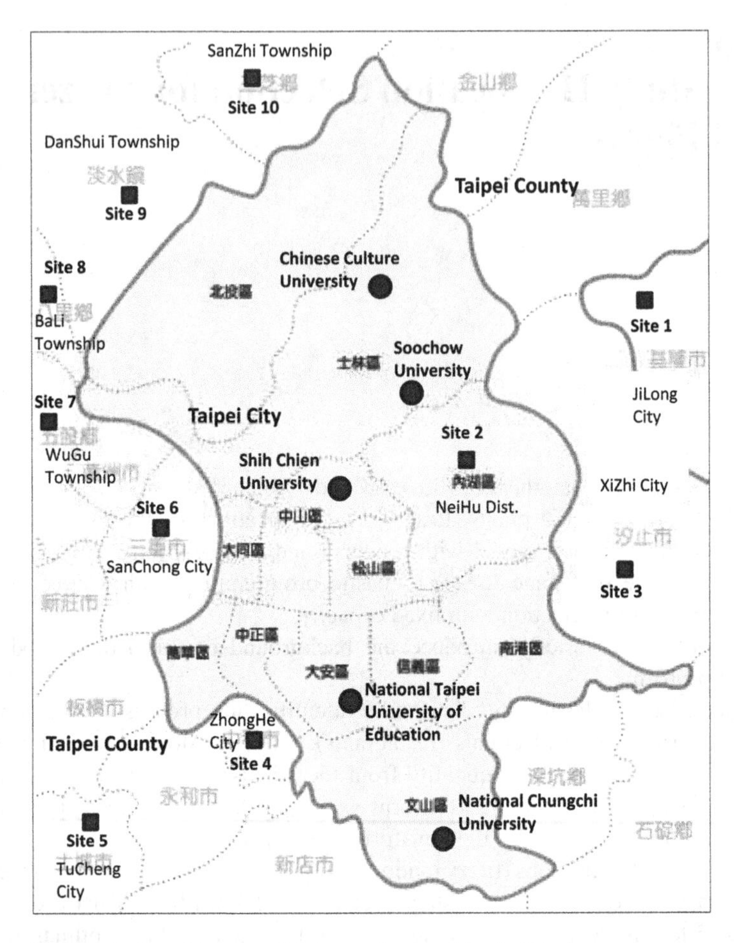

Fig. 9.1 The distribution of the potential sites and universities in Taipei

The chain restaurants mainly serve various kinds of take-out set meals, e.g., beef set meal, pork set meal, and so on, and all the meat materials will be supplied by the frozen food plants, whose ten potential sites are plotted in Fig. 9.1 and their locations as well as the area sizes are listed in Table 9.1. So our task is to find the optimal sites for the meat suppliers (frozen food plants) from all the ten potential alternatives.

9.2 Distribution Identification for Client Demand

As we have discussed in Chap. 5, the client demand is a critical uncertain parameter in the location selection problems. In this case study, since the students are the actual consumers, the amount of meat demand of the five chain restaurants in five

Table 9.1 The location with area size for each potential site of frozen food plants

Site i	Location	Area size
1	Malingkang Road, QiDu Dist., JiLong City	383 m^2
2	Technological Region, LeiHu Dist., Taipei City	150 m^2
3	Datong Road, XiZhi City, Taipei Country	175 m^2
4	Lide Street, ZhongHe City, Taipei Country	127 m^2
5	Central Road, TuCheng City, Taipei Country	100 m^2
6	Chongxin Road, SanChong City, Taipei Country	321 m^2
7	Wugong Road, Wugu Township, Taipei Country	350 m^2
8	Beitou Road, BaLi Township, Taipei Country	172 m^2
9	Yuhuadian Road, DanShui Township, Taipei Country	380 m^2
10	101 Avenue, SanZhi Township, Taipei Country	432 m^2

universities can be roughly and reasonably equated with the expected consumption by the students. However, note that the five chain restaurants were newly opened; not much historical demand data are available. In order to obtain the distributions of meat demand, we do questionnaires and acquire the demand information directly from the students. The whole distribution identification procedure is doing the questionnaires, processing the data collected, and determining the distribution.

9.2.1 Questionnaire

With the menu of the chain restaurant, we poll 100 students randomly in each university. To each student, we capture the average demand for each week in four different time periods in one academic year, i.e., the first semester, winter vacation, the second semester, and summer vacation. The questionnaire is designed as Table E.1 (Appendix E).

In Appendix E, the collected answers of students (Nos. 1–100) in National Chengchi University are given in Figs. E.1–E.3. Here we explain the answer sheet a little: in Fig. E.1, intervals $[a_j, c_j]$ for $j = 1,2,3,4$ represent the answers to the Q.1 in Table E.1 for periods of the first semester , winter vacation, the second semester, and summer vacation, respectively. Furthermore, $b_j, j = 1,2,3,4$ represent the answers to the Q.2 in Table E.1 in different corresponding periods. From those demand data gathered from the students, it is easy to see that most of them are imprecise data, which reflects the imprecision or fuzziness in nature of demand of the customers when confronted with the decision making to new products.

9.2.2 Fuzzy Data Process and Analysis

From the answers collected, we can see that the demand of student i for period j can be represented as $(a_j, b_j, c_j)^i$, for $j = 1,2,3,4$, and $i = 1,2,\cdots,100$. Now we merge the average demands per week of four different periods into that of one academic year by the following equation:

$$\left(a^i,b^i,c^i\right) = \sum_{j=1}^{4} w_j \cdot (a_j,b_j,c_j)^i, \tag{9.1}$$

for $i = 1,2,\cdots,100, j = 1,2,3,4$, where w_j is the proportion of period j in one academic year that takes on values of $w_1 = w_3 = 0.34, w_2 = 0.10$, and $w_4 = 0.22$.

Since each fuzzy data with the form of

$$(a^i,b^i,c^i) \quad \text{with} \quad a^i \le b^i \le c^i$$

can be written equivalently as

$$(b^i,b^i-a^i,c^i-b^i)$$

for $i = 1,2,\cdots,100$, where b^i, b^i-a^i, and c^i-b^i can be any (independent) positive real numbers. Therefore, b^i, b^i-a^i, and c^i-b^i can be treated essentially as realizations of three independent random variables X^C, X^L, and X^R, respectively. Furthermore, we employ the $\mathbb{R}$ statistical software to analyze all the transformed data of b^i, b^i-a^i, and c^i-b^i so as to make the statistic inference for the distributions of X^C, X^L, and X^R, respectively, for each university (see Fig. 9.2).

After the statistic analysis and inference with $\mathbb{R}$ for all the demand data of students in five universities, we obtain the distribution results of random parameters for individual (student) demand in Table 9.2 in which $\Gamma(\nu,\theta)$, $\mathscr{W}(\omega,\upsilon)$, and $\mathscr{N}(\mu,\sigma^2)$ represent Gamma distribution, Weibull distribution, and Normal distribution, respectively.

9.2.3 Determining Distributions

From the theory of statistics, each student can be regarded as a random sample, which is independent and identically distributed individual of the population, i.e., all the students in the university. From the above procedure of data process and analysis, we can identify the distribution of the frequency of buying meals per week in one academic year for each student, i.e.,

$$(X_k^C, X_k^L, X_k^R)$$

for university $k, k = 1,2,3,4,5$.

Since each set meal served by the chain restaurant contains averagely 80 g meat, and 1 year consists of 52 weeks, the demand (for meat) of each student for the whole year is $4.16\left(X_k^C, X_k^L, X_k^R\right)$ kg. Furthermore, denoting the population of the students in university k be N_k, the demand $\mathscr{D}_k$ of the university k can be expressed as

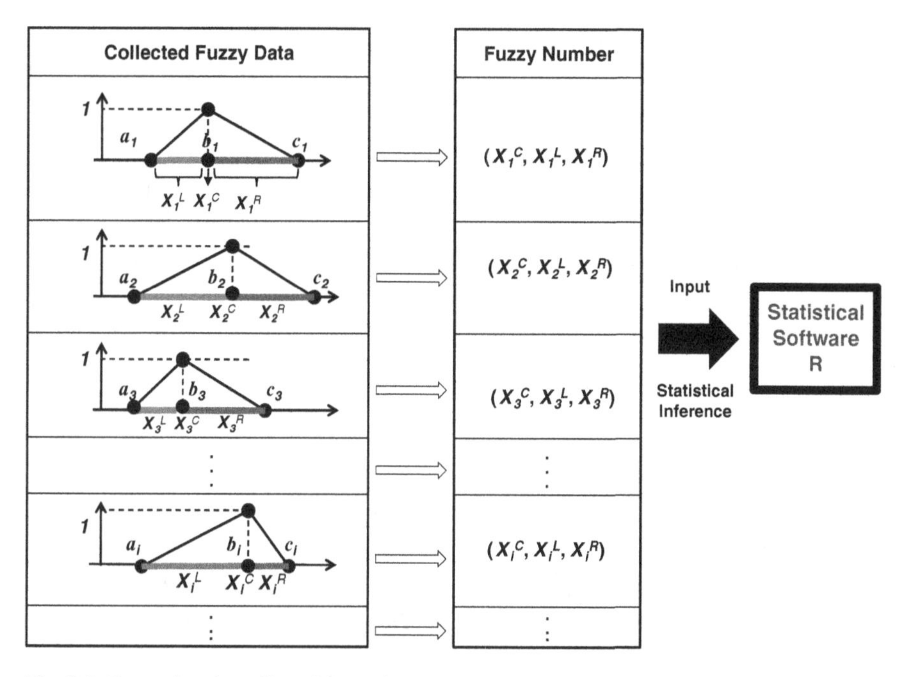

Fig. 9.2 Processing the collected fuzzy data

Table 9.2 The probability distributions for random parameters X^C, X^L, and X^R of individuals

Individual (university k)	X_k^C	X_k^L	X_k^R
1	$\mathscr{N}(2.45, 0.63^2)$	$\mathscr{W}(2.50, 0.89)$	$\Gamma(3.04, 0.42)$
2	$\Gamma(4.70, 0.56)$	$\mathscr{N}(0.86, 0.41^2)$	$\mathscr{W}(2.19, 1.48)$
3	$\mathscr{W}(2.94, 2.63)$	$\mathscr{N}(1.03, 0.49^2)$	$\mathscr{N}(1.02, 0.45^2)$
4	$\Gamma(3.13, 0.68)$	$\Gamma(1.94, 0.26)$	$\mathscr{N}(2.18, 1.28^2)$
5	$\mathscr{W}(2.34, 2.99)$	$\mathscr{W}(1.84, 1.07)$	$\Gamma(2.71, 0.60)$

$$\mathscr{D}_k = \sum_{i=1}^{N_k} 4.16\left(X_k^C, X_k^L, X_k^R\right)$$
$$= 4.16\left(\sum_{i=1}^{N_k} X_k^C, \sum_{i=1}^{N_k} X_k^L, \sum_{i=1}^{N_k} X_k^R\right), \tag{9.2}$$

for $k = 1,2,3,4,5$. Furthermore, from the relationship between the probability distributions of individual and that of the population with size N (refer to [23]), we have

$$\sum_{i=1}^{N} \Gamma(v, \theta) \sim \Gamma\Big(v, N \cdot \theta\Big), \tag{9.3}$$

Table 9.3 The probability distributions of parameters $\mathscr{X}^C, \mathscr{X}^L$, and $\mathscr{X}^R$ for each university

University k	$\mathscr{X}_k^C$	$\mathscr{X}_k^L$	$\mathscr{X}_k^R$
1	$\mathscr{N}(38220, 9828^2)$	$\mathscr{W}(39000, 0.89)$	$\Gamma(3.04, 6552)$
2	$\Gamma(4.70, 8848)$	$\mathscr{N}(13588, 6478^2)$	$\mathscr{W}(34602, 1.48)$
3	$\mathscr{W}(39342, 2.63)$	$\mathscr{N}(16274, 7742^2)$	$\mathscr{N}(16116, 7110^2)$
4	$\Gamma(3.13, 18292)$	$\Gamma(1.94, 6994)$	$\mathscr{N}(58642, 34432^2)$
5	$\mathscr{W}(13806, 2.99)$	$\mathscr{W}(10856, 1.07)$	$\Gamma(2.71, 3540)$

$$\sum_{i=1}^{N} \mathscr{W}(\omega, \upsilon) \sim \mathscr{W}\left(N \cdot \omega, \upsilon\right), \tag{9.4}$$

and

$$\sum_{i=1}^{N} \mathscr{N}(\mu, \sigma^2) \sim \mathscr{N}\left(N \cdot \mu, N^2 \cdot \sigma^2\right). \tag{9.5}$$

Hence, given the student number N_k for each university k, i.e., $N_1 = 15{,}600$, $N_2 = 15{,}800$, $N_3 = 15{,}800$, $N_4 = 26{,}900$, and $N_5 = 5{,}900$, if we denote

$$\mathscr{X}_k^C = \sum_{i=1}^{N_k} X_k^C, \mathscr{X}_k^L = \sum_{i=1}^{N_k} X_k^L, \text{ and } \mathscr{X}_k^R = \sum_{i=1}^{N_k} X_k^R,$$

then the demand $\mathscr{D}_k$ of the university k can be rewritten as

$$\begin{aligned} \mathscr{D}_k &= 4.16\left(\sum_{i=1}^{N_k} X_k^C, \sum_{i=1}^{N_k} X_k^L, \sum_{i=1}^{N_k} X_k^R\right) \\ &= 4.16\left(\mathscr{X}_k^C, \mathscr{X}_k^L, \mathscr{X}_k^R\right), \end{aligned} \tag{9.6}$$

where the probability distributions for $\mathscr{X}_k^C, \mathscr{X}_k^L$, and $\mathscr{X}_k^R$ based on (9.3)–(9.5) are provided in following Table 9.3.

From (9.6), apparently, the meat demands $\mathscr{D}_k, k = 1,2,3,4,5$ of clients (restaurants in universities) are fuzzy random variables, and for each outcome of random parameters $\mathscr{X}_k^C, \mathscr{X}_k^L$, and $\mathscr{X}_k^R$ from the probability distributions in Table 9.3, the realization of $\mathscr{D}_k$ is a triangular fuzzy number. The process of determining the distributions for the fuzzy random demands is also illustrated in Fig. 9.3.

9.3 Expected Value Model and Solution

As for the other model parameters, the fixed cost for the frozen food plant roughly consists of land cost, workshop establishment, freezing storehouse, processing equipment, thermal insulation, worker employment, electricity, water,

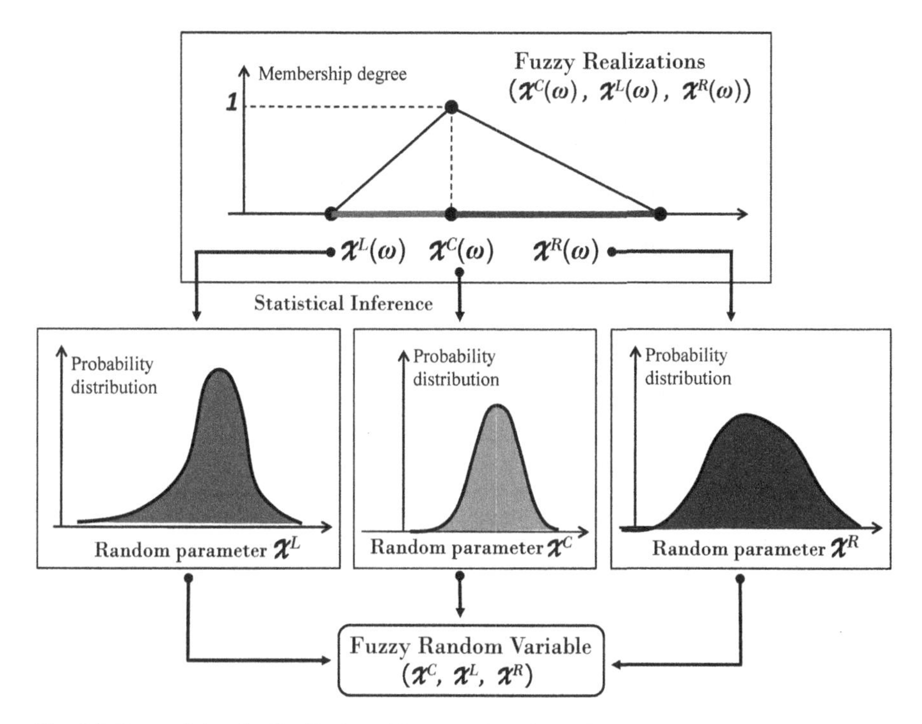

Fig. 9.3 Determining the distribution of fuzzy random demand

and maintenance. The unit variable cost is mainly the expense on the purchase as well as process of unit raw materials, which is fluctuating and depending on the sales situation. As an instance, for Site 1 with 380 m^2 in Jilong City, the land cost is 12,000,000 New Taiwan Dollar (NTD), workshop establishment costs 12,000,000 NTD, the freezing storehouse costs 3,000,000 NTD, the processing equipments cost 7,600,000 NTD, the expense for worker employment is 7,200,000 NTD (1 year), the cost for electricity and water consumption is 1,800,000 NTD, and the maintenance cost is 250,000 NTD. Meanwhile, the unit variable cost for raw material is estimated as the possible values in a range between 90,000 and 110,000 NTD per ton (T), and the productivity capacity for Site 1 is around 4,000 T per year.

The gross fixed cost and unit variable cost, as well as the productivity capacity of each plant, are listed in the following Table 9.4.

What's more, since all the sites as well as universities are located within the area of Taipei (including Taipei City and Taipei County), the average transportation cost of 1 T frozen foods from each site to each university is around $t_{ik} = 1,600$ NTD, equally for $i = 1,2,\cdots,10, k = 1,2,3,4,5$. Finally, the unit price r_k of 1 T meat charged to client k (chain restaurant) is 150,000 NTD, equally for $k = 1,2,3,4,5$.

With the obtained fuzzy random demand $\mathscr{D}_k$ of each university for $k = 1,2,3,4,5$, interval values [90, 100] or [100, 120] for unit variable cost V_i of each plant, and the other parameter values mentioned above, the frozen food plant location problem

Table 9.4 The fixed cost, variable cost, and capacity for each potential site

Site i	Capacity s_i (T)	Fixed costc_i (thousand NTD)	Unit variable cost V_i (thousand NTD)
1	4,000	43,850	$90 \sim 110$
2	1,500	105,620	$100 \sim 120$
3	1,800	53,120	$100 \sim 120$
4	1,300	101,800	$100 \sim 120$
5	1,000	29,300	$100 \sim 120$
6	3,400	113,720	$90 \sim 110$
7	3,700	44,550	$90 \sim 110$
8	1,800	44,120	$100 \sim 120$
9	4,000	41,850	$90 \sim 110$
10	4,200	51,600	$90 \sim 110$

can be formulated by using the two-stage fuzzy random facility location model with recourse and fixed capacity (FR-FLM-RFC) in Chap. 5 as follows:

$$\left.\begin{array}{ll}\max & \mathcal{Q}(x) - \sum_{i=1}^{10} c_i x_i \\ \text{subject to } x_i \in \{0,1\}, i = 1,2,\cdots,10, & \end{array}\right\} \tag{9.7}$$

where $\mathcal{Q}(x) = E\left[Q(x,\xi)\right]$, and for each realization (ω,γ), the value function for the second-stage problem is

$$\left.\begin{array}{rl}Q\Big(x,\xi(\omega,\gamma)\Big) = \max & \sum_{i=1}^{10}\sum_{k=1}^{5}\big(r_k - V_i(\gamma) - t_{ik}\big)y_{ik} \\ \text{subject to} & \\ & \sum_{i=1}^{10} y_{ik} \le \mathcal{D}_k(\omega,\gamma), k = 1,2,\cdots,5, \\ & \sum_{k=1}^{5} y_{ik} \le s_i x_i, i = 1,2,\cdots,10, \\ & y_{ik} \ge 0, i = 1,2,\cdots,10, k = 1,2,\cdots,5. \end{array}\right\} \tag{9.8}$$

Solving the above problem with the hybrid MBACO approach (Algorithm 5.3 in Chap. 5), we obtain the optimal solution, i.e., fuzzy random solution (FRS), as

$$FRS = x^* = (0,0,0,0,0,0,0,0,1,0),$$

with objective value (expected return) $\mathcal{Q}(x^*) = 3417{,}800$ NTD. This means the Site 9 in DanShui Township is optimal potential site for the new frozen food plant. Testing this solution by using exhausted research which compares all $2^{10} = 1024$ possible solutions, it shows that the above FRS is the real best solution of the problems (9.7)–(9.8).

Table 9.5 Solution comparison for location problem (9.7)–(9.8)

Approach	Optimal solution	Expected profit
FRS	(0,0,0,0,0,0,0,0,1,0)	3,417,800 NTD
RS	(0,0,0,0,1,0,0,0,0,0)	2,129,600 NTD

Similarly as we have done in Case Study I, we further discuss on the value loss (profit loss) in the case that we simplify the real imprecise probability distributions of the demands (fuzzy random vector) to a "reasonable" random demands (random vector), and treat the original problem as an easier stochastic optimization problem.

With the simplification, all the fuzzy random distributions for demands are replaced with some random variables as follows:

$$\mathscr{D}'_k = 4.16\mathscr{X}_k^C \Longleftarrow \mathscr{D}_k = 4.16\left(\mathscr{X}_k^C, \mathscr{X}_k^L, \mathscr{X}_k^R\right)$$

for $k = 1,2,3,4,5$. As for the fuzzy unit variable costs $V_i = \left[v_i^L, v_i^U\right]$, apparently, the mean value is the most "suitable" value chosen to replace the original interval, i.e.,

$$V_i' = \frac{v_i^U - v_i^L}{2} \Longleftarrow V_i = \left[v_i^L, v_i^U\right],$$

for $i = 1,2,\cdots,10$.

Based on the above simplification, inputting all the modified parameters $\mathscr{D}'_k$ and V_i' to the problems (9.7)–(9.8), the original problem has been changed into a task of stochastic programming problem. Solving this stochastic problem and utilizing the "best" solution obtained into the original real fuzzy stochastic optimization problem, we get the random solution (RS). A comparison is showed in the following Table 9.5.

The above Table 9.5 shows that if we ignore the fuzziness and use the solution $x = (0,0,0,0,1,0,0,0,0,0)$ obtained through the simplified stochastic model as the decision which selects Site 5 in Tucheng City to build the new frozen food plant, it causes a loss in expected profit with amount of 1,288,200 NTD, compared with the original approach that considers fully the fuzzy random information.

9.4 VaR Model and Solution

Now we remodel this frozen food plant location problem with fixed capacity using the two-stage fuzzy stochastic programming with Value-at-Risk (FSP-VaR) discussed in Chap. 6. The objective

$$\mathrm{VaR}_{1-\beta}(x) = \sup\left\{\lambda \mid \mathrm{Ch}\left\{\sum_{i=1}^{10} c_i x_i - \mathscr{R}(x,\xi) \geq \lambda\right\} \geq \beta\right\},$$

is to minimize the largest loss at the confidence $1-\beta$ produced by the location decision $x=(x_1,x_2,\cdots,x_{10})$. Here $\sum_{i=1}^{10} c_i x_i$ represents the fixed cost at each location decision x, and $\mathscr{R}(x,\xi)$ is the variable return based on the fuzzy random parameter ξ.

Making use of the two-stage FSP-VaR in (6.12)–(6.14), the above location problem can be built as follows with confidence level $1-\beta$:

$$\left.\begin{array}{ll} \min & \mathrm{VaR}_{1-\beta}(x) \\ \text{subject to } x_i \in \{0,1\}, i=1,2,\cdots,10, & \end{array}\right\} \tag{9.9}$$

where

$$\mathrm{VaR}_{1-\beta}(x)=\sup\left\{\lambda \mid \mathrm{Ch}\left\{\sum_{i=1}^{10} c_i x_i-\mathscr{R}(x,\xi)\geq\lambda\right\}\geq\beta\right\}, \tag{9.10}$$

and the second-stage value function

$$\mathscr{R}\Big(x,\xi(\omega,\gamma)\Big)=Q\Big(x,\xi(\omega,\gamma)\Big) \tag{9.11}$$

which is the same as (9.8) for each scenario (ω,γ).

Combining the approximate algorithm of fuzzy random VaR (Algorithm 6.1 in Chap. 6) with the binary ACO algorithm, we can obtain easily the optimal solution for problem (9.9)–(9.11):

$$FRS_{1-\beta}=x^*=(0,0,0,0,1,0,0,0,0,0)$$

with objective values 5,862,300 NTD, 3,050,700 NTD, and 2,260,800 NTD, for different confidence levels $1-\beta$ of 0.9, 0.85, and 0.8, respectively. This optimal solution indicates when we deal with the frozen food plant location problem with VaR model, the optimal choice of the location should be the Site 5 other than the Site 1 which is the best solution to expected value model.

Such different solutions to the same problem exactly reflect the distinction of modeling appetites between the expected value model and VaR model: the expected value model holds the neutral attitude toward the risk and focuses on the long-term average return based on the decision made, whereas the VaR model stresses risk aversion and minimizes the largest possible loss at some confidence. So when coping with the same location problem by VaR model, the optimal solution changes from Site 9 to Site 5, which implies that although the decision Site 5 is unable to produce the maximum long-term average profit, it does reduce the risk (VaR) to the minimum level, while the Site 9 is the best choice for the converse situation.

Finally, the following Table 9.6 shows the comparison of the FRS solution and RS solution to problem (9.9)–(9.11) with VaR objective. Here the RS solution

Table 9.6 Solution comparison for location problem (9.9)–(9.11)

Approach	Confidence level	Optimal solution	Largest possible loss(VaR)
FRS	0.9	(0,0,0,0,1,0,0,0,0,0)	5,862,300 NTD
FRS	0.85	(0,0,0,0,1,0,0,0,0,0)	3,050,700 NTD
FRS	0.8	(0,0,0,0,1,0,0,0,0,0)	2260,800 NTD
RS	0.9	(0,0,0,0,0,0,0,0,1,0)	10,305,100 NTD
RS	0.85	(0,0,0,0,0,0,0,0,1,0)	7,205,500 NTD
RS	0.8	(0,0,0,0,0,0,0,0,1,0)	5,736,900 NTD

is obtained via the same simplification as described in the preceding case of expected value model. The comparison shows that in VaR model, ignoring fuzziness which exists intrinsically in the original parameter distributions increases the largest possible loss at different confidence levels, i.e., VaR.

Appendix A
Semicontinuity of a Real Function

This appendix gives some basic mathematical concepts on semicontinuity of a real valued function that have been used in the book.

Let $f(x)$ be a function defined on the set $\Re$ of real values. Then we say $f(x)$ is upper semicontinuous at x_0 if for any $\varepsilon > 0$, there exists $\delta > 0$ such that $f(x) < f(x_0) + \varepsilon$ for all x with $|x - x_0| < \delta$. We say $f(x)$ is lower semicontinuous at x_0 if for any $\varepsilon > 0$, there exists $\delta > 0$ such that $f(x) > f(x_0) - \varepsilon$ for all x with $|x - x_0| < \delta$.

In addition, the limit superior of function $f(x)$ at x_0 is defined as

$$\limsup_{x \to x_0} f(x) = \inf_{\delta > 0} \sup_{0 < |x - x_0| < \delta} f(x),$$

the right limit superior of $f(x)$ at x_0 is defined by

$$\limsup_{x \to x_0 +} f(x) = \inf_{\delta > 0} \sup_{0 < x - x_0 < \delta} f(x),$$

and the left limit superior of $f(x)$ at x_0 is given as

$$\limsup_{x \to x_0 -} f(x) = \inf_{\delta > 0} \sup_{-\delta < x - x_0 < 0} f(x).$$

Furthermore, for the upper semicontinuity of a real function, we have the following result.

Proposition A.1. *Let $f(x)$ be a real function, $f(x_0) < \infty$, $x_0 \in \Re$. Then $f(x)$ is upper semicontinuous at x_0 if and only if*

$$\limsup_{x \to x_0} f(x) \leq f(x_0).$$

More detailed properties about the upper and lower semicontinuity can be found in [112, 126].

S. Wang and J. Watada, *Fuzzy Stochastic Optimization: Theory, Models and Applications*, DOI 10.1007/978-1-4419-9560-5,

Appendix
Semicontinuity of a Real Function

[illegible]

Appendix B
Some Necessary Results on Fuzzy Variables

In this appendix, we present some properties of distribution functions of fuzzy random variables and some results on the operations of fuzzy variables.

Let $(\Gamma, \mathscr{A}, \mathrm{Pos})$ be a possibility space, X a fuzzy variable defined on Γ, and Cr be a credibility measure on $\mathscr{A}$. The credibility distribution functions of fuzzy variable X are defined as follows.

Definition B.1 ([87]). Let X be a fuzzy variable defined on Γ. The credibility distribution functions of X are defined by

$$\Phi_L(x) = \mathrm{Cr}\{\gamma \in \Gamma \mid X(\gamma) \leq x\}, \quad \text{and}$$
$$\Phi_U(x) = \mathrm{Cr}\{\gamma \in \Gamma \mid X(\gamma) \geq x\}$$

for every $x \in \Re$.

As for the credibility distribution functions, we have the following results:

Theorem B.1 ([145]). *If fuzzy variable X is left continuous and lower semicontinuous, then the credibility distribution function Φ_L is left continuous.*

Theorem B.2 ([145]). *If fuzzy variable X is right continuous and lower semicontinuous, then the credibility distribution function Φ_U is right continuous.*

Theorem B.3 ([145]). *Let X be a fuzzy variable. We have*

$$\Phi_L(x) = \mathrm{Cr}\{X < x\} \tag{B.1}$$

and

$$\Phi_U(x) = \mathrm{Cr}\{X > x\} \tag{B.2}$$

hold except on an at most countable subset of $\Re$. Moreover, if X is a lower semicontinuous fuzzy variable, then (B.1) and (B.2) hold for every $x \in \Re$.

As for the minimum and maximum t-norm operations of convex fuzzy variables, we have the following results:

S. Wang and J. Watada, *Fuzzy Stochastic Optimization: Theory, Models and Applications*, DOI 10.1007/978-1-4419-9560-5, © Springer Science+Business Media New York 2012

Theorem B.4 ([61]). *Suppose that fuzzy variables $X_1, \cdots, X_n$ have membership functions $\mu_1, \cdots, \mu_n$, respectively. If μ_i is nondecreasing on $(-\infty, m_i]$, nonincreasing on $[m_i, \infty)$ with $\mu_i(m_i) = 1$ for $1 \le i \le n$, and $m_1 \le \cdots \le m_n$, then we have the membership function of $\bigvee_{i=1}^n X_i$ is*

$$\mathrm{Pos}\left\{\bigvee_{i=1}^{n} X_i = z\right\} = \begin{cases} \bigwedge_{i=1}^n \mu_i(z), & z < m_1 \\ \bigwedge_{i=k+1}^n \mu_i(z), & m_k \le z < m_{k+1}, 1 \le k \le n-1 \\ \bigvee_{i=1}^n \mu_i(z), & z \ge m_n, \end{cases} \tag{B.3}$$

where $\bigwedge_{i=n}^n \mu_i(z) = \mu_n(z)$.

Theorem B.5 ([61]). *Under the same condition as in Theorem B.4, we have the membership of $\bigwedge_{i=1}^n X_i$ is*

$$\mathrm{Pos}\left\{\bigwedge_{i=1}^{n} X_i = z\right\} = \begin{cases} \bigvee_{i=1}^n \mu_i(z), & z < m_1 \\ \bigwedge_{i=1}^k \mu_i(z), & m_k \le z < m_{k+1}, 1 \le k \le n-1 \\ \bigwedge_{i=1}^n \mu_i(z), & z \ge m_n, \end{cases} \tag{B.4}$$

where $\bigwedge_{i=1}^1 \mu_i(z) = \mu_1(z)$.

Appendix C
Binary Particle Swarm Optimization

The binary particle swarm optimization (BPSO) was introduced by Kennedy and Eberhart [63] to deal with binary-coded optimization problems. It is different from the standard PSO which is a real-coded algorithm.

In the BPSO, an n-dimensional potential solution to a problem is represented as a particle i having current position x_i which is an integer vector in $\{0,1\}^n$ and the current velocity v_{id} of the particle which represents the probability of x_i taking on the value 1. Each particle i maintains a record of the position of its previous best performance in a vector called p_{id}. The variable g is the index of the particle with best performance so far in the population.

An iteration comprises *evaluation* of each particle: let's say that all the particles are evaluated via the value of the fitness function, and the *updating process* for all the binary particles: first of all, it needs to determine the best particle p_{gd} (with the highest fitness) in the entire population, and for each particle x_i, find the p_{id} with the best previous performance, where $i = 1,2,\cdots,P_{size}$; here P_{size} is the population size. Then, for each i, the velocity vector v_{id} is computed through following formula:

$$v_{id} = \mathscr{W} * v_{id} + c_1 * \text{rand}() * (p_{id} - x_i) + c_2 * \text{rand}() * (p_{gd} - x_i), \tag{C.1}$$

where c_1 and c_2 are learning rates which are usually generated in the interval [0,4], rand() is a uniform random number in the interval [0,1], and $\mathscr{W}$ is the inertia weight whose value decreases linearly as the number of iterations of the algorithm increases. After that, each particle $x_i, i = 1,2,\cdots,P_{size}$ can be updated according to the following rule

$$\texttt{if}\Big(\text{rand}() < S(v_{id,j})\Big)\ \texttt{then}\ x_{ij} = 1; \texttt{else}\ x_{ij} = 0 \tag{C.2}$$

for $j = 1,2,\cdots,n$, where $v_{id,j}$ and x_{ij} are the components of the vector v_{id} and x_i, respectively, and $S(\cdot)$ is a sigmoid function

S. Wang and J. Watada, *Fuzzy Stochastic Optimization: Theory, Models and Applications*, DOI 10.1007/978-1-4419-9560-5, © Springer Science+Business Media New York 2012

$$S(x) = \frac{1}{1+e^{-x}}.$$

Through formulas (C.1) and (C.2), a new generation of particles $x'_1, x'_2, \cdots, x'_{P_{size}}$ can be generated.

For more detailed knowledge on BPSO, see [63, 64, 136, 152].

Appendix D
Tabu Search

Tabu search (TS) was introduced by Glover [40], which is a metaheuristic algorithm based on extension of local search. The TS makes use of a flexible structure of storage (Tabu List) to avoid the local recycle, and on the other hand, the involved aspiration criteria can make it able to pick up some "good" solutions which are even tabu.

In the TS algorithm, we let N_{size} be the size of the neighborhood, TT the tabu tenure, which is a random integer number $(\text{int})\text{rand}(a,b)$ from a specified interval [a,b], and x be an n-dimensional potential integer-vector solution, i.e., $x \in \mathscr{K}^n$, where $\mathscr{K}$ is some predetermined set of integers. A TS algorithm consists of the following critical portions.

Neighborhood structure: To produce the neighborhood $\mathbf{N}(x_{\text{current}})$ of the current solution x_{current}, two different positions i and j are generated randomly from $\{1,2,\cdots,n\}$, and then exchange the values of the ith and jth positions in x_{current} (as for binary optimization problems, it picks up randomly two different positions i and j from $\{1,2,\cdots,n\}$, and then change the values of x_i and x_j between 0 and 1). Repeat the above operations N_{size} times, and the $\mathbf{N}(x_{\text{current}})$ can be generated.

Update process: Evaluate the current solution x_{current} and all of its neighbors by computing the fitness function $\mathbf{F}(x)$, and find the best solution x^* of x_{current}. For the case that x^* is nontabu, if $\mathbf{F}(x^*) > \mathbf{F}(x_{\text{best}})$, then let $x_{\text{current}} = x^*$ and $x_{\text{best}} = x^*$, and update the tabu list; otherwise, $x_{\text{current}} = x^*$. For the case that x^* is tabu, if it satisfies the aspiration criteria, then let $x_{\text{current}} = x^*$ and $x_{\text{best}} = x^*$; otherwise, find the best solution x^{**} of the neighbors that are nontabu in $N(x_{\text{current}})$, and set $x_{\text{current}} = x^{**}$, and then update the tabu list.

Tabu moves: To further avoid local recycle, if the values of the ith and jth positions in x have been exchanged at iteration M (for the binary version, if x_i and x_j have changed their values between 0 and 1 at iteration M), then they are not allowed to change again until the iteration $M+TT$

Aspiration criteria: If the objective value of a solution x is better than that of the best solution x_{best} known so far, then it can be selected even it is tabu.

More discussions on TS can be found in [40–42, 167, 168].

S. Wang and J. Watada, *Fuzzy Stochastic Optimization: Theory, Models and Applications*, DOI 10.1007/978-1-4419-9560-5, © Springer Science+Business Media New York 2012

Appendix E
Questionnaire Sheet and Some Answer Samples in Case Study II

Table E.1 The questionnaire sheet

Semester 1
Q.1. Around ____ times will you buy meals on the menu per week? (Answer could be INTERVALs or VALUEs)
Q.2. MOST POSSIBLY, around ____times per week will you buy meals on the menu?
Winter Vacation
Q.1. Around ____ times will you buy meals on the menu per week? (Answer could be INTERVALs or VALUEs)
Q.2. MOST POSSIBLY, around ____times per week will you buy meals on the menu?
Semester 2
Q.1. Around ____ times will you buy meals on the menu per week? (Answer could be INTERVALs or VALUEs)
Q.2. MOST POSSIBLY, around ____times per week will you buy meals on the menu?
Summer Vacation
Q.1. Around ____ times will you buy meals on the menu per week? (Answer could be INTERVALs or VALUEs)
Q.2. MOST POSSIBLY, around ____times per week will you buy meals on the menu?

S. Wang and J. Watada, *Fuzzy Stochastic Optimization: Theory, Models and Applications*,
DOI 10.1007/978-1-4419-9560-5,

Student number	Semester 1			Winter Vacation			Semester 2			Summer Vacation		
	a1	c1	b1	a2	c2	b2	a3	c3	b3	a4	c4	b4
1	2	9	4	2	4	2	2	4	2	2	4	2
2	3	5	3	2	4	2	3	4	3	3	4	3
3	1	3	2	0	0	0	1	3	2	0	0	0
4	3	7	3	2	3	2	3	7	3	1	3	2
5	2	4	2	1	2	1	2	4	2	1	2	1
6	2	3	2	1	2	1	2	3	2	1	2	1
7	5	10	5	0	10	5	5	10	5	5	10	5
8	1	3	2	1	2	1	1	3	2	0	1	0
9	1	3	2	1	3	2	1	3	2	1	3	2
10	2	5	3	1	3	1	2	5	3	1	2	1
11	2	3	2	2	3	2	2	3	2	2	3	2
12	1	10	3	0	0	0	2	3	2	0	0	0
13	1	3	2	2	3	2	2	4	2	1	5	2
14	2	4	3	1	2	2	1	4	2	2	3	2
15	2	2	2	0	0	0	2	2	2	0	0	0
16	3	5	5	1	3	2	3	5	3	2	4	3
17	3	4	3	2	4	4	2	3	3	2	4	3
18	2	5	5	0	0	0	1	2	2	0	0	0
19	2	5	3	1	3	2	2	2	2	2	4	3
20	2	5	2	2	4	3	2	4	3	1	2	2
21	2	5	5	2	4	3	2	4	2	1	4	2
22	3	5	5	1	2	2	2	6	2	2	3	2
23	2	10	3	0	0	0	2	5	4	0	0	0
24	1	5	3	0	0	0	3	6	4	0	0	0
25	3	3	3	1	2	2	2	4	3	2	3	2
26	2	3	3	1	2	1	1	4	2	3	4	3
27	2	3	2	2	3	2	1	2	2	2	4	2
28	2	4	3	1	3	2	2	4	3	2	3	3
29	1	4	2	1	3	2	1	6	2	2	5	4
30	2	3	3	1	2	2	2	5	3	0	0	0
31	2	4	3	0	0	0	2	4	3	0	0	0
32	2	3	3	1	2	2	3	5	4	1	3	2
33	2	5	3	0	0	0	2	4	3	0	0	0
34	2	5	3	1	3	2	2	4	3	1	2	2
35	2	5	5	2	4	3	3	5	4	3	4	3
36	2	7	3	1	3	2	2	5	3	1	3	2
37	3	4	3	2	3	2	1	4	2	2	3	2
38	2	3	2	1	3	2	2	4	3	3	5	4
39	1	4	3	2	3	2	3	5	3	3	6	5
40	2	3	2	0	0	0	1	4	2	0	0	0

Fig. E.1 Answers of students (Nos. 1–40) in National Chengchi University

Student number	Semester 1			Winter Vacation			Semester 2			Summer Vacation		
	a1	c1	b1	a2	c2	b2	a3	c3	b3	a4	c4	b4
41	2	4	3	2	3	3	2	3	3	2	5	2
42	2	5	3	2	3	2	2	5	3	2	3	3
43	2	6	3	2	4	3	3	5	4	2	3	3
44	2	5	4	1	2	1	2	4	3	1	2	2
45	3	8	3	1	3	2	2	3	2	4	5	5
46	2	4	2	0	0	0	1	2	2	0	0	0
47	3	5	3	2	3	2	1	3	2	1	2	1
48	2	3	3	1	4	3	2	6	3	2	5	4
49	3	4	3	2	3	3	2	5	3	2	2	2
50	2	3	3	0	0	0	1	2	1	0	0	0
51	2	4	3	1	3	2	1	4	2	2	5	4
52	1	4	2	2	4	2	1	4	2	2	3	3
53	2	4	3	1	5	2	1	3	2	1	4	2
54	2	3	2	2	3	2	1	2	1	2	5	2
55	3	5	5	2	3	2	2	4	3	1	3	2
56	2	3	2	0	0	0	2	3	2	0	0	0
57	2	4	3	0	0	0	1	2	1	0	0	0
58	1	2	2	2	3	2	1	2	2	1	4	2
59	3	5	4	1	3	2	1	2	2	2	2	2
60	1	3	2	5	5	5	2	4	3	2	4	2
61	2	6	2	2	3	2	2	3	2	1	2	2
62	3	8	3	2	3	3	2	4	3	1	3	2
63	2	10	3	0	0	0	1	2	2	0	0	0
64	3	8	3	2	2	2	2	3	2	1	2	1
65	1	2	2	2	4	3	2	4	2	1	3	2
66	2	3	3	2	3	2	2	3	2	1	2	2
67	2	4	3	2	4	3	2	4	3	1	2	2
68	2	4	2	2	4	2	2	3	4	1	2	2
69	2	6	5	1	3	2	1	3	2	1	2	2
70	2	8	2	2	5	3	1	3	2	1	2	2
71	2	6	3	0	0	0	2	2	2	0	0	0
72	3	4	4	3	3	3	2	4	2	1	3	2
73	1	3	2	2	2	2	2	5	3	2	2	2
74	3	3	3	1	1	1	3	3	3	1	1	1
75	3	6	4	2	3	2	2	4	4	2	3	2
76	2	3	3	4	4	4	2	5	3	1	2	2
77	2	3	3	1	2	1	2	5	5	2	3	2
78	3	5	3	0	0	0	1	3	2	0	0	0
79	1	4	2	2	3	3	2	3	2	1	2	2
80	2	4	3	2	2	2	2	5	3	1	2	2

Fig. E.2 Answers of students (Nos. 41–80) in National Chengchi University

Student	Semester 1			Winter Vacation			Semester 2			Summer Vacation		
number	a1	c1	b1	a2	c2	b2	a3	c3	b3	a4	c4	b4
81	2	3	2	1	2	1	1	3	2	2	4	4
82	2	5	3	2	4	3	2	5	4	1	2	2
83	3	5	3	2	5	3	1	3	2	2	3	2
84	2	2	2	2	4	2	2	5	3	1	4	2
85	1	3	2	1	5	2	1	2	2	1	4	2
86	2	6	2	0	0	0	1	3	2	0	0	0
87	1	3	3	2	4	3	1	2	2	2	5	2
88	2	5	4	2	3	3	1	3	2	2	4	2
89	2	7	5	1	1	1	3	6	3	2	3	3
90	3	5	3	0	0	0	2	3	3	0	0	0
91	1	4	4	2	3	2	2	4	3	1	2	2
92	3	5	3	2	3	2	2	4	3	1	2	1
93	2	3	2	1	2	2	1	2	1	2	3	2
94	1	5	4	2	3	2	2	4	2	1	2	2
95	2	3	3	0	0	0	1	4	2	0	0	0
96	2	4	2	2	5	5	1	3	2	2	5	2
97	1	10	5	2	3	2	2	4	3	1	5	2
98	2	4	3	1	2	2	2	4	3	1	2	1
99	3	5	3	3	5	4	1	3	2	1	4	1
100	2	3	3	0	0	0	1	2	2	0	0	0

Fig. E.3 Answers of students (Nos. 81–100) in National Chengchi University

References

1. P.T. Agraz, On Borel measurability and large deviations for fuzzy random variables, *Fuzzy Sets and Systems*, vol. 157, no. 19, pp. 2558–2568, 2006.
2. E.E. Ammar, On fuzzy random multiobjective quadratic programming, *European Journal of Operational Research*, vol. 193, no. 2, pp. 329–341, 2009.
3. U. Akinc and B.M. Khumawala, An efficient branch and bound algorithm for the capacitated warehouse location problem, *Management Science*, vol. 23 no.6, pp. 585–594, 1977.
4. G. Aletti, E.G. Bongiorno and V. Capasso, Statistical aspects of fuzzy monotone set-valued stochastic processes, *Application to birth-and-growth processes Fuzzy Sets and Systems,* vol. 160, no. 21, pp. 3140–3151, 2009.
5. M.A. Badri, Combining the analytic hierarchy process and goal programming for global facility location-allocation problem, *International Journal of Production Economics*, vol. 62, no. 3, pp. 237–248, 1999.
6. S. Bag, D. Chakraborty and A.R. Roy, A production inventory model with fuzzy random demand and with flexibility and reliability considerations, *Computers & Industrial Engineering*, vol. 56, no. 1, pp. 411–416, 2009.
7. R.E. Barlow and F. Proschan, *Statistical Theory of Reliability and Life Testing*, Holt, Rinehart, Winston, 1975.
8. O. Berman and Z. Drezner, A probabilistic one-centre location problem on a network, *Journal of the Operational Research Society*, vol. 54, no. 88, pp. 871–877, 2003.
9. M. Benrejeb, A. Sakly, K.B. Othman and P. Borne, Choice of conjunctive operator of TSK fuzzy systems and stability domain study, *Mathematics & Computers in Simulation*, vol. 76, nos. 5-6, pp. 410–421, 2008.
10. U. Bhattacharya, J.R. Rao, and R.N. Tiwari, Fuzzy multi-criteria facility location problem, *Fuzzy Sets and Systems*, vol. 51, no. 3, pp. 277–287, 1992.
11. R.B. Bhat, *Modern Probability Theory*, John Wiley & Sons, New York, 1985.
12. R.P. Bigelow, *Computers and the Law: An Introductory Handbook*, Commerce Clearing House, New York, 3rd Edition, 1981.
13. J.R. Birge and F.V. Louveaux, *Introduction to Stochastic Programming*, Springer-Verlag, New York, 1997.
14. C. Blum and M. Dorigo, The hyper-cube framework for ant colony optimization, *IEEE Transactions on Systems, Man, and CyberneticsPart B*, vol. 34, no. 2, pp. 1161–1172, 2004.
15. I. Bloch and H. Maitre, Fuzzy mathematical morphologies: a comparative study, *Pattern Recognition*, vol. 28, no. 9, pp. 1341–387, 1995.
16. Y.-H.O. Chang, Hybrid fuzzy least-squares regression analysis and its reliability measures, *Fuzzy Sets and Systems*, vol. 119, no. 2, pp. 225–246, 2001.

S. Wang and J. Watada, *Fuzzy Stochastic Optimization: Theory, Models and Applications*,
DOI 10.1007/978-1-4419-9560-5,

17. C. Chantrapornchai, W. Surakampontorn and E.H.-M. Sha, Design exploration with imprecise latency and register constraints, *IEEE Transactions on Computer-Aided Design of Integrated Circuits and Systems*, vol. 25, no. 12, pp. 2650–2662, 2006.
18. Y.J. Chen and Y.-K. Liu, A strong law of large numbers in credibility theory, *World Journal of Modeling and Simulation*, vol. 2, no. 5, pp. 331–337, 2006.
19. S.-M. Chen and J.-Y. Wang, Document retrieval using knowledge-based fuzzy information retrieval techniques, *IEEE Transactions on Systems, Man and Cybernetics*, vol. 25, no. 5, pp. 793–803, 1995.
20. W.-N. Chen and J. Zhang, An ant colony optimization approach to a grid workflow scheduling problem with various QoS requirements, *IEEE Transactions on Systems, Man, and Cybernetics, Part C: Applications and Reviews*, vol. 39, no. 1, pp. 29–43, 2009.
21. V. Chvatal, *Linear Programming*, W.H. Freeman and Company, New York, 1983.
22. M. Clerc and J. Kennedy, The particle swarm — Explosion, stability, and convergence in a multidimensional complex space, *IEEE Transactions on Evolutionary Computaion*, vol. 6, no. 1, pp. 58–73, 2002.
23. H. Cramer, *Mathematical Methods of Statistics*, Princeton University Press, Princeton, 1966.
24. K. Das, T.K. Roy and M. Maiti, Multi-item stochastic and fuzzy-stochastic inventory models under two restrictions, *Computers & Operations Research*, vol. 31, no. 11, pp. 1793–1806, 2004.
25. G. De Cooman, Possibility theory III, *International Journal of General Systems*, vol.25, no.4, pp.291–371, 1997.
26. T.Q. Deng and H.J.A.M. Heijmans, Gray-scale morphology based on fuzzy logic, *Journal of Mathematical Imaging and Vision*, vol. 16, no. 2, pp. 155–171, 2002.
27. M. Dorigo, *Optimization, Learning and Natural Algorithms,* Ph.D. dissertation, Dipartimento di Elettronica, Politecnico di Milano, Milan, Italy, 1992.
28. M. Dorigo, V. Maniezzo and A. Colorni, Ant System: Optimization by a colony of cooperating agents, *IEEE Transactions on Systems, Man, and CyberneticsPart B*, vol. 26, no. 1, pp. 29–41, 1996.
29. D. Dubois and H. Prade, *Possibility Theory*, Plenum Press, New York, 1988.
30. D. Duffie and J. Pan, An overview of value-at-risk, *Journal of Derivatives*, vol. 4, no. 3, pp. 7–49, 1997.
31. L. Dupont, Branch and bound algorithm for a facility location problem with concave site dependent costs, *International Journal of Production Economics*, vol. 112, no. 1, pp. 245–254, 2008.
32. P. Dutta, D. Chakraborty and A.R. Roy, A single-period inventory model with fuzzy random variable demand, *Mathematical and Computer Modelling*, vol. 41, nos. 8-9, pp. 915–922, 2005.
33. A.O.C. Elegbede, C. Chu, K.H. Adjallah and F. Yalaoui, Reliability allocation through cost minimization, *IEEE Transactions on Reliability*, vol. 52, no. 1, pp. 106–111, 2003.
34. A.T. Ernst and M. Krishnamoorthy, Solution algorithms for the capacitated single allocation hub location problem, *Annals of Operations Research*, vol. 86, no. 1-4, pp. 141–159, 1999.
35. X. Feng and Y. K. Liu, Measurability criteria for fuzzy random vectors, *Fuzzy Optimization and Decision Making*, vol. 5, no. 3, pp. 245–253, 2006.
36. G.S. Fishman, *Monte Carlo: Concepts, Algorithms, and Applications,* Springer-Verlag, New York, 1996.
37. J.M. Garibaldi and T. Ozen, Uncertain fuzzy reasoning: A case study in modelling expert decision making, *IEEE Transactions on Fuzzy Systems*, vol. 15, no. 1, pp. 16–30, 2007.
38. M.A. Gil, L.D. Miguel and D.A. Ralescu, Overview on the development of fuzzy random variables, *Fuzzy sets and systems*, vol. 157, no. 19, pp. 2546–2557, 2006.
39. D. E. Goldberg, *Genetic Algorithms in Search, Optimization and Machine Learning*, Addison-Wesley, Reading, MA, 1989.
40. F. Glover, Tabu search-Part I, *ORSA Journal on Computing*, vol. 1, no. 3, pp. 190–206, 1989.
41. F. Glover, Tabu search-Part II, *ORSA Journal on Computing*, vol. 2, no. 1, pp. 4–32, 1990.

42. F. Glover, Parametric tabu-search for mixed integer programs, *Computers & Operations Research*, vol. 33, no. 9, pp. 2449–2494, 2006.
43. P. Halmos, *Measure Theory*, Van Nostrand, Princeton, 1950.
44. F.F. Hao and Y.K. Liu, Mean-variance models for portfolio selection with fuzzy random returns, *Journal of Applied Mathematics and Computing*, vol. 30, nos. 1-2, pp. 9–38, 2009.
45. F.F. Hao and R. Qin, Variance formulas for trapezoidal fuzzy random variables, *Journal of Uncertain Systems*, vol. 3, no. 2, pp. 145–160, 2009.
46. S.L. Ho and S. Yang, A computationally efficient vector optimizer using ant colony optimizations algorithm for multiobjective designs, *IEEE Transactions on Magnetics*, vol. 44, no. 6, pp. 1034–1037, 2008.
47. N.V. Hop, Fuzzy stochastic goal programming problems, *European Journal of Operational Research*, vol. 176, no. 1, pp. 77–86, 2007.
48. J. Holland, *Adaptation in Natural and Artificial Systems*, University of Michigan Press, Ann Arbor, 1975.
49. D.H. Hong and P.I. Ro, The law of large numbers for fuzzy numbers with unbounded supports, *Fuzzy Sets and Systems*, vol. 116, no. 2, pp. 269–274, 2000.
50. X. Huang, Two new models for portfolio selection with stochastic returns taking fuzzy information, *European Journal of Operational Research*, vol. 180, no. 1, pp. 396–405, 2007.
51. H. Huang, C.-G. Wu and Z.-F. Hao, A pheromone-rate-based analysis on the convergence time of ACO algorithm, *IEEE Transactions on Systems, Man, and Cybernetics, Part B: Cybernetics,* vol. 39, no. 4, pp. 910–923, 2009.
52. C.-M. Hwang and J.-S. Yao, Independent fuzzy random variables and their application, *Fuzzy Sets and Systems*, vol. 82, no. 3, pp. 335–350, 1996.
53. C.W. Hwang, A theorem of renewal process for fuzzy random variables and its application, *Fuzzy Sets and Systems* vol. 116, no. 2, pp. 237–244, 2000.
54. S.-T. Hsieh, T.-Y. Sun, C.-C. Liu and S.-J. Tsai, Efficient population utilization strategy for particle swarm optimizer, *IEEE Transactions on Systems, Man, and Cybernetics, Part B: Cybernetics,* vol. 39, no. 2, pp. 444– 456, 2009.
55. H. Ishii, Y.L. Lee, and K.Y. Yeh, Fuzzy facility location problem with preference of candidate sites, *Fuzzy Sets and Systems*, vol. 158, no. 17, pp. 1922–1930, 2007.
56. L.S. Iliadis, S. Spartalis and S. Tachos, Application of fuzzy T-norms towards a new Artificial Neural Networks' evaluation framework: A case from wood industry, *Information Sciences*, vol. 178, no. 20, pp. 3828–3839, 2008.
57. S. Janson, D. Merkle and M. Middendorf, Molecular docking with multi-objective particle swarm optimization, *Applied Soft Computing*, vol. 8, no. 1, pp. 666–675, 2008.
58. S.Y. Joo, Y.K. Kim, J.S. Kwon and G.S. Choi, Convergence in distribution for level-continuous fuzzy random sets, *Fuzzy Sets and Systems,* vol. 157, no. 2, pp. 243–255, 2006.
59. P. Jorion, *Value at Risk: The New Benchmark for Controlling Market Risk*, McGraw-Hill, New York, 2000.
60. P. Kall and S.W. Wallace, *Stochastic Programming,* John Wiley & Sons, Chichester, 1994.
61. N.N. Karnik and J.M. Mendel, Operations on type-2 fuzzy sets, *Fuzzy Sets and Systems*, vol. 122, no. 2, pp. 327–348, 2001.
62. J. Kennedy and R.C. Eberhart, Particle swarm optimization, *In: Proceedings of the 1995 IEEE International Conference on Neural Networks*, IV, pp. 1942–1948, 1995.
63. J. Kennedy and R.C. Eberhart, A discrete binary version of the particle swarm algorithm, *In: Proceedings of the 1997 IEEE International Conference on Systems, Man, and Cybernetics*, Orlando, pp. 4104–4108, 1997.
64. J. Kennedy, R.C. Eberhart and Y. Shi, *Swarm Intelligence,* Morgan Kaufmann Publishers, San Francisco, 2001.
65. I. Khindanova, S. Rachev and E. Schwartz, Stable modeling of value at risk, *Mathematical and Computer Modelling,* vol. 34, nos. 9-11, pp. 1223–1259, 2001.
66. P. Klement, R. Mesiar and E. Pap, *Triangular Norms*, Kluwer Academic Publishers, Netherlands, 2000.

67. R. Korner, On the variance of fuzzy random variables, *Fuzzy Sets and Systems,* vol. 92, no. 1, pp. 83–93, 1997.
68. V. Krätschmer, Limit theorems for fuzzy-random variables, *Fuzzy Sets and Systems,* vol. 126, no. 2, pp. 253–263, 2002.
69. R. Kruse and K.D. Meyer, *Statistics with Vague Data,* D. Reidel Publishing Company, Dordrecht, 1987.
70. W. Kuo, V.R. Prasad, F.A. Tillman and C. L. Hwang, *Optimal Reliability Design*, Cambridge University Press, 2001.
71. W. Kuo and R. Wan, Recent Advances in Optimal Reliability Allocation, *IEEE Transactions on Systems, Man, and Cybernetics – Part A: Systems and Humans*, vol. 37, no. 2, pp. 143–156, 2008.
72. H. Kwakernaak, Fuzzy random variables–I. Definitions and theorems, *Information Sciences*, vol. 15, no. 1, pp. 1–29, 1978.
73. H. Kwakernaak, Fuzzy random variables–II. Algorithm and examples, *Information Sciences*, vol. 17, no. 3, pp. 253–278, 1979.
74. K. J. Laidler, *The World of Physical Chemistry*, Oxford University Press, Oxford, 1993.
75. H.K. Lam and M. Narimani, Stability analysis and performance design for fuzzy-model-based control system under imperfect premise Matching, IEEE Transactions on Fuzzy Systems, vol. 17, no. 4, pp. 949–961, 2009.
76. G. Laporte, F.V. Louveaux and L.V. Hamme, Exact solution to a location problem with stochastic demands, *Transportation Science*, vol. 28, no. 2, pp. 95–103, 1994.
77. S. Lee, S. Soak, S. Oh, W. Pedrycz and M. Jeon, Modified binary particle swarm optimization, *Progress in Natural Science*, vol. 18, no. 9, pp. 1161–1166, 2008.
78. S. Li and L. Guan, Fuzzy set-valued Gaussian processes and Brownian motions, *Information Sciences,* vol 177, no. 16, pp. 3251–3259, 2007.
79. S. Li, Y. Ogura, F.N. Proske and M.L. Puri, Central limit theorems for generalized set-valued random variables, *Journal of Mathematical Analysis and Applications*, vol. 285, no. 1, pp. 250–263, 2003.
80. S. Li, R. Zhao and W. Tang, Fuzzy random delayed renewal process and fuzzy random equilibrium renewal process, *Journal of Intelligent & Fuzzy Systems*, vol. 18, no. 2, pp. 149–156, 2007.
81. B. Li, M. Zhu and K. Xu, A practical engineering method for fuzzy reliability analysis of mechanical structures, *Reliability Engineering & System Safety*, vol. 67, no. 3, pp. 311–315, 2000.
82. L. Lin and H.M. Lee, A fuzzy decision support system for facility site selection of multinational enterprises, *International Journal of Innovative Computing, Information & Control*, vol. 3, no. 1, pp. 151–162, 2007.
83. F. Liu, An efficient centroid type-reduction strategy for general type-2 fuzzy logic system, *Information Sciences*, vol. 178, no. 9, pp. 2224–2236, 2008.
84. B. Liu, Fuzzy random chance-constrained programming, *IEEE Transactions on Fuzzy Systems*, vol. 9, no. 5, pp. 713–720, 2001.
85. B. Liu, Fuzzy random dependent-chance programming, *IEEE Transactions on Fuzzy Systems*, vol. 9, no. 5, pp. 721–726, 2001.
86. B. Liu, *Theory and Practice of Uncertain Programming*, Physica-Verlag, Heidelberg, 2002.
87. B. Liu, *Uncertainty Theory: An Introduction to Its Axiomatic Foundations*, Springer-Verlag, Berlin, 2004.
88. B. Liu and Y.K. Liu, Expected value of fuzzy variable and fuzzy expected value models, *IEEE Transaction on Fuzzy Systems*, vol. 10, no. 4, pp. 445–450, 2002.
89. Y.K. Liu and B. Liu, Fuzzy random variable: A scalar expected value operator, *Fuzzy Optimization and Decision Making*, vol. 2. no. 2, pp. 143–160, 2003.
90. Y.K. Liu and B. Liu, A class of fuzzy random optimization: expected value models, *Information Sciences,* vol. 155, nos. 1-2, pp. 89–102, 2003.
91. Y.K. Liu and B. Liu, On minimum-risk problems in fuzzy random decision systems, *Computers & Operations Research*, vol. 32, no. 2, pp. 257–283, 2005.

92. Y.K. Liu, Convergent results about the use of fuzzy simulation in fuzzy optimization problems, *IEEE Transactions on Fuzzy Systems*, vol. 14, no. 2, pp. 295–304, 2006.
93. Y. K. Liu and S. Wang, A credibility approach to the measurability of fuzzy random vectors, *International Journal of Natural Sciences & Technology*, vol. 1, no. 1, pp. 111–118, 2006.
94. Y. K. Liu, Fuzzy programming with recourse, International Journal of Uncertainty, Fuzziness & Knowledge-Based Systems 13 (4) (2005) 381–413.
95. Y.K. Liu and J. Gao, The independence of fuzzy variables with applications to fuzzy random optimization, *International Journal of Uncertainty, Fuzziness & Knowledge-Based Systems*, vol.15, no.2, pp.1–19, 2007.
96. Y.K. Liu, The approximation method for two-stage fuzzy random programming with recourse, *IEEE Transactions on Fuzzy Systems*, vol. 15, no. 6, pp. 1197–1208, 2007.
97. Y.K. Liu, The convergent results about approximating fuzzy random minimum risk problems, *Applied Mathematics & Computation*, vol. 205, no. 2, pp. 608–621, 2008.
98. Y.K. Liu, Z.Q. Liu and J. Gao, The modes of convergence in the approximation of fuzzy random optimization problems, *Soft Computing*, vol. 13, no. 2, pp. 117–125, 2009.
99. R. Logendran and M.P. Terrell, Uncapacitated plant location-allocation problems with price sensitive stochastic demands, *Computers and Operations Research*, vol. 15, no. 2, pp. 189–198, 1988.
100. F.V. Louveaux and D. Peeters, A dual-based procedure for stochastic facility location, *Operations Research*, vol. 40, no. 3, pp. 564–573, 1992.
101. M. López-Diaz and M. A. Gil, Constructive definitions of fuzzy random variables, *Statistics and Probability Letters*, vol. 36, no. 2, pp. 135–143, 1997.
102. S. Lozano, F. Guerrero, L. Onieva and J. Larrañeta , Kohonen maps for solving a class of location-allocation problems, *European Journal of Operational Research*, vol. 108, no. 1, pp. 106–117, 1998.
103. M.K. Luhandjula, On possibilistic linear programming, *Fuzzy Sets and Systems*, vol. 18, pp. 15–30, 1986.
104. M.K. Luhandjula, Fuzziness and randomness in an optimization framework, *Fuzzy Sets and Systems*, vol. 77, no. 3, pp. 291–297, 1996.
105. M.K. Luhandjula, Optimization under hybrid uncertainty, *Fuzzy Sets and Systems*, vol. 146, no. 2, pp. 187–203, 2004.
106. M.K. Luhandjula, Fuzzy stochastic linear programming: Survey and future research directions, *European Journal of Operational Research*, vol. 174, no. 3, pp. 1353–1367, 2006.
107. G.S. Mahapatra and T.K. Roy, Fuzzy multi-objective mathematical programming on reliability optimization model, *Applied Mathematics and Computation*, vol. 174, no. 1, pp. 643–659, 2006.
108. D. Merkle, M. Middendorf and H. Schmeck, Ant colony optimization for resourceconstrained project scheduling, *IEEE Transactions on Evolutionary Computation*, vol. 6, no. 4, pp. 333–346, 2002.
109. Z. Michalewicz, *Genetic Algorithms + Data Stractures = Evolution Programs*, Springer, New York, 1994.
110. B.M. Mohan and A. Sinha, Mathematical models of the simplest fuzzy PI/PD controllers with skewed input and output fuzzy sets, *ISA Transactions*, vol. 47, no. 3, pp. 300–310, 2008.
111. A.W. Mohemmed, N.C. Sahoo and T. K. Geok, Solving shortest path problem using particle swarm optimization, *Applied Soft Computing*, vol. 8, no. 4, pp. 1643–1653, 2008.
112. A. Mukherjea, and K. Pothoven, *Real and Functional Analysis*, Plenum Press, New York, 1984.
113. T.M. Mullin, Experts' estimation of uncertain quantities and its implications for knowledge acquisition, *IEEE Transactions on Systems, Man and Cybernetics*, vol. 19, no. 3, pp. 616–625, 1989.
114. S. Nahmias, Fuzzy variable, *Fuzzy Sets and Systems*, vol. 1, no. 2, pp. 97–101, 1978.
115. W. Nather, Regression with fuzzy random data, *Computational Statistics & Data Analysis*, vol. 51, no. 1, pp. 235–252, 2006.
116. H.T. Nguyen and B. Wu, *Foundations of Statistics with Fuzzy Data*, Springer, New York, 2006.

117. W. Pedrycz and F. Gomide, *An Introduction to Fuzzy Sets; Analysis and Design*, MIT Press, Cambridge, MA, 1998.
118. W. Pedrycz and F. Gomide, *Fuzzy Systems Engineering: Toward Human-Centric Computing*, J. Wiley, Hoboken, NJ, 2007.
119. W. Pedrycz, B.J. Park and N.J. Pizzi, Identifying core sets of discriminatory features using particle swarm optimization, *Expert Systems with Applications*, vol. 36, no. 3, pp. 4610–4616, 2009.
120. H. Pham, Optimal design of parallel-series systems with competing failure modes, *IEEE Transactions on Reliability*, vol. 41, no. 4, pp. 583–587, 1992.
121. E. Popova and H.C. Wu, Renewal reward processes with fuzzy rewards and their applications to T-age replacement policies, *European Journal of Operational Research*, vol. 117, no. 3, pp. 606–617, 1999.
122. M.L. Puri and D.A. Ralescu, Fuzzy random variables, *Journal of Mathematical Analysis & Applications*, vol. 114, no. 2, pp. 409–422, 1986.
123. V.R. Prasad and M. Raghavachari, Optimal allocation of interchangeable components in a series-parallel system, *IEEE Transactions on Reliability*, vol. 47, no. 3, pp. 255–260, 1998.
124. R. Qin and Y.K. Liu, A new data envelopment analysis model with fuzzy random inputs and outputs, *Journal of Applied Mathematics & Computing*, in press (DOI: 10.1007/s12190-009-0289-7).
125. A. Ratnweera, S.K. Halgamuge and H.C. Watson, Self-organizing hierarchical particle swarm optimizer with time-varying acceleration coefficients, *IEEE Transactions on Evolutionary Computation*, vol. 8, no. 3, pp. 240–255, 2004.
126. H.L. Royden, *Real Analysis*, Prentice-Hall, New Jersey, 1988.
127. S.M. Ross, *Stochastic Processes*, 2nd Edition, John Wiley & Sons, New York, 1996.
128. M. Sadeghi and S. Shavvalpour, Energy risk management and value at risk modeling, *Energy Policy*, vol. 34, no. 18, pp. 3367–3373, 2006.
129. B. Schweizer and A. Sklar, Associative functions and abstract semigroups, *Publicationes Mathematicae Debrecen*, vol. 10, pp. 69–81, 1963.
130. R. Schutz, L. Stougie and A. Tomasgard, Stochastic facility location with general long-run costs and convex short-run cost, *Computers & Operations Research*, vol. 35, no. 9, pp. 2988–3000, 2008.
131. A.F. Shapiro, Fuzzy random variables, *Insurance: Mathematics and Economics*, vol. 44, no. 2, pp. 307–314, 2009.
132. M. Shen and Y. Tsai, CPW-fed monopole antenna characterized by using particle swarm optimization incorporating decomposed objective functions, *International Journal of Innovative Computing, Information & Control*, vol. 4, no. 8, pp. 1897–1919, 2008.
133. K. Smimou, C.R. Bector and G. Jacoby, Portfolio selection subject to experts' judgments, *International Review of Financial Analysis*, vol. 17, no. 5, pp. 1036–1054, 2008.
134. W.E. Stein and K. Talati, Convex fuzzy random variables, *Fuzzy Sets and Systems*, vol. 6, no. 3, pp. 271–283, 1981.
135. T. Stützle and H.H. Hoos, $\mathcal{MAX}-\mathcal{MIN}$ Ant System, *Future Generation Computer Systems*, vol. 16, no. 8, pp. 889–914, 2000.
136. M.F. Tasetiren and Y. C. Liang, A binary particle swarm optimization algorithm for lot sizing problem, *Journal of Economic and Social Research*, vol. 5, no. 2, pp. 1–20, 2003.
137. R. Tavakkoli-Moghaddam, J. Safari and F. Sassani, Reliability optimization of series-parallel system with a choice of redundancy strategies using a genetic algorithm, *Reliability Engineering & System Safety*, vol. 93, no. 4, pp. 550–556, 2008.
138. P. Terán, Probabilistic foundations for measurement modelling using fuzzy random variables, *Fuzzy Sets and Systems*, vol. 158, no. 9, pp. 973–986, 2007.
139. J.V. Tiel, *Convex Analysis: An Introductory Text*, John Wiley & Sons, 1984.
140. M.K. Urbanski and J. Wasowski, Fuzzy approach to the theory of measurement inexactness, *Measurement*, vol. 34, no. 1, pp. 67–74, 2003.
141. R. Viertl, On reliability estimation based on fuzzy lifetime data, *Journal of Statistical Planning and Inference*, vol. 139, no. 5, pp. 1750–1755, 2009.

142. G. Wang and Z. Qiao, Linear programming with fuzzy random variable coefficients, *Fuzzy Sets and Systems*, vol. 57, no. 3, pp. 295–311, 1993.
143. B. Wang, S. Wang and J. Watada, Real options analysis based on fuzzy random variables, International Journal of Innovative Computing, Information & Control, vol.6, no.4, pp.1689–1698, 2010.
144. B. Wang, S. Wang and J. Watada, Fuzzy portfolio selection models with Value-at-Risk, IEEE Transactions on Fuzzy Systems, to appear
145. S. Wang, Y. Liu and X. Dai, On the continuity and absolute continuity of credibility functions, *Journal of Uncertain Systems*, vol.1, no.3, pp.185–200, 2007.
146. S. Wang and J. Watada, T-independence condition for fuzzy random vector based on continuous triangular norms, *Journal of Uncertain Systems*, vol. 2, no. 2, pp. 155–160, 2008.
147. S. Wang and J. Watada, Studying distribution functions of fuzzy random variables and its applications to critical value functions, *International Journal of Innovative Computing, Information & Control*, vol. 5, no. 2, pp. 279–292, 2009.
148. S. Wang, Y.-K. Liu and J. Watada, Fuzzy random renewal process with queueing applications, Computers & Mathematics with Applications, vol.57, no.7, pp. 1232–1248, 2009.
149. S. Wang and J. Watada, Modelling redundancy allocation for a fuzzy random parallel-series system, *Journal of Computational & Applied Mathematics,* vol. 232, no. 2, pp. 539–557, 2009.
150. S. Wang and J. Watada, Fuzzy random renewal reward process and its applications, *Information Sciences*, vol. 179, no. 23, pp. 4057–4069, 2009.
151. S. Wang and J. Watada, Reliability optimization of a series-parallel system with fuzzy random lifetimes, International Journal of Innovative Computing, Information & Control, vol.5, no.6, pp.1547–1558, 2009.
152. S. Wang, J. Watada and W. Pedrycz, Value-at-Risk-based two-stage fuzzy facility location problems, *IEEE Transactions on Industrial Informatics*, vol. 5, no. 4, pp. 465–482, 2009.
153. S. Wang, and J. Watada, T-norm-based limit theorems for fuzzy random variables, Journal of Intelligent & Fuzzy Systems, vol.21, no.4, pp.233–242, 2010.
154. S. Wang, J. Watada and W. Pedrycz, Recourse-based facility location problems in hybrid uncertain environment, *IEEE Transactions on Systems, Man, and Cybernetics, Part B: Cybernetics*, vol. 40, no. 4, pp. 1176–1187, 2010.
155. S. Wang and J. Watada, Value of information and solution under VaR criterion for fuzzy random optimization problems, *Proc. of the 2010 IEEE World Congress on Computational Intelligence*, IEEE Publisher, Barcelona, Spain, pp. 3290–3295, 2010.
156. S. Wang and J. Watada, Two-stage fuzzy stochastic programming with Value-at-Risk criteria, *Applied Soft Computing*, vol. 11, no. 1, pp. 1044–1056, 2011.
157. S. Wang and J. Watada, Some properties of T-independent fuzzy variables, Mathematical & Computer Modelling, vol.53, nos.5-6, pp.970–984, 2011.
158. S. Wang and J. Watada, A hybrid modified PSO approach to VaR-based facility location problems with variable capacity in fuzzy random uncertainty, Information Sciences, to appear, doi:10.1016/j.ins.2010.02.014.
159. J. Watada and W. Pedrycz, A fuzzy regression approach to acquisition of linguistic rules, *Handbook on Granular Computation*, In: W. Pedrycz, A. Skowron, V. Kreinovich (eds), Chapter 32, John Wiley & Sons Ltd, pp. 719–740, 2008.
160. J. Watada and S. Wang, Regression model based on fuzzy random variables, Seising Rodulf (Ed.), Chapter 26, *Views on Fuzzy Sets and Systems from Different Perspectives,* Spring-Verlag, Berlin, 2009.
161. J. Watada, S. Wang and W. Pedrycz, Building confidence-interval-based fuzzy random regression models, *IEEE Transactions on Fuzzy Systems,* vol. 17, no. 6, pp. 1273–1283, 2009.
162. P. Wang, Fuzzy contactability and fuzzy variables, *Fuzzy Sets and Systems*, vol. 8, no. 1, pp.81–92, 1982.
163. M. Wen and K. Iwamura, Fuzzy facility location-allocation problem under the Hurwicz criterion, *European Journal of Operational Research*, vol. 184, no. 2, pp. 627–635, 2008.

164. H.-C. Wu, Probability density functions of fuzzy random variables, *Fuzzy Sets and Systems*, vol. 105, no. 1, pp. 139–158, 1999.
165. H.-C. Wu, The central limit theorems for fuzzy random variables, *Information Sciences*, vol. 120, nos. 1-4, pp. 239–256, 1999.
166. J. Xu and Y.G. Liu, Multi-objective decision making model under fuzzy random environment and its application to inventory problems, *Information Sciences*, vol. 178, no. 15, pp. 2899–2914, 2008.
167. L. Yang and L. Liu, Fuzzy fixed charge solid transportation problem and algorithm, *Applied Soft Computing*, vol. 7, no. 3, pp. 879–889, 2007.
168. L. Yang, X. Ji, Z. Gao and K. Li, Logistics distribution centers location problem and algorithm under fuzzy environment, *Journal of Computational and Applied Mathematics*, vol. 208, no. 2, pp. 303–315, 2007.
169. L. Yang, Chance-constrained methods for optimization problems with random and fuzzy parameters, *International Journal of Innovative Computing, Information & Control*, vol. 5, no. 2, pp. 413–422, 2009.
170. H. Yu, C. Chu, E. Chatelet and F. Yalaoui, Reliability optimization of a redunant system with failure dependencies, *Reliability Engineering & System Safety*, vol. 92, no. 12, pp. 1627–1634, 2007.
171. L.A. Zadeh, Fuzzy sets, *Information and Control*, vol. 8, no. 3, pp. 338–353, 1965.
172. L.A. Zadeh, Fuzzy sets as a basis for a theory of possibility, *Fuzzy Sets and Systems*, vol. 1, no. 1, pp. 3–28, 1978.
173. L.A. Zadeh, Fuzzy probabilities, *Information Processing & Management*, vol. 20, no. 3, pp. 363–372, 1984.
174. L.A. Zadeh, Toward a perception-based theory of probabilistic reasoning with imprecise probabilities, *Journal of Statistical Planning and Inference*, vol. 105, no. 1, pp. 233–264, 2002.
175. L.A. Zadeh, Toward a generalized theory of uncertainty (GTU) – an outline, *Information Science*, vol. 172, no. 1-2, pp. 1–40, 2005.
176. L.A. Zadeh, Generalized theory of uncertainty (GTU) – principal concepts and ideas, *Computational Statistics & Data Analysis*, vol. 51, no. 1, pp. 15–46, 2006.
177. L.A. Zadeh, Is there a need for fuzzy logic?, *Information Sciences*, vol. 178, no. 13, pp. 2751–2779, 2008.
178. T. Zeng and J. Ward, The stochastic location-assignment problem on a tree, *Annals of Operations Research*, vol. 136, no. 1, pp. 81–97, 2005.
179. R. Zhao and B. Liu, Standby redundancy optimization problems with fuzzy lifetimes, *Computers & Industrial Engineering*, vol. 49, no. 2, pp. 318–338, 2005.
180. R.Q. Zhao and W. Tang, Some properties of fuzzy random renewal processes, *IEEE Transactions on Fuzzy Systems*, vol. 14, no. 2, pp. 173–179, 2006.
181. J. Zhou and B. Liu, Modeling capacitated location-allocation problem with fuzzy demands, *Computers & Industrial Engineering*, vol. 53, no. 3, pp. 454–468, 2007.

Index

S. Wang and J. Watada, *Fuzzy Stochastic Optimization: Theory, Models and Applications*, DOI 10.1007/978-1-4419-9560-5, © Springer Science+Business Media New York 2012